Photoelectric Measurement and Sensing: New Technology and Applications—2nd Edition

Photoelectric Measurement and Sensing: New Technology and Applications—2nd Edition

Guest Editors

Qibo Feng
Jiakun Li
Qixin He

MDPI

Basel • Beijing • Wuhan • Barcelona • Belgrade • Novi Sad • Cluj • Manchester

Guest Editors

Qibo Feng
MOE Key Lab of Luminescence and Optical Information
Beijing Jiaotong University
Beijing
China

Jiakun Li
MOE Key Lab of Luminescence and Optical Information
Beijing Jiaotong University
Beijing
China

Qixin He
MOE Key Lab of Luminescence and Optical Information
Beijing Jiaotong University
Beijing
China

Editorial Office
MDPI AG
Grosspeteranlage 5
4052 Basel, Switzerland

This is a reprint of the Special Issue, published open access by the journal *Sensors* (ISSN 1424-8220), freely accessible at: https://www.mdpi.com/journal/sensors/special_issues/O70446R258.

For citation purposes, cite each article independently as indicated on the article page online and as indicated below:

Lastname, A.A.; Lastname, B.B. Article Title. *Journal Name* **Year**, *Volume Number*, Page Range.

ISBN 978-3-7258-4495-1 (Hbk)
ISBN 978-3-7258-4496-8 (PDF)
https://doi.org/10.3390/books978-3-7258-4496-8

Contents

sensors

MDPI

Article

A Rail Profile Measurement Method Based on Polarization Fusion Imaging

Qiang Han, Xinxin Zhao, Jing Shi, Shengchun Wang, Peng Dai, Ning Wang and Le Wang *

Infrastructure Inspection Research Institute, China Academy of Railway Sciences Corporation Limited, Beijing 100081, China
* Correspondence: 13001988989@163.com

Abstract: The smooth area on the rail surface causes abnormal exposure in the laser section image, resulting in measurement errors of the rail profile. To address this issue, a novel rail profile measurement technique based on polarization fusion imaging is proposed. A polarized camera is utilized to acquire the four-directional polarization component images, Stokes parameter images, linear polarization angle images, and linear polarization degree images of the rail laser section. A polarization image data fusion algorithm based on Segmented Random Sample Consensus (S-RANSAC) is designed using these images as data sources, and the optimal rail profile fitting curve is obtained. Experimental results demonstrate that the proposed method can still obtain accurate and effective rail profile data in regions where traditional methods fail to capture profile data. Compared with the traditional method, the measurement error of the rail profile is reduced from 0.137 mm to 0.081 mm, and the measurement accuracy is improved by 40.9%. Evidently, this method avoids data loss in key areas of the rail profile caused by local underexposure, thus significantly enhancing the measurement accuracy. This method can provide a valuable reference for high-precision measurement of the rail profile under complex working conditions.

Keywords: polarization fusion; underexposure; rail profile; image fusion

1. Introduction

It is widely acknowledged that changes in the rail profile directly impact the safe operation of railways. As a vital means of railway operation and maintenance, rail profile detection is instrumental in understanding the service condition of the rail. Based on this understanding, more effective rail grinding can be carried out [1–3]. Rail profile detection refers to the process of comparing the measured rail profile data with the standard rail profile data. This comparison helps to obtain parameters such as vertical wear and side wear of the rail [4]. Currently, rail profile detection mainly falls into two categories: contact-type detection and non-contact-type detection. Specifically, contact-type detection has disadvantages such as low detection efficiency and high labor costs. This is because it requires probes to be in contact with the rail. In contrast, non-contact-type detection aims to extract rail profile data from the intensity information of the reflected light on the rail surface. The line-structured light profile detection technology is a typical non-contact-type detection technology. The line-structured light rail profile detection technology, based on the triangulation measurement principle enables real-time acquisition of rail profile information. Due to its high speed, high precision, and non-contact nature, it is extensively used for dynamic rail profile detection at home and abroad [5–7].

However, influenced by harsh working conditions, the rail will undergo some changes after a period of operation. These changes include increased surface roughness, the presence of foreign matter on the surface, and rust on the rail head and the light stripe area of the rail. These surface changes will affect the energy distribution of the reflected light on the rail surface, leading to abnormal energy distribution. For example, the light stripe area of the rail head has a relatively smooth surface, strong specular reflection ability, and very weak diffuse reflection ability. Most of the incident light energy is distributed near the specular reflection direction. Only a small amount of diffuse reflection light is collected by the camera. The traditional line-structured light rail profile detection technology employs the diffuse reflection light as the measurement signal. The specular reflection light is considered the interference signal to obtain the intensity information of the reflected light on the rail surface. For the smooth light stripe area of the rail, when the traditional light rail profile detection method is used to acquire the light intensity image of the rail profile, the phenomenon of underexposure occurs.

Accurately extracting the center of the light stripe is crucial for ensuring the accuracy of rail profile detection [8]. Many scholars have conducted extensive research on the innovation and improvement of light stripe center extraction algorithms [9–11]. In summary, light stripe center extraction algorithms are mainly divided into geometric center extraction algorithms and energy center extraction algorithms [12]. Geometric center extraction algorithms are based on edge information, threshold information, or refinement techniques. They are applicable to simple working environments and situations where the requirements for measurement accuracy are not high. As prevailing light stripe center extraction algorithms, energy center extraction algorithms are based on the center-of-gravity, directional template, or maximum point. They are suitable for harsh working conditions, objects with complex shapes, and scenarios where high-precision measurement is required. The underexposed areas of rail profile light stripe images are characterized by weak light stripe energy, low contrast, and low confidence in the light stripe center. In such underexposed areas, even with more accurate light stripe center extraction algorithms like the center-of-gravity algorithm and the Steger algorithm, accurate rail profile information cannot be obtained. Moreover, when the light stripe energy is extremely low, the light stripe cannot be detected. As a result, it gives rise to partial loss of profile data and reduces the accuracy of profile detection [13–16]. Although the underexposure problem of the light stripe area can be mitigated to some extent by prolonging the exposure time, it will cause overexposure of light stripes in the normal area of the same image. This, in turn, affects the overall profile detection accuracy.

Traditional imaging technology can only acquire light intensity information. In contrast, the information obtained by polarization imaging detection technology is expanded from three dimensions (light intensity, spectrum, and space) to seven dimensions (light intensity, spectrum, space, polarization degree, polarization angle, polarization ellipticity, and rotation direction). The additional polarization information is often used to improve the imaging quality of the measured object. Many scholars have conducted extensive research on polarization imaging technology. For example, Wolff [17] developed a polarization imaging system composed of a polarization splitting prism and two CCD cameras to analyze the polarization state of specular reflection light on the object surface. In terms of enhancing contrast, Li [18] explored the potential of active polarization imaging technology in various underwater applications. He fully utilized the polarization characteristics of the target reflected light. Exponential functions were introduced to reconstruct cross-polarized backscatter images. The proposed method demonstrated an improvement for high-polarization objects under various turbidity conditions. Mo [19] proposed a method to calculate the polarization characteristics image that can reflect the differences in polarization characteristics of different materials. They fused the multi-angle orthogonal differential polarization characteristics (MODPC) image with the intensity image. The fused polarization image effectively enhanced the object detection information. This provided a basis for object classification, recognition, and segmentation. Umeyama [20] captured images of the measured object from different polarization angles by rotating the polarizer and separated the specular reflection components from the diffuse reflection components through independent component analysis. Overall, polarization imaging technology has certain advantages in reducing the impact of specular reflection light on the object surface, enhancing contrast, and improving imaging quality. Le Wang et al. [21] proposed a line-structured light imaging method for rail profiles based on polarization fusion. They proposed obtaining polarization component images and total intensity images of the rail laser section from multiple angles using a polarized camera. They solved the problem of local overexposure of laser images caused by specular reflection on the rail grinding surface through polarization fusion imaging technology. However, they did not discuss the existing problem of local underexposure.

Statistic features of multiple images are effectively extracted based on information complementarity after the image fusion. This compensate for the information insufficiency problems of single images, such as image information interference by noise, and too little image information. The result is more accurate information about the measured object. In consideration of the information correlation and complementarity between polarization images, researchers have proposed a variety of polarization image fusion methods, mainly including the frequency domain image fusion and spatial domain image fusion. In terms of frequency domain image fusion, Zhang Jingjing et al. proposed decomposing an image into low-frequency and high-frequency parts of different scales using the discrete wavelet transform (DWT). The wavelet coefficients of the fusion image are determined based on the wavelet coefficients of the low-frequency and high-frequency images as the statistical features [22]. Qiao Juan put forward a polarization image fusion algorithm based on the two-dimensional DWT, to enhance image details and improve the visual effect of images [23]. In terms of spatial domain image fusion, Yin Haining et al. proposed a polarization image fusion method based on feature analysis. Through this method, the fusion weight of an image can be calculated according to its gray feature, texture feature and shape feature. In addition, image fusion is used to solve the problem of detail loss that occurs when the polarization parameter image is calculated [24]. Recently, some image fusion methods based on deep learning, such as DPFN [25] and Gan [26], have become research hotspots. However, they are usually specific to natural scenes with rich color and texture features, while laser stripe data are relatively insufficient and lacks rich texture and color features. Therefore, these methods are not suitable for fusion of laser polarized stripe images.

To solve the above problems, based on previous research results, this paper proposes an improved rail profile measurement method based on polarization imaging. Specifically, a polarized camera is first used to capture the polarization component images of the rail laser section from multiple angles. Then, the polarization information of the rail laser section is extracted, and the Stokes parameter images, linear polarization angle images, and linear polarization degree images are calculated. Based on the S-RANSAC algorithm, the rail profile data corresponding to multiple polarization component images are fused. The fused data are used as the final rail profile measurement result. This method effectively improves the accuracy of rail profile measurement results under complex working conditions.

The structure of this paper is as follows: Section 2 introduces the traditional rail profile measurement methods and the exposure anomaly problems they face. Section 3 describes the rail profile measurement model based on polarization fusion. This includes the basic principle of polarization imaging, rail polarization component images, and the polarization data fusion method. Section 4 presents the experimental results, including laboratory static experiments and field dynamic experiments, and makes comparisons with other methods. Finally, Section 5 summarizes the conclusion of this paper.

2. The Structured Light Measurement Model and Exposure Anomaly Problem

2.1. The Structured Light Measuring Model

Figure 1 shows the perspective projection geometric model of the line-structured light profile measurement technology. The line-structured light profile measurement components are composed of the line laser, lens and camera. The line-structured light incident on the measured object surface is modulated into a light stripe reflecting the profile information of the measured object. The laser section image of the measured object is captured by photographing the light stripe. The actual profile of the measured object is calculated based on the pixel coordinates of image light stripe center, pixel coordinates of light stripe center, and system calibration parameters.

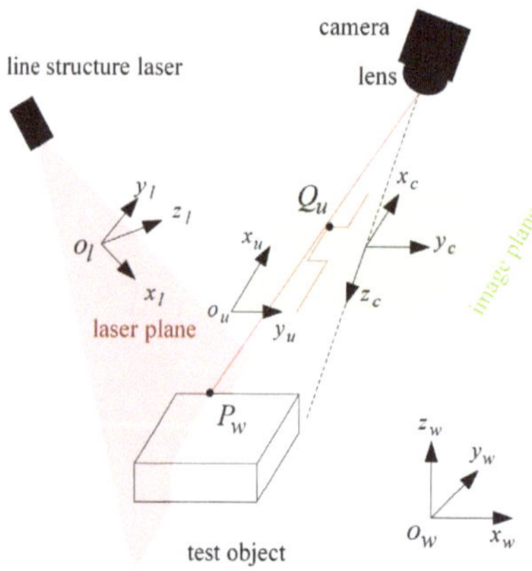

Figure 1. The geometric model of line-structured light perspective projection.

Before the rail profile measurement, the measurement system needs to be calibrated to obtain the internal parameters of cameras on both sides. It is also important to determine the parameters of laser planes on both sides. As shown in Figure 1, $o_w x_w y_w z_w$ is the world coordinate system, $o_c x_c y_c z_c$ represents the camera coordinate system, $o_l x_l y_l z_l$ denotes the laser plane coordinate system, and $o_u x_u y_u$ indicates the pixel coordinate system. $\boldsymbol{P_w} = (x_w, y_w, z_w, 1)^T$ refers to the coordinate (in the world coordinate system) of $\boldsymbol{P_l}$ in the laser plane, and $Q_u = (x_u, y_u, 1)^T$ represents the image point corresponding to $\boldsymbol{P_l}$. The expression based on the pinhole imaging model is as follows.

$$sQ_u = A[R, t]P_w \tag{1}$$

where s is the scale factor, matrix $\boldsymbol{A}$ refers to the internal parameter matrix of the camera, and $[\boldsymbol{R}, \boldsymbol{t}]$ denotes the external parameter matrix of the camera, representing the rotation matrix and translation vector from the world coordinate system to the camera coordinate system respectively. In addition, $\boldsymbol{P_w}$ in the laser plane meets the requirements of the following laser plane equation.

$$ax_w + by_w + cz_w + d = 0 \tag{2}$$

where a, b, c, d represent the laser plane parameters. The mathematical model of line-structured light profile measurement can be obtained through combining Equations (1) and (2).

$$\begin{cases} sQ_u = A[R, t]P_w \\ ax_w + by_w + cz_w + d = 0 \end{cases} \quad (3)$$

Calculate the internal parameter matrix, external parameter matrix and laser plane parameters of the camera according to the method specified in the literature [27] first, then calculate P_w according to the Equation (3), and finally obtain the actual profile of the measured rail.

2.2. Abnormal Exposure Issues

Figure 2 shows the intensity image of the rail laser section captured by the traditional profile measurement method based on an unpolarized camera. The light stripe area of the rail head and the area near the rail gauge point are prone to low contrast and underexposure. This is because of the smooth surfaces and curvature changes in these areas, which result in a lack of diffuse reflection components in the camera. These factors lead to insufficient light reflection, making it difficult for the camera to capture clear images. This issue is highlighted in the dotted rectangular box in Figure 2c. Figure 3 shows the results of light stripe center extraction from the area in the dotted rectangular box in Figure 2. Here, (a), (b), and (c) represent the maximum value algorithm [28], the center-of-gravity algorithm [29], and the Steger algorithm [30] respectively, while (d) indicates the result of mapping the measurement data of the Miniprof profilometer to the image coordinate system. The Miniprof profilometer can achieve high-accuracy measurement because it directly contacts the test object. Its measurement results can be used to compare the effects of different light stripe center extraction algorithms. The comparison results show that in the normal intensity image area of the rail laser section, the rail profiles obtained by different light stripe center extraction algorithms are basically the same as those obtained by the Miniprof profilometer. In the underexposed area of the image, due to weak light stripe energy and low contrast, the light stripe center obtained by any light stripe center extraction algorithm is interrupted to varying degrees. The interrupted light stripe center cannot reflect the actual profile of the rail, resulting in the loss of rail profile data and affecting the rail profile registration. Severe underexposure will lead to the loss of data near the rail gauge point. Since the rail gauge point is a characteristic point for rail wear measurement, rail wear measurement may fail when this characteristic point is missing. The imaging quality and accuracy of traditional rail profile measurement based on intensity information need to be improved. Prolonging the exposure time can alleviate local underexposure. However, it causes overexposure of light stripes in the normal area. This undermines the accuracy of light stripe center positioning in the normal area. Therefore, the problem of local underexposure of rail laser section images cannot be effectively solved only by adjusting the exposure time.

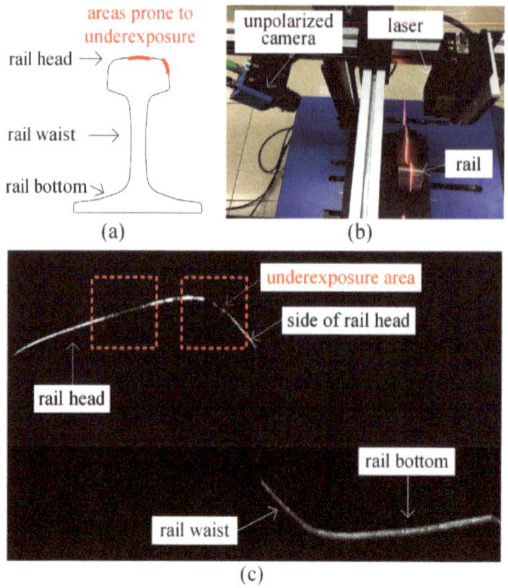

Figure 2. Local underexposure of the rail laser section image. (**a**) Rail areas prone to underexposure; (**b**) image acquisition device; (**c**) local underexposure image.

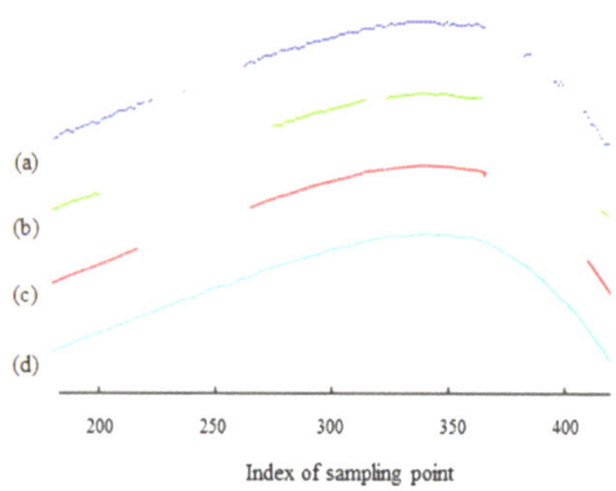

Figure 3. The center-of-light stripe in the underexposed area of the rail laser section image. (a) The maximum value algorithm; (b) the center-of-gravity algorithm; (c) the Steger algorithm; (d) the Miniprof profilometer.

3. The Measurement Principle

3.1. Polarization Imaging

Through polarization optical imaging, multiple intensity images of the measured object in different analyzer directions can be captured. Polarization information of the measured object can also be acquired. In other words, intensity information and polarization information of the measured object can be collected simultaneously. In contrast, traditional imaging technology is mainly used to capture intensity images of the measured object, without polarization information. For convenience, the cameras through which both intensity information and polarization information of the measured object can be acquired are collectively referred to as polarized cameras. Cameras through which only intensity information can be acquired are referred to as unpolarized cameras.

A polarized camera is equipped with four polarization filters in different directions, which can simultaneously capture the polarization component images of the rail laser section from four directions. These images are abbreviated as four-directional polarization component images. Figure 4 shows the polarization filters and pixel distribution of the polarized camera. The four polarization filters of the polarized camera are arranged in a 2 × 2 configuration. Sub-pixels in the 2 × 2 template correspond to nanowire grating polarization filters at 0 degrees, 135 degrees, 45 degrees, and 90 degrees, respectively. The polarized light whose vibration direction is perpendicular to the nanowire grating will pass through the filter, while the polarized light whose vibration direction is parallel to the nanowire grating will be filtered out. All sub-pixels of the 2 × 2 template in the same polarization direction constitute a polarization component image. The gray values of all the sub-pixels at the same position in the 2 × 2 template are extracted to obtain four polarization component images that are 1/2 of original images in width and height. The four images are marked as I_0, I_{45}, I_{90} and I_{135} respectively. According to the Stokes representation method of polarized light, total intensity image I_t is expressed below:

$$I_t = I_0 + I_{90} = I_{45} + I_{135} \tag{4}$$

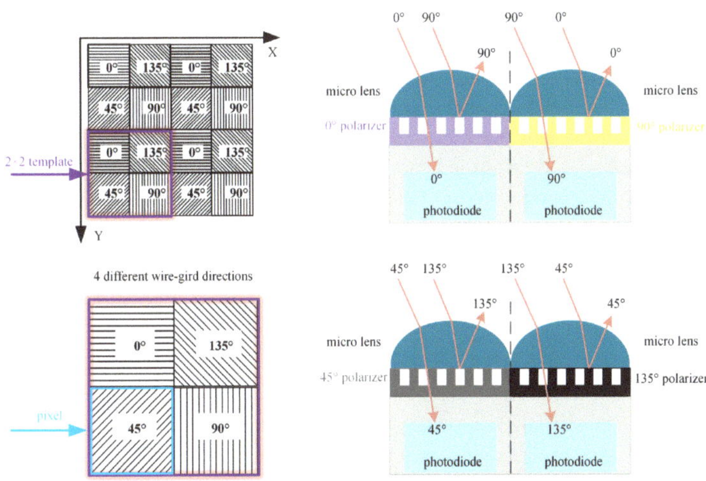

Figure 4. Polarization filters and pixel distribution of polarized camera.

In fact, the total intensity image is an intensity image captured through the traditional method and an ordinary intensity camera. Each pixel in a polarization component image is derived from the same 2 × 2 template. Polarization component images are pixel-aligned. Therefore, the total intensity image, generated by the superposition of polarization component images, is also pixel-aligned with the polarization component images.

According to reference [16], both the normal area and the overexposed area of the light stripe have partially polarized light with the same polarization angle. The interference part of the overexposed area has a high degree of polarization, while the non-interference part of the overexposed area and the normal area have a relatively low degree of polarization. Therefore, the four polarizers of the polarized camera in the orthogonal transmission direction can be used to filter out the interference components with a higher degree of polarization in the overexposed area. However, because the four-directional polarization component images have the effect of polarization filtering, this method cannot solve the problem of local underexposure of the rail laser section image. In this case, the polarization information of the rail laser section needs to be further extracted.

According to electromagnetic theory, light is defined as a transverse wave, and its electric field direction and magnetic field direction are perpendicular to the direction of propagation. In a plane perpendicular to the direction of light propagation, the electric vector may have different vibration states, which are also known as the polarization states of light. According to the polarization state, light can be further divided into natural light, partially polarized light, and fully polarized light. Fully polarized light is further divided into elliptically polarized light, linearly polarized light, and circularly polarized light. The Stokes vector S can be used to describe the polarization state of any light. The relationships between each component of the Stokes vector and the amplitude components E_x, E_y of the light's electric vector and the phase difference are shown in Equation (5).

$$\begin{bmatrix} S_0 \\ S_1 \\ S_2 \\ S_3 \end{bmatrix} = \begin{bmatrix} E_x^2 + E_y^2 \\ E_x^2 - E_y^2 \\ 2E_xE_y\cos\delta \\ 2E_xE_y\sin\delta \end{bmatrix} \tag{5}$$

where S_0 represents the total intensity of light, S_1 denotes the light intensity difference between the linear polarization component of light wave in the x direction and the linear polarization component in the y direction, S_2 refers to the light intensity difference between the linear polarization component of light wave in the 45 degree direction and the linear polarization component in the 135 degree direction, and S_3 represents the light intensity difference between the left-handed circular polarization component and the right-handed circular polarization component. Natural light is usually partially polarized light, while partially polarized light can be regarded as a combination of fully polarized light and natural light. The first three components of the Stokes vector S can be expressed as follows:

$$\begin{bmatrix} S_0 \\ S_1 \\ S_2 \end{bmatrix} = \begin{bmatrix} I_0 + I_{90} \\ I_0 - I_{90} \\ I_{45} - I_{135} \end{bmatrix} \tag{6}$$

The linear polarization degree ***DoLP*** can represent the proportion of linearly polarized light in the partially polarized light. The linear polarization degree is represented in the Equation (7). The linear polarization angle ***AoP*** is the angle between the long axis of polarization ellipse and the x axis, namely the angle between the strongest vibration direction and the x axis. The expression of ***AoP*** is shown in the Equation (8).

$$DoLP = \frac{\sqrt{S_1^2 + S_2^2}}{S_0^2} \tag{7}$$

$$AoP = \frac{1}{2}\tan^{-1}\frac{S_2}{S_1} \tag{8}$$

Polarization information of the measured object mainly involves the linear polarization components in various directions, Stokes vector, linear polarization degree ***DoLP***, and linear polarization angle ***AoP***. Through a traditional camera, only the intensity information of the measured object, namely the first component of the Stokes vector, can be acquired. In contrast, a polarized camera based on the polarization imaging technology makes it possible to obtain all the above polarization information. Therefore, rail profile information collected through a polarized camera is much more than that obtained through a traditional camera. Being capable to collect both polarization information and intensity information, a polarized camera is often used to enhance the contrast and reduce the impact of specular reflection light.

3.2. Polarization Component Images of Rail

The same position of the rail shown in Figure 2 was photographed with a polarized camera, to obtain four polarization component images of the rail laser section, as shown in Figure 5, the red box represents the same position as the red box in Figure 2. Then, the Stokes parameter images S_0, S_1 and S_2, linear polarization degree image I_{DoLP}, and linear polarization angle image I_{AoP} were obtained based on the results of calculation according to the Equations (6), (7) and (8) respectively, as shown in Figures 6 and 7. The red box represents the same position as the red box in Figure 2.

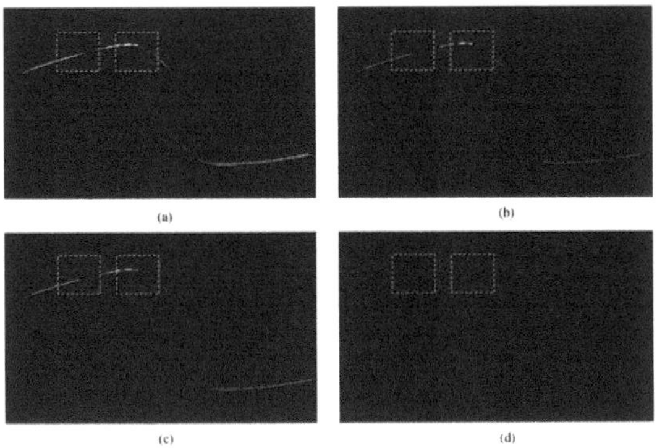

Figure 5. Four-directional polarization component image of the rail laser section. (**a**) 0 degree; (**b**) 135 degree; (**c**) 45 degree; (**d**) 90 degree.

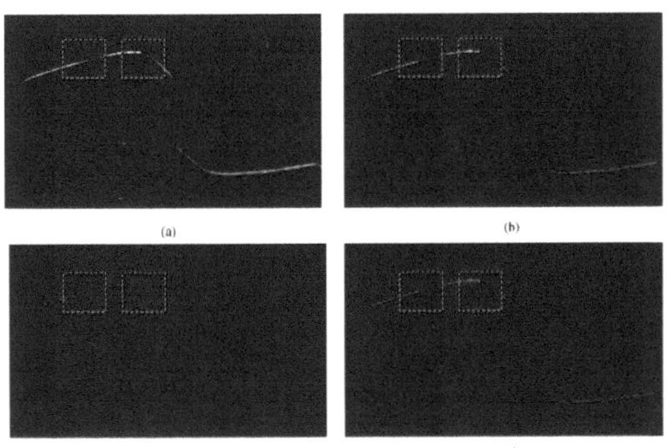

Figure 6. Stokes parameter image and linear polarization angle image. (**a**) S_0; (**b**) S_1; (**c**) S_2; (**d**) I_{AoP}.

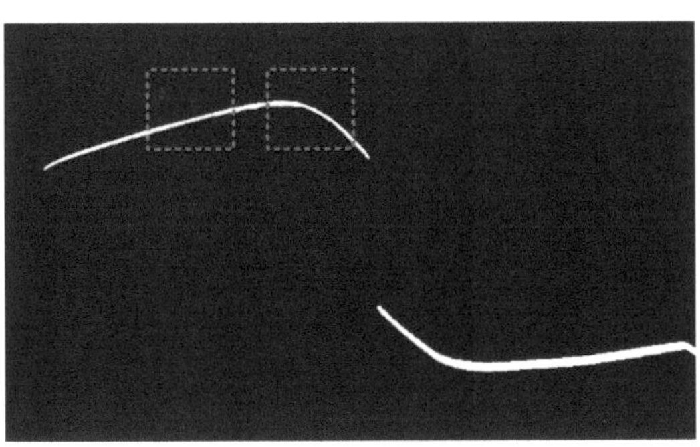

Figure 7. Linear polarization degree image.

Polarization component images are pixel-aligned. Therefore, the synthesized Stokes parameter image, linear polarization degree image, and linear polarization angle image are also pixel-aligned. These images can reflect the rail profile information to a certain extent.

There is a certain information correlation between them. Stokes parameter image S_0 is an intensity image captured through the traditional rail profile measurement method, and is equivalent to that in Figure 2c. The low gray value of the underexposed area of Stokes parameter image S_0 leads to the low confidence of the light stripe center. In contrast, the corresponding area of the linear polarization degree image is characterized by the strong light stripe energy. It also exhibits high imaging contrast and light stripe center confidence. This shows a certain degree of information complementarity.

The four-directional polarization component image, Stokes parameter image, linear polarization degree image, and linear polarization angle image are pixel-aligned with each other. They are correlated and complementary in rail profile information. Therefore, fusion of the aforesaid images to obtain the rail laser section images on the principle of reducing the fusion weight of the underexposed area and improving the fusion weight of the normal area. This approach can better solve the problem of underexposure of light stripe images captured through the traditional rail profile detection technology.

Based on the above analysis results, this paper proposes a rail profile measurement method based on polarization imaging, as shown in Figure 8. Specifically, the laser is equipped with a linear polarizer to obtain the linearly polarized light in the laser plane in the vibration direction, and the polarization direction of the laser is shown in the figure. In addition, a polarized camera is used to capture the polarization component images of the rail laser section from multiple polarization angles. Stokes parameter images S_0, S_1 and S_2, linear polarization degree image I_{DoLP}, and linear polarization angle image I_{AoP} are synthesized from such polarization component images. The aforesaid images are fused through the image fusion algorithm to obtain rail laser section images, which lays a foundation for light stripe center extraction, calibration, profile stitching, profile registration, and final rail profile measurement.

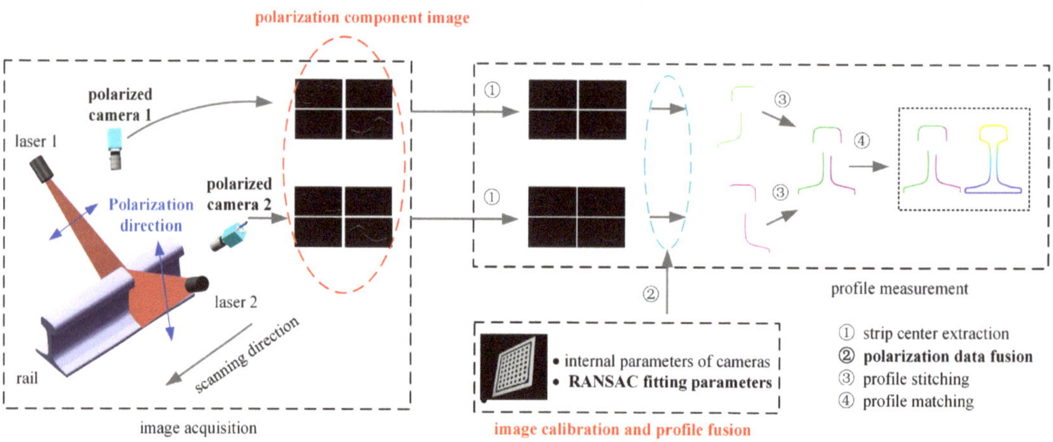

Figure 8. Schematic diagram of the rail profile measurement based on polarization imaging.

3.3. The Polarization Data Fusion Method

The determination of fusion weights is crucial for improving the quality of multi-polarized light image fusion. Wang [21] introduced a light stripe reliability evaluation mechanism to determine the fusion weights of source images. For light stripe reliability evaluation, statistical features such as light stripe width, gray value, and average residual sum of squares were used as evaluation indicators. For each source image, the light stripe reliability in each column needed to be calculated. The total pixel intensity or light stripe width was separately selected as an evaluation index to calculate the reliability of stripe polarization imaging, and the weights of each component image were continuously adjusted to achieve the optimal fusion effect. Additionally, these two evaluation indexes could also be used comprehensively to calculate the light stripe reliability and obtain image fusion weights. This method effectively overcomes the problem of rail surface reflection. However, when applied to the dynamic measurement of the profile system, there are still some issues to be resolved:

(1) The fusion strategy and weight calculation lack systematic quantitative analysis, overly depending on qualitative analysis results and artificial experience thresholds.

(2) Light stripes change during train operation, and unpredictable changes may occur due to factors such as rail wear, sunlight, and foreign matter interference. Thus, using the light stripe width and brightness as criteria for determining fusion weights makes it difficult to adapt to the complex and ever-changing conditions of an entire railway line.

(3) The fusion calculation of multiple polarization images is time-consuming, thus affecting the real-time performance of the measurement system.

To solve the above problems, and a data fusion algorithm for segmented RANSAC polarization point cloud is proposed in this paper, as shown in Figure 9.

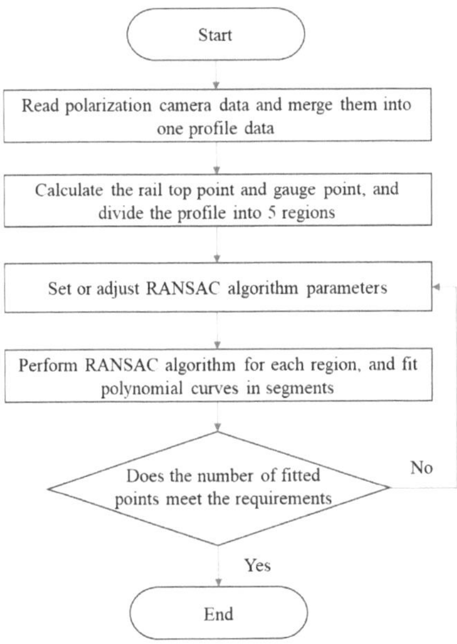

Figure 9. Flowchart of the polarization profile data fusion algorithm based on S-RANSAC.

The algorithm process is described as follows:

(1) For nine polarization component images represented by I_0, I_{45}, I_{90}, I_{135}, I_{AoP}, I_{DoP}, S_0, S_1 and S_2, solve the light stripe centers to obtain the corresponding polarization profile data, denoted as P_0, P_{45}, P_{90}, P_{135}, P_{AoP}, P_{DoP}, S_0', S_1' and S_2';

(2) According to point coordinates, merge nine pieces of polarization profile data into one piece of profile data denoted as

$$P = \cup(P_0, P_{45}, P_{90}, P_{135}, P_{AoP}, P_{DoP}, S_0', S_1', S_2') \tag{9}$$

where $\cup(\cdot)$ represents the fusion of profile coordinate data.

(3) Based on the top and gauge points, divide these profiles into five areas denoted as

$$(P_t, P_{S_1}, P_{S_2}, P_{w_1}, P_{w_2}) = \text{segment}(P, T_{x,y}, G_{x,y}) \tag{10}$$

where segment$(\cdot)$ is a region-based segmentation, $T_{x,y}$ is the coordinate of the rail top, and $G_{x,y}$ is the coordinate of the rail gauge point. The division of the five regions incorporates strong prior information about rail shapes. By utilizing the rail top and gauge points, as well as the rail model, the rails can be precisely divided into the rail top region, rail head transition region, rail head side region, rail web region, and rail bottom region according to Formula (10). Each region possesses a relatively fixed curvature, ensuring consistent reflection of light.

(4) Implement the random sample consensus (RANSAC) for each region to obtain the best-fit polynomial equation, denoted as

$$(C_t, C_{S_1}, C_{S_2}, C_{w_1}, C_{w_2}) = \begin{cases} \text{RANSAC}(P_t, \xi_t, m_t, n_t) \\ \text{RANSAC}(P_{S_1}, \xi_{S_1}, m_{S_1}, n_{S_1}) \\ \text{RANSAC}(P_{S_2}, \xi_{S_2}, m_{S_2}, n_{S_2}) \\ \text{RANSAC}(P_{w_1}, \xi_{w_1}, m_{w_1}, n_{w_1}) \\ \text{RANSAC}(P_{w_2}, \xi_{w_2}, m_{w_2}, n_{w_2}) \end{cases} \tag{11}$$

where $C_t, C_{S_1}, C_{S_2}, C_{w_1}, C_{w_2}$ represent the polynomial curves of RANSAC fitting for five segments; $\xi_t, \xi_{S_1}, \xi_{S_2}, \xi_{w_1}, \xi_{w_2}$ are interior point thresholds, the samples with values less than such thresholds are used for fitting, while other samples with values greater than such thresholds are removed as noise; $m_t, m_{S_1}, m_{S_2}, m_{w_1}, m_{w_2}$ represent the times of sampling iterations; $n_t, n_{S_1}, n_{S_2}, n_{w_1}, n_{w_2}$ represent the optimal polynomial fitting powers for five segments, which can be obtained by using the least square method on the basis of selecting multiple pieces of typical profile data from actual railway line, and constructing a global optimization model.

(5) Stitch the fitting curves $C_t, C_{S_1}, C_{S_2}, C_{w_1}, C_{w_2}$ of the five segments into a complete half section profile of steel rail denoted as $C = \text{stitching}(C_t, C_{S_1}, C_{S_2}, C_{w_1}, C_{w_2})$, where C is the optimal profile fitting curve formed after fusion of multiple polarization point cloud data.

4. Experimental Results

4.1. Laboratory Static Experiments

The rail profile measurement device was constructed, as shown in Figure 10. The rail shown in Figure 2 was placed on the electronic control translation platform. The Genie Nano M2450 polarized camera manufactured by Teledyne Dalsa, a Canadian company headquartered in Waterloo, Ontario, was selected to photograph the same position of the rail in Figure 2. The laser cross-sectional images of the rail were acquired at equal intervals of 2 mm. The main parameters are shown in Table 1.

Table 1. The main parameters in laboratory static experiments.

Index	Item	Parameter
1	Camera	Resolution: 2464 × 056, Pixel Size: 3.45 μm × 3.45 μm
2	Lens	Focus: 12 mm
3	Laser	Power: 500 mw

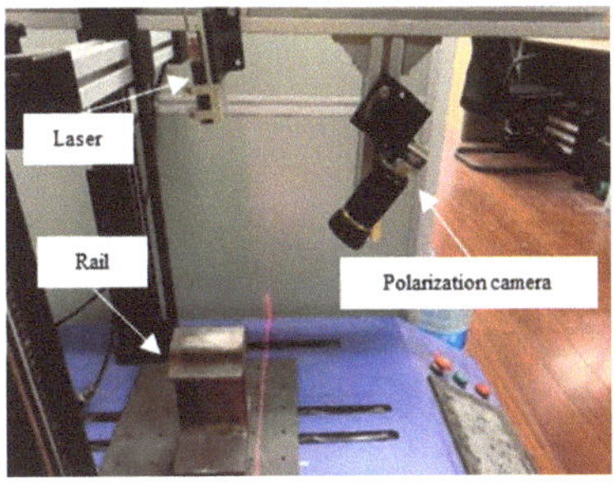

Figure 10. Experimental setup.

This process enabled the acquisition of the four-directional polarization component images of the rail laser section. Then, Stokes parameter images S_0, S_1 and S_2, linear polarization degree image I_{DoLP}, and linear polarization angle image I_{AoP} are calculated and obtained respectively according to the Equations (6), (7) and (8), as shown in Figures 6 and 7. We extracted the light stripe center using the gray centroid method [29]. The proposed S-RANSAC algorithm was then employed to fuse each polarization component, and the outcomes are illustrated in Figure 11. Data 1 to data 5 represent the partition fitting results, which are respectively the rail top region, rail head transition region, rail head side region, rail web region, and rail bottom region. By comparing with Figure 3, it becomes evident that the traditional method incurs data loss in the rail head area. In contrast, the proposed method is capable of obtaining effective contour data in the rail head area.

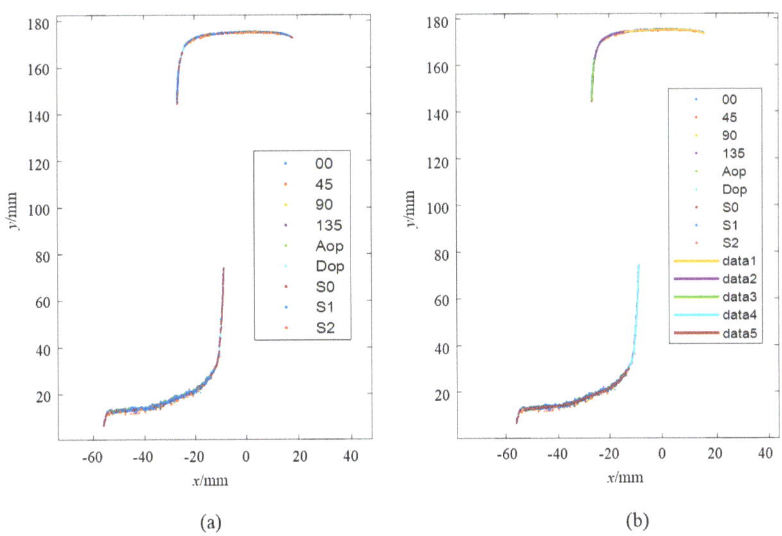

Figure 11. Experimental result of S-RANSAC: (**a**) fused point cloud points; (**b**) segmented polyfit result.

To further verify the effectiveness of the proposed method for the three-dimensional reconstruction of rails, the proposed method was utilized to perform 3D reconstruction of the measured rail. Meanwhile, the profile of the rails obtained from the Stokes parameter image S_0 was regarded as the measurement result of the traditional method. Figure 12 presents the 3D reconstruction results of the steel rails using the traditional method and the proposed method respectively. It can be clearly observed that the traditional method led to partial data loss in the 3D-reconstructed steel rails due to abnormal exposure. Conversely, the proposed method did not exhibit any data loss in the 3D-reconstructed steel rails, and the reconstruction results still accurately reflected the true condition of the steel rails.

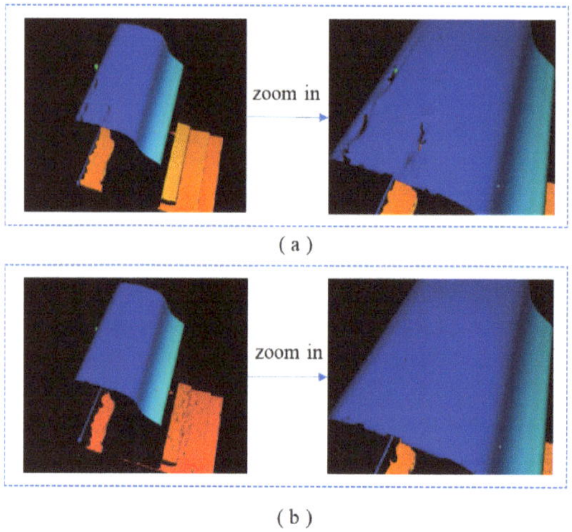

Figure 12. 3D reconstruction of rail: (**a**) the traditional method; (**b**) the proposed method.

4.2. The On-Site Dynamic Test

The on-site dynamic test was carried out near Qinghecheng Station at K 346 on the Beijing–Kowloon Line. The Beijing–Kowloon Line is a trunk railway with shared passenger and freight traffic. Due to its high traffic volume and diverse cargo types, the on-site operational conditions are relatively complex: the rail surface is often relatively shiny and mixed with foreign objects, which affects image quality. The profiles of the tested railway line include situations such as web burial of the rail, ambient light interference, and the polished rails. Miniprof, a contact-type rail profile measurement device with a measurement accuracy of 0.02 mm, was used as a reference. The rail profile obtained from the Stokes parameter image was considered as the measurement result of the traditional method. Given the difference in the number of points collected by the measurement system and the number of points obtained by Miniprof, it was necessary to perform smoothing processing before discretizing the profile. This ensured that corresponding point pairs could be compared, thereby enabling a genuine evaluation of the profile difference. Figure 13 shows a comparison diagram of the rail profile at one typical position. The figure contains two enlarged areas, namely the underexposed area at the top of the rail (in the upper left corner of the figure) and the overexposed area on the side of the rail head (on the right side of the figure). In the underexposed area, partial data loss occurred (the red dots are interrupted) when using the traditional method. However, since the proposed method acquires more polarization information through the polarization imaging technology and adopts a fusion method based on divided regions, continuous and valid contour data can still be obtained in this area, and there is a high degree of consistency with the contour data obtained by Miniprof. Similarly, in the overexposed area, data distortion and deformation occurred (the red dots deviate significantly from the green dots) with the traditional method. Nevertheless, the contour data obtained by the proposed method still show a high degree of consistency with that obtained by Miniprof. Thus, it can be demonstrated that the proposed method can simultaneously address the issues of both underexposure and overexposure.

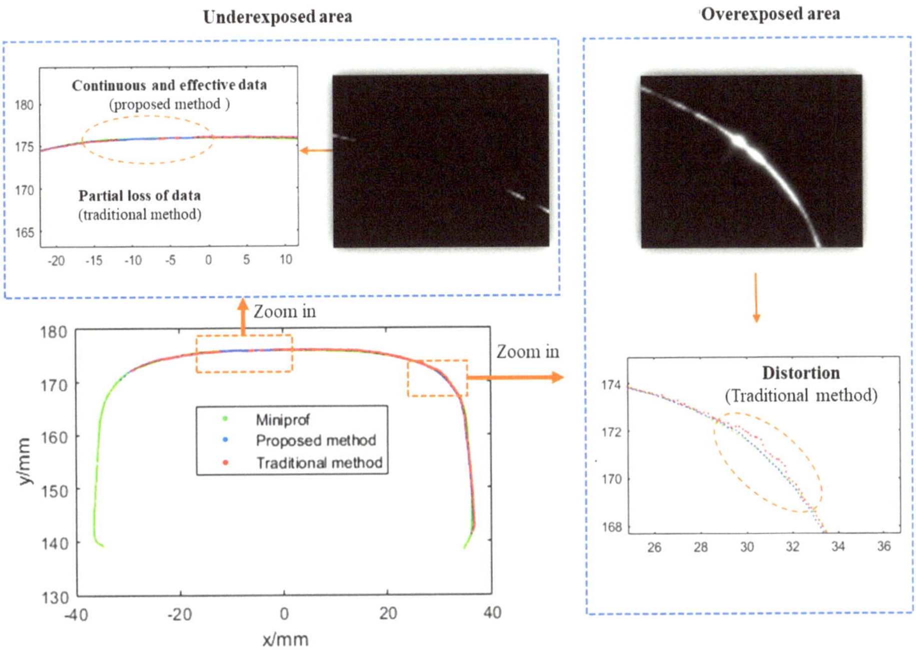

Figure 13. Comparison of rail profile measurement results between the proposed and traditional methods.

To quantitatively compare the accuracy of the rail profile measurement of the proposed method, 100 sets of rail profile data were collected using Miniprof. For each sampling point, the Miniprof measurement data were taken as the true value, and the differences between the rail profile data obtained by the traditional method and the proposed method and the Miniprof data were statistically analyzed. The statistical parameters included the maximum error (**ME**), average error (**AE**), and 95th percentile error (**PE**). The results are presented in Figure 14, and Table 2 shows the average values of the three statistical measures mentioned above. Compared with the traditional method, the maximum, average, and 95th percentile values of the rail profile measurement errors have all decreased to varying degrees. Taking the 95th percentile value of the rail profile measurement error as an example, it has significantly declined from 0.137 mm, as recorded in conventional methodologies, to 0.081 mm. This represents a remarkable reduction of 40.9%. This clearly demonstrates the enhanced precision of the rail profile data obtained through the proposed method, which now more accurately represents the true profile of the rail.

Table 2. Statistical results of 100 sampling points (mm).

Measurement Error	Max Value	Average Value	95th Percentile Value
Traditional method	0.254	0.084	0.137
Proposed method	0.092	0.042	0.081

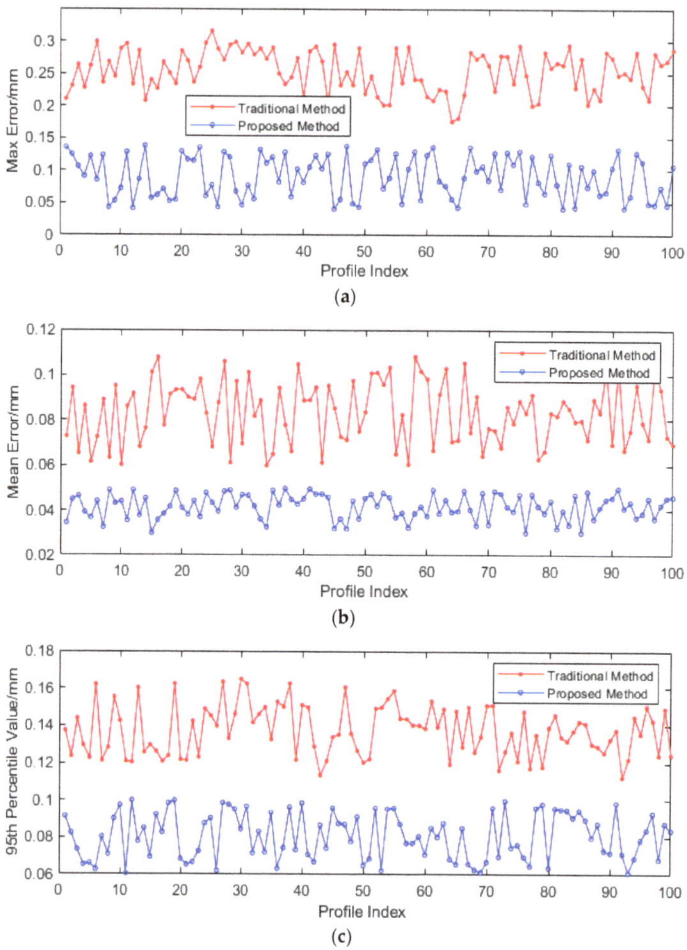

Figure 14. The measurement error of rail profile between the proposed and traditional methods. (a) **ME**, (b) **AE**, and (c) **PE**.

To verify the execution efficiency of the S-RANSAC algorithm, typical profile data were selected for testing. The computer CPU parameters were Intel(R) Core(TM) i7-10700K CPU @ 3.80 GHz, and the programming language used was C++ (version 20). To improve the efficiency of algorithm execution, a programming architecture for data concurrency was designed. The 4000 sets of profile data collected per kilometer were divided into 10 concurrent queues, with each queue buffering 400 sets of profile data. The relationship between algorithm execution efficiency and data size is shown in Figure 15. Due to the randomness of RANSAC, the time consumption may vary slightly each time, but it can process approximately 200 m of data per second.

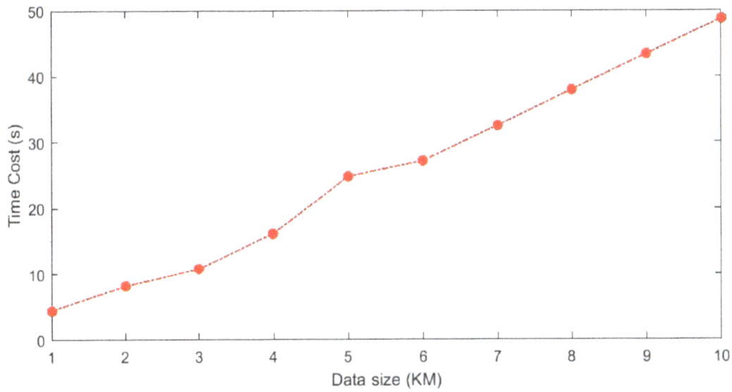

Figure 15. The relationship between algorithm execution efficiency and data size.

4.3. Comparison with Other Research Methods

To evaluate the performance of the proposed method, 550 sets of polarization component images of steel rails were randomly selected, and the proposed method, along with the methods in references Huang [31], Hayat [32], Qu [33], and Wang [34], were used for fusion analysis. **SSIM** and **PSNR** are the most commonly used indicators for evaluating multi-exposure fusion algorithms in dynamic scenes. The **PSNR** indicator measures the similarity between the fused image and the source images in terms of image gray levels. A larger **PSNR** means that the fused image is close to the source images and has less distortion. Therefore, the larger the **PSNR** value, the better the fusion performance. **SSIM** is used to model image losses and distortions, to which the human visual system is sensitive. It consists of three parts, namely correlation, luminance, and contrast distortion. The **SSIM** between the source image and the fused image is defined as the product of these three parts. **SSIM** reflects the degree to which the fused image preserves the local structural details of the input images. The larger the value of this indicator, the greater the degree of structural preservation, and the better the fusion effect. **ET** represents the algorithm execution time. As shown in Figures 16–18, it can be observed that the proposed method achieved the highest scores in both **PSNR** and **SSIM**, which were 37.57 and 0.99 respectively. Compared with the **PSNR** of 36.44 and the **SSIM** of 0.928 of the method proposed by Qu, who ranked second, and the execution time of 536.84 ms, the proposed method is not only better in maintaining data quality and structural integrity, but also more efficient in algorithm execution, with the time being only 175.86 ms, reflecting its efficient processing ability and good fusion effect.

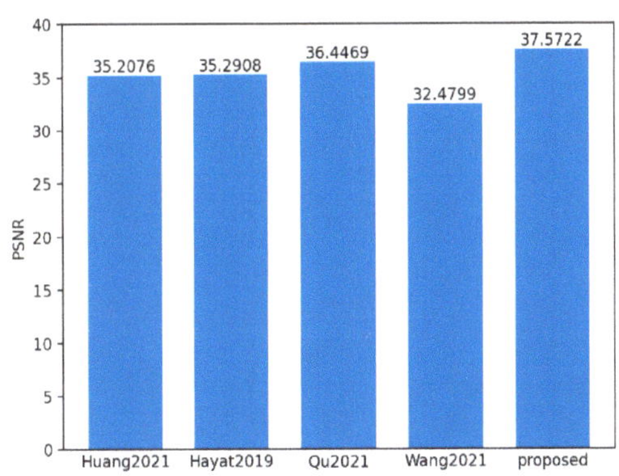

Figure 16. Mean value histogram of **PSNR** indicators obtained by the five fusion methods.

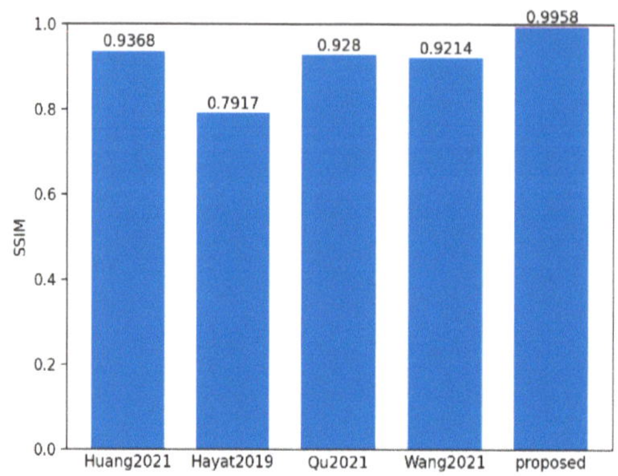

Figure 17. Mean value histogram of **SSIM** indicators obtained by the five fusion methods.

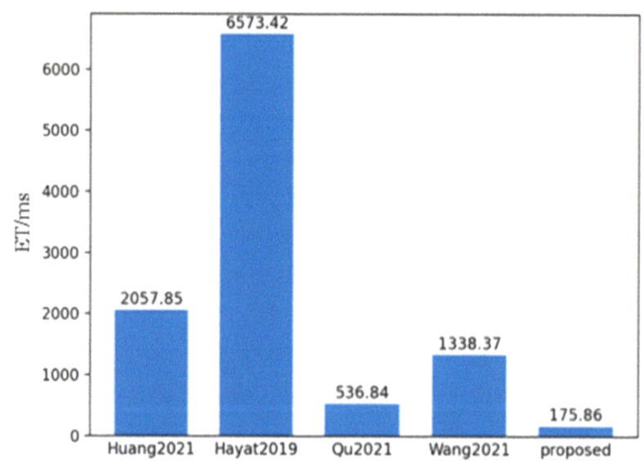

Figure 18. Mean value histogram of algorithm time obtained by the five fusion methods.

On this basis, we compared the proposed S-RANSAC method with the above-mentioned mainstream fusion methods before and after the sub-area division. The experimental results are shown in Table 3. In line with Table 3, S-RANSAC has a more competitive performance under **PSNR**, **SSIM** and time efficiency. In consequence, it is selected as the for polarized component fusion task.

Table 3. Comparisons on polarization fusion algorithms.

Algorithm	PSNR (Single/Multi-Region)	SSIM (Single/Multi-Region)	Time (Single/Multi-Region)
Huang	34.21/35.15	0.94/0.98	2057.85 ms/3125.14 ms
Hayat	34.29/34.98	0.79/0.91	6573.42 ms/7402.27 ms
Qu	36.45/37.05	0.93/0.97	536.84 ms/1026.33 ms
Wang	32.48/33.93	0.9210.97	1338.37 ms/1579.43 ms
S-RANSAC	36.57/37.91	0.98/0.99	175.86 ms/341.10 ms

5. Conclusions

A novel rail profile measurement method founded on multi-polarization fusion has been presented to resolve the issue of insufficient local exposure in laser cross-section images, which is a common hurdle in traditional rail profile measurement techniques. This advance involves the creation of a profile data fusion algorithm that utilizes the S-RANSAC algorithm, specifically designed for four-directional polarization component images, Stokes parameter images, linear polarization angle images, and linear polarization degree images. This approach effectively alleviates the problem of local underexposure in the laser cross-section of steel rails, securing a comprehensive and accurate depiction of the rail profile. Following three-dimensional reconstruction, the method guarantees that the steel rail no longer suffers from data loss, which is a crucial improvement over traditional methods. This innovation surmounts the exposure insufficiency in key areas of laser cross-section images of steel rails, which can significantly influence the extraction of light strip centers. By preserving the integrity of profile data in critical areas, the method boosts the accuracy and stability of profile detection under complex working conditions. This not only ensures the effectiveness of profile analysis, comparison, and evaluation but also facilitates the expansion of rail profile detection application scenarios.

Future research will probe into alternative image fusion methods, such as frequency-domain fusion or deep-learning-based fusion, with the objective of further enhancing algorithm efficiency and robustness. This could potentially result in more accurate and reliable rail profile measurements, even in the most demanding operating environments. Additionally, efforts will be made to optimize the existing S-RANSAC algorithm to reduce its computational complexity and improve its real-time performance, making it more suitable for practical applications in railway infrastructure inspection. Through these continuous improvements, the proposed method is expected to play an increasingly significant role in ensuring the safety and reliability of railway transportation systems.

Author Contributions: Conceptualization, Q.H. and S.W.; methodology, Q.H. and L.W.; software, X.Z.; validation, Q.H., N.W. and X.Z.; formal analysis, L.W.; investigation, P.D.; resources, Q.H. and J.S.; data curation, X.Z.; writing—original draft preparation, Q.H., X.Z. and L.W.; writing—review and editing, S.W.; visualization, L.W.; supervision, S.W.; project administration, P.D.; funding acquisition, Q.H. All authors have read and agreed to the published version of the manuscript.

Funding: Research and Development Plan of China Academy of Railway Sciences Co., Ltd. (2023YJ041); Science and Technology Research and Development Plan of China National Railway Group Co., Ltd. (K2022T006).

Institutional Review Board Statement: Not applicable.

Informed Consent Statement: Not applicable.

Data Availability Statement: Data underlying the results presented in this paper are not publicly available at this time but may be obtained from the authors upon reasonable request.

Conflicts of Interest: Authors Qiang Han, Xinxin Zhao, Jing Shi, Shengchun Wang, Peng Dai, Ning Wang and Le Wang were employed by the company China Academy of Railway Sciences Corporation Limited. The remaining authors declare that the research was conducted in the absence of any commercial or financial relationships that could be construed as a potential conflict of interest.

References

1. Shang, D.; Su, S.; Sun, Y.; Wang, F.; Cao, Y.; Liu, H.; Yang, W.; Zhou, J. Fault Diagnosis for Rail Profile Data Using Refined Dispersion Entropy and Dependence Measurements. *IEEE Trans. Intell. Transp. Syst.* **2024**, *25*, 1689–11700. [CrossRef]
2. Xiong, Z.; Li, Q.; Mao, Q.; Zou, Q. A 3D Laser Profiling System for Rail Surface Defect Detection. *Sensors* **2017**, *17*, 1791. [CrossRef] [PubMed]
3. Zhang, J.; Yao, X.; Xing, Z.; Cai, C.; Chen, Y. A Multiline Structured Light Sensor-based System for Measuring the Geometries of Train Wheelsets. *IEEE Sens. J.* **2024**, *24*, 24609–24621. [CrossRef]
4. Li, Y.; Yang, X.; Chen, L.; Zhi, Y.; Liu, H. Robust Registration of Rail Profile and Complete Detection of Outliers in Complex Field Environment. *IEEE Trans. Intell. Transp. Syst.* **2022**, *23*, 20098–20109. [CrossRef]

5. Gazafrudi, S.; Younesian, D.; Torabi, M. A High Accuracy and High Speed Imaging and Measurement System for Rail Corrugatio Inspection. *IEEE Trans. Ind. Electron.* **2020**, *68*, 8894–8903. [CrossRef]
6. Millara, F.; Molleda, J. Profile Measurement of the Rails in a Rolling Mill: Implementing and Evaluating Autonomic Computin Capabilities. *IEEE Trans. Ind. Appl.* **2019**, *55*, 5466–5475. [CrossRef]
7. Zhang, Z.; Zhang, H.; Wu, Y.; Zhang, F. Simple and Precise Calibration of the Line-Structured Light Vision System using a Plana Target. *J. Opt. Soc. Am. A* **2023**, *40*, 1397–1408. [CrossRef]
8. Liu, J.; Zhang, J.; Ma, Z.; Zhang, H.; Zhang, S. Three-Dimensional Measurement of Full Profile of Steel Rail Cross-Section Base on Line-Structured Light. *Electronics* **2023**, *12*, 3194. [CrossRef]
9. Zhang, D.; Lingamanaik, S.; Chung, H. Image-based 3D Reconstruction for Rail Profile Measurement. *Proc. Inst. Mech. Eng. Par F J. Rail Rapid Transit* **2023**, *237*, 309–321. [CrossRef]
10. Ye, J.; Stewart, E.; Zhang, D.; Chen, Q.; Thangaraj, K.; Roberts, C. Integration of Multiple Sensors for Noncontact Rail Profile Measurement and Inspection. *IEEE Trans. Instrum. Meas.* **2021**, *70*, 1–12. [CrossRef]
11. Hauck, J.; Gniado, P. Laser Scan Compression for Rail Inspection. *Sensors* **2024**, *24*, 6722. [CrossRef] [PubMed]
12. Zheng, Q.; Shi, H.; Zhu, P.; Liu, Q.; Ma, W.; Li, X. Centerline Extraction of Structured Light Stripe based on Principal Componen Analysis. *Univ. Politeh. Buchar. Sci. Bull. Ser. C-Electr. Eng. Comput. Sci.* **2023**, *85*, 91–104.
13. Huang, Y.; Kang, W.; Lu, Z. Improved Structured Light Centerline Extraction Algorithm Based on Unilateral Tracing. *Photonics* **2024**, *11*, 723. [CrossRef]
14. Li, S.; Gao, X.; Xie, Z. Underwater Structured Light Stripe Center Extraction with Normalized Grayscale Gravity Method. *Sensors* **2023**, *23*, 9839. [CrossRef]
15. Liu, F.; He, C.; Shen, A.; Li, J.; Lai, J. Optimized Compensation Method of Divisional Projection for Saturated Region of Structured Light. *Acta Opt. Sin.* **2018**, *38*, 0612001.
16. Yang, P.B.; Deng, L.J.; Chen, Y. Three-Dimensional Shape Measurement of Highly Reflective Objects based on Structured Light *Chin. J. Laser* **2019**, *46*, 0204004. [CrossRef]
17. Wolff, B. Polarized Camera for Computer Vision with a Beam Splitter. *J. Opt. Soc. Am.* **1994**, *11*, 2935–2945. [CrossRef]
18. Li, H.; Zhu, J.; Deng, J.; Guo, F.; Zhang, N.; Sun, J.; Hou, X. Underwater Active Polarization Descattering based on a Single Polarized Image. *Opt. Express* **2023**, *31*, 21988–22000. [CrossRef]
19. Mo, S.; Duan, J.; Zhang, W.; Wang, X.; Liu, J.; Jiang, X. Multi-Angle Orthogonal Differential Polarization Characteristics and Application in Polarization Image Fusion. *Appl. Opt.* **2022**, *61*, 9737–9748. [CrossRef]
20. Umeyama, S.; Godin, G. Separation of Diffuse and Specular Components of Surface Reflection by Use of Polarization and Statistical Analysis of Images. *IEEE Trans. Pattern Anal. Mach. Intell.* **2004**, *26*, 639–647. [CrossRef]
21. Wang, L.; Fang, Y.; Wang, S.C.; Wang, H.; Ren, S.W.; Li, H.Q.; Wang, F. Line-Structured Light Imaging Method of the rail Profile Based on Polarization Fusion. *Acta Opt. Sin.* **2020**, *40*, 2211001. [CrossRef]
22. Zhang, J.; Fang, Y. Algorithm and Evaluation for Polarization Image Fusion based on Edge Information. *Opto-Electron. Eng.* **2007**, *34*, 78–81.
23. Qiao, J. Research on Image Fusion Technology Based on Polarization. Ph.D. Thesis, Changchun University of Technology, Changchun, China, 2017.
24. Yin, H.; Hu, L.; Fan, Z. Fusion Method for Polarization Images based on Analysis of Features. *J. Appl. Opt.* **2015**, *36*, 220–226.
25. Lin, B.; Fan, X.; Peng, P.; Guo, Z. Dynamic Polarization Fusion Network (DPFN) for Imaging in Different Scattering Systems. *Opt. Express* **2024**, *32*, 511–525. [CrossRef]
26. Liu, J.; Duan, J.; Hao, Y.; Chen, G.; Zhang, H. Semantic-Guided Polarization Image Fusion Method based on a Dual-Discriminator GAN. *Opt. Express* **2022**, *30*, 43601–43621. [CrossRef]
27. Li, W.; Wang, P.; Chen, Y. Calibration of Binocular Vision Measurement System by Line Structured Light for Rail Full Profile. *Chin. J. Sci. Instrum.* **2019**, *40*, 204–211.
28. Wan, Z.; Lai, L.; Mao, J.; Zhu, L. Extraction and Segmentation Method of Laser Stripe in Linear Structured Light Scanner. *Opt. Eng.* **2021**, *60*, 046104. [CrossRef]
29. Wang, J.; Wu, J.; Jiao, X.; Ding, Y. Research on the Center Extraction Algorithm of Structured Light Fringe based on an Improved Gray Gravity Center Method. *J. Intell. Syst.* **2023**, *32*, 1597–1603. [CrossRef]
30. Steger, C. An Unbiased Detector of Curvilinear Structures. *IEEE Trans. Pattern Anal. Mach. Intell.* **1998**, *20*, 113–125. [CrossRef]
31. Huang, L.; Li, Z.; Xu, C.; Feng, B. Multi-Exposure Image Fusion based on Feature Evaluation with Adaptive Factor. *IET Image Process.* **2021**, *15*, 3211–3220. [CrossRef]
32. Hayat, N.; Imran, M. Multi-Exposure Image Fusion Technique using Multi-Resolution Blending. *IET Image Process.* **2019**, *13*, 2554–2561. [CrossRef]

33. Qu, Z.; Huang, X.; Liu, L. An Improved Algorithm of Multi-Exposure Image Fusion by Detail Enhancement. *Multimed. Syst.* **2021**, *27*, 33–44. [CrossRef]
34. Wang, Q.; Chen, W.; Wu, X.; Li, Z. Detail-Enhanced Multi-Scale Exposure Fusion in YUV Color Space. *IEEE Trans. Circuits Syst. Video Technol.* **2019**, *30*, 2418–2429. [CrossRef]

sensors

MDPI

Article

High-Precision Measurement Method for Small Angles Based on the Defect Spot Mode of the Position-Sensitive Detector

Yusheng Zhai [1,*], Guorong Wang [1], Yiheng Zhao [1], Rongxin Wu [1], Lin Zhang [2], Zhan Su [3], Zhifeng Zhang [1,*], Peng Yang [1] and Ruiliang Zhang [1]

[1] School of Electronics and Information, Zhengzhou University of Light Industry, Zhengzhou 450000, China; 332211030780@zzuli.edu.cn (G.W.); 332011030614@zzuli.edu.cn (Y.Z.); 332311030847@zzuli.edu.cn (R.W.); 2013005@zzuli.edu.cn (P.Y.); 2017073@zzuli.edu.cn (R.Z.)

[2] White Pigeon Abrasive Tools Co., Ltd., Zhengzhou 450000, China; zl8082zl@163.com

[3] Henan Academy of Sciences Institute of Applied Physics Co., Ltd., Zhengzhou 450000, China; suzhan0813@163.com

* Correspondence: 2013001@zzuli.edu.cn (Y.Z.); 2009041@zzuli.edu.cn (Z.Z.); Tel.: +86-13503839747 (Y.Z.); +86-13838318673 (Z.Z.)

Abstract: The paper proposes and verifies a small-angle measurement method based on the defect spot mode of the position-sensitive detector (PSD). With the output characteristics of the PSD in the defect spot mode and the size transformation properties of a focused beam, the measurement sensitivity can be significantly improved. Calibration experiments with the piezoelectric transducer (PZT) indicate that compared with the current PSD-based autocollimation method, the proposed method can improve the sensitivity of small-angle measurement by 57 times, and the measurement sensitivity of the proposed method can be further improved by optimizing the system parameters, while the proposed method has the advantages of a simple system and high real-time performance. Therefore, the proposed method is expected to be used in high-precision motion error detection, as well as in shape and position measurement.

Keywords: photoelectric technology; angle measurement; PSD; defect spot working mode

1. Introduction

High-precision angle measurement is important in many fields, such as motion error detection, shape and position measurement, etc. Due to their non-contact, real-time applicability, and high sensitivity, many optical methods have been proposed and applied. Among them, the methods based on the polarization characteristics of light [1,2], the Sagnac effect [3,4], and vortex beam transformation [5] have a larger measurement range, while the methods based on internal reflection effect [6,7], laser interference [8,9], and autocollimation [10–21] have higher sensitivity in small-angle measurement.

The sensitivity of the internal reflection methods can reach 0.015 μrad. Such methods are based on polarized light intensity or phase detection, which is easily disturbed by factors such as the change in polarization state and stray light. The sensitivity of the laser interferometer can reach 0.03 μrad (Renishaw Corp., London, UK, XM-60). The interferometry methods represented by a laser interferometer are based on phase detection; the change in light intensity does not influence measurement results, but the influences of environmental stability need to be considered, and the interferometry system is complex and expensive. The autocollimator is the most widely used high-precision angle measuring instrument at present. Laser collimation measurement methods represented by the autocollimator are based on the detection of the spatial position of the beam. The accuracy of this kind of method is mainly determined by the spatial stability of the beam and the sensitivity of the position detection device. There are many types of position-sensitive detection devices, including the PSD, the quadrant detector (QD), and other photodiode

(PD) combinations based on photocurrent detection, as well as the CCD and the CMOS based on image recognition. The sensitivity of the current PSD-based photoelectric autocollimator is approximately 0.05 μrad [10,11]. The method proposed by Gao W.'s research group, which used QD or PD photosensitive zone edges for focused spot position detection, can further improve the measurement sensitivity, reaching 0.005 μrad. The principle is to use the edge of the photosensitive area of a PD to convert the displacement of the focused spot into the energy of the partial spot entering its photosensitive area and combine it with the energy of the complete spot measured by another PD to convert it into the degree of spot defect. The way to improve the sensitivity of this method is to reduce the size of the focused spot to increase the degree of spot defect. Obviously, in order to further improve sensitivity, obtaining a smaller focused spot that breaks through the diffraction limit is a problem that needs to be solved. Meanwhile, directly utilizing the edge area of the detector for energy detection is an unconventional use of the detection device, and the performance of the edge area of the device needs to be considered [12,13]. The sensitivity of the CCD or CMOS-based digital autocollimator can even reach 0.0005 μrad (Taylor Hobson Corp., Leicester, UK, Ultra HP 142-204); however, the high sensitivity of the digital autocollimator is mainly due to the image processing algorithm [14–19]; therefore, the system is relatively complex, and the real-time responsiveness needs to be considered.

In summary, for high-precision measurement of small angles, when comprehensively considering the measurement sensitivity, system structure, and cost, laser collimation measurement methods have more advantages. The further improvement of the sensitivity of this kind of method is mainly limited by the sensitivity of the position detection device. Although the sensitivity of the digital autocollimator is greatly improved by the image processing algorithm, the system based on image processing is relatively complex, and the real-time response is reduced.

In this study, a simple and high-precision small-angle measurement method is proposed. For the first time, the innovative PSD in defect spot working mode has been used for angle measurement, achieving ultra-high sensitivity improvement through a simple projection relationship. The proposed method is an almost entirely analog system, which has natural advantages in system complexity and real-time responsiveness compared to digital autocollimators. Meanwhile, compared with the current PSD-based autocollimation methods, the measurement sensitivity of the proposed method improved by 57 times. Compared with the method of using the detector edge to detect the degree of spot defect [12,13], the proposed method is more flexible, easier to generalize, and has greater potential for sensitivity improvement.

2. Detection Principle

2.1. The Conventional PSD-Based Autocollimation Methods and the Defect Spot Working Mode of the PSD

The PSD can identify the position of the spot energy center in its photosensitive area. In the conventional working mode of the PSD, the complete light spot moves on the photosensitive area of the PSD; the corresponding displacement of the complete spot energy center is output linearly by the PSD [10,11].

The autocollimation method based on the normal working mode of the PSD is shown in Figure 1. The collimated beam is incident on the mirror M. For any change in the angle of M, the beam deflection will be twice the rotation angle of the M, and the focused spot of the reflected light on the PSD moves accordingly. The angular change α of M can be expressed as follows:

$$\alpha = -\Delta X_a/2f = k_1\Delta X_a, \tag{1}$$

where f is the focal length of the lens, ΔX_a denotes the displacement of the light spot on the PSD caused by α, and k_1 represents the constant determined by system parameters.

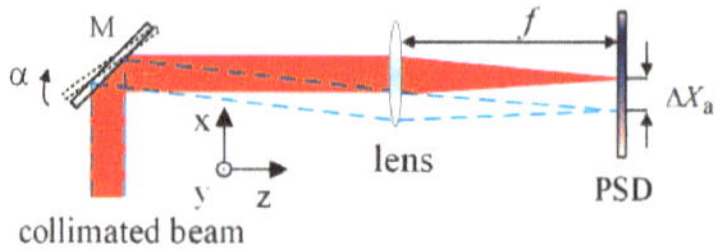

Figure 1. Measuring optical path of autocollimator based on conventional working mode of the PSD.

The defect of the spot can also lead to the change of the spot energy center, but there are few related studies [12,22]. An innovative defect working mode [23] of the PSD is shown in Figure 2. The spot received by the PSD is a circular Gaussian spot, which is the most common shape and pattern of light spots, and its center is located at the center of the PSD. When the beam is occluded, the light spot is no longer complete; the energy center of the defect spot output from PSD changes according to the defect of light spot.

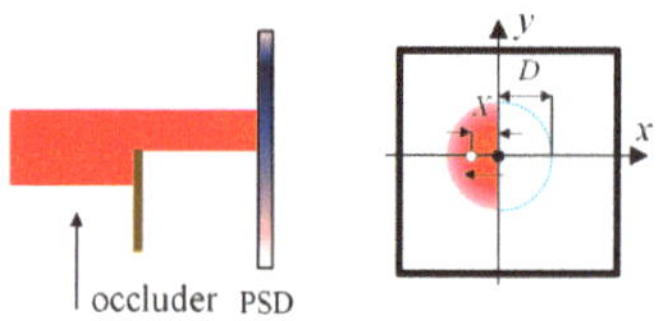

Figure 2. The defect spot working mode of the PSD. The black and white circles represent the spot center before and after the change of the measured beam, respectively.

As shown in Figure 2, assuming that the defect is along the x direction, and the width of the defect is D, and the spot energy center positions is X. The spot energy center position output characteristics of the PSD in the defect spot mode are shown in Figure 3.

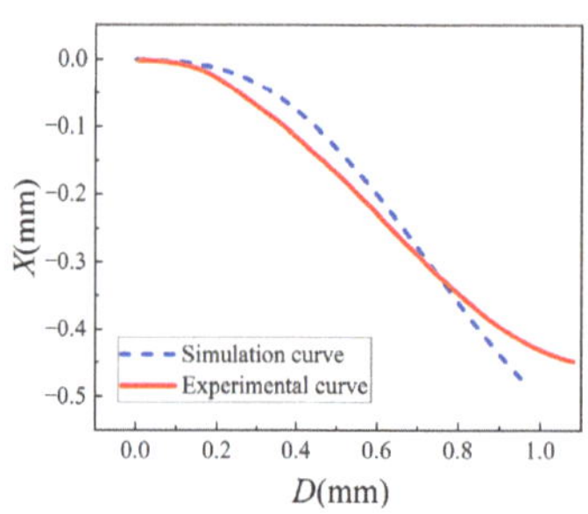

Figure 3. Spot energy center position output characteristics of the PSD in the defect spot mode.

It can be observed from Figure 3 that in the middle of the curve, the PSD is more sensitive, and the output is approximately linear, although for light-displacement detection, the sensitivity in the defect spot mode is lower than that in conventional working mode of the PSD shown in Figure 1. However, the defect spot mode of the PSD, combined with certain optical structures, can provide new features, which will be discussed in the next section.

2.2. Small-Angle Measurement Principle Based on the Defect Spot Mode of the PSD

We propose a high-precision small-angle measurement method based on the defect spot mode of the PSD. The optical path structure is shown in Figure 4. A single-mode fiber-coupled semiconductor laser was used as the system light source, and the beam was collimated using a collimating objective (CO). The collimated beam was reflected by the mirror (target to be detected) mounted on the piezoelectric transducer (PZT) rotation stage and focused by a lens with focal length f. A rectangular plate (RP) that partially (about half) occluded the beam was placed near the focus point of the lens. The PSD was placed behind the focal point of the lens, and the direction in which the beam was occluded was consistent with the x-axis direction of the PSD.

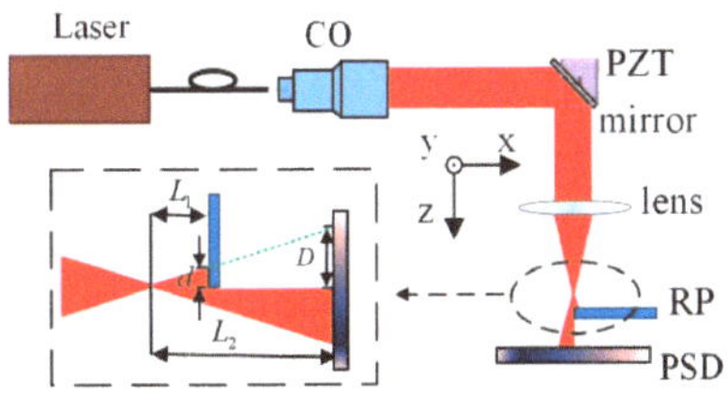

Figure 4. Optical path of high-precision small-angle measurement method based on defect spot mode of the PSD.

Assuming that the distance between RP and focal point of the lens is L_1, and the distance between the PSD and focal point of the lens is L_2, the relationship between the defect width D of the spot on the PSD and the defect width d of the beam occluded by RP can be expressed as follows:

$$D = \left(\frac{L_2}{L_1}\right)d. \tag{2}$$

When the mirror rotates around the y-axis by an angle α, the focused spot moves along the x-axis. According to the optical characteristics of the lens, as shown in Figure 1, the collimated beam reflected by the mirror is focused on the intersection of the main ray and the focal plane. The overall light spot displacement of the beam on the RP plane can be expressed by the corresponding main ray displacement Δd. So, the relationship between Δd and α can be expressed as follows:

$$\Delta d = 2(f + L_1)\alpha. \tag{3}$$

From Equation (2), the size of the spot defect on PSD changed according to α.

$$\Delta D = (L_2/L_1)2(f + L_1)\alpha \tag{4}$$

It is known from Figure 3 that PSD has approximately linear output characteristics within a certain range near half defect point; therefore, the change of the spot position output from PSD caused by the spot defect can be expressed as follows:

$$\Delta X(D) = X'(D)\left(\frac{L_2}{L_1}\right)2(f + L_1)\alpha. \tag{5}$$

In practice, the spot displacement output from the PSD should be expressed as follows:

$$S = \Delta X(D) + \Delta X_a, \tag{6}$$

where ΔX_a is the same as defined in Equation (1). From Figure 1, the overall light spot displacement of the beam on the PSD can be expressed by the corresponding main ray displacement ΔX_α:

$$\Delta X_a = -2(f + L_2)\alpha. \tag{7}$$

So, Equation (6) also can be expressed as follows:

$$S = 2\left[X'(D)(L_2/L_1)(f + L_1) - (f + L_2)\right]\alpha. \tag{8}$$

The angle change α of mirror can be expressed by the following relation:

$$\alpha = S/2\left[X'(D)(L_2/L_1)(f + L_1) - (f + L_2)\right] = k_2 S, \tag{9}$$

where k_2 represents a constant determined by system parameters.

Considering the size of the PSD sensitive area, L_2 is about same as f. From Equations (5) and (7), as long as L_1 is sufficiently small, the magnification of the spot energy center displacement $\Delta X(D)$ will be very large; $\Delta X(D)$ will be much greater than ΔX_a. Therefore, compared with the method shown in Figure 1, from Equations (1) and (9), the angle measurement sensitivity of the proposed system can be significantly improved. This amplification property is derived from the defect spot mode of the PSD, which can amplify the displacement of the spot energy center at the same magnification as the size transformation of the asymmetric defect spot.

The proposed method can be used for biaxial rotation measurements as long as the orthogonal double RP are used as the shield. In order to express this succinctly, follow-up analysis and experiments are only carried out for the single-axis rotation.

Because the light beam reaching the RP is not collimated, and the RP has a certain thickness, before and after the center of the light spot passes through the edge of the RP, the position of the edge of the RP that blocks the light will change. As shown in Figure 5a, the RP is placed behind the focal point, and before and after being obstructed by the RP edge at the center of the light spot, the rear and front surfaces of the RP edge respectively block the light. This implies that L_1 changes before and after the center of the spot passes through the edge of the RP. From Equation (9), it can be observed that the measurement sensitivity changes abruptly. Therefore, the edge of the RP should be selected as a knife-edge (KE), as shown in Figure 5b.

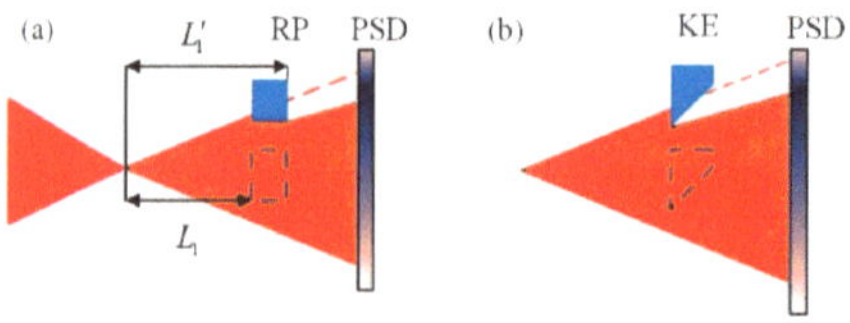

Figure 5. Influence of the thickness of RP. (**a**) Before adopting KE; (**b**) After adopting KE.

According to Equation (9), the greater the value of f, the smaller the value of L_1, and the higher the sensitivity of the system; at the same time, the influence of light drifts [21,24] as noise on the measurement accuracy of the system is correspondingly greater. Therefore, the parameter settings of f and L_1 need to be considered comprehensively, and the experiments described in Section 3.2 can be used as a reference.

3. Experiments and Analysis

The system structure and experimental setup for high-precision small-angle measurement are shown in Figures 4 and 6, respectively. The light source of the experimental system was a single-mode fiber-coupled semiconductor laser with a wavelength of 635 nm and a power of 5 mW (Xilong Optoelectronics Technology Co., Ltd., Shanghai, China, FC-635-005-SM). The laser power supply of this model is equipped with voltage and temperature stabilization modules. The diameter of the collimated beam was approximately 3.6 mm. The focal length f of the focusing lens was 50 mm, and the value of L_2 was approximately 50 mm. The beam was partially (about half) occluded by a KE, and the defect spot energy center position is detected by a PSD with a photosensitive area of 4 × 4 mm^2 (First Senor Corp., Berlin, Germany, DL16-7-PCBA3). The AD module adopts the ADS1256 module, which has eight channels, 24 bits, and a maximum sampling frequency of 30 K, and the data are transmitted to the computer through a USB. All experimental equipment was fixed on a vibration isolation optical platform, and the experiments were conducted in a laboratory environment.

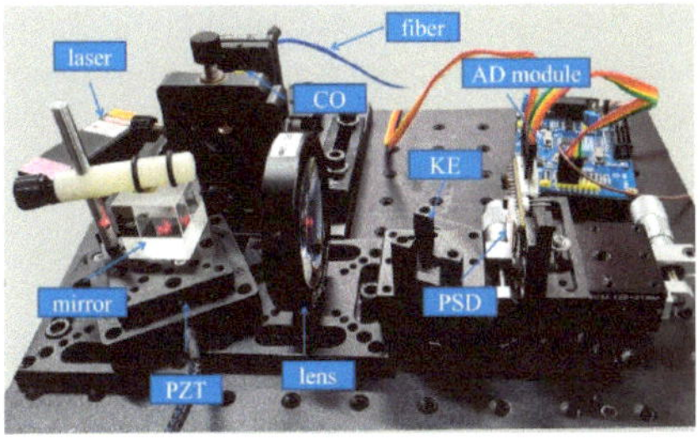

Figure 6. Experimental setup.

3.1. System Stability Experiment

In order to test the influence of electronic noise and environmental disturbance, the stability experiment was performed. In the experiment, the value of L_1 was approximately 3.5 mm, and the test point was chosen near the midpoint of the approximate linear working region of the PSD output characteristic curve, where the spot defect is about half and the relative position of the light spot on the PSD is 0.2327; the sampling frequency was 500 Hz, and testing time was 15 min. The experimental results after the sliding average algorithm (n = 10) for each sampling point are shown in Figure 7.

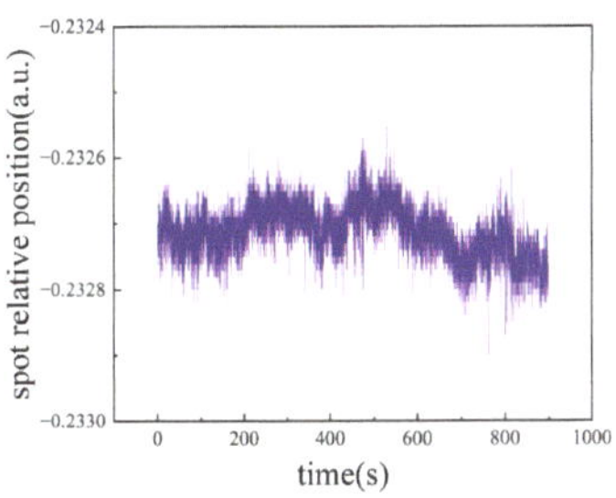

Figure 7. Stability experimental results.

As shown in Figure 7, the relative position of the spot on the PSD fluctuated between −0.2329 and −0.2325. Therefore, according to the principle of significant figures in errors theory, the output values of the experimental system can be read up to 1/10,000 bit. For the selected PSD, the width of the photosensitive region is 4 mm, so a relative position change of 0.0001 corresponds to a displacement of approximately 0.2 μm.

3.2. Calibration Experiments

We performed several sets of small-angle measurement calibration experiments with different values of L_1 using the PZT rotation stage (Core Morrow Technology Co., Ltd., Harbin, China, S21.R7S) as the standard meter. Corresponding to Figure 8a–c, the values of L_1 are 3.5 mm, 1.5 mm, and 0.5 mm, respectively. The resolution, repeatability, and measuring range of the PZT rotation stage was 0.2 μrad, 0.2%, and 6 mrad, respectively. According to Equation (8) and the output characteristics of the PSD shown in Figure 3, the piecewise linear fitting method was used for the analysis and processing of the experimental data. The experimental results are shown in Figure 8.

As shown in Figure 8a, the slops of the three piecewise fitting lines S_1, S_2, and S_3 in the range of 0 mrad to 0.08 mrad, 0.08 mrad to 0.5 mrad, and 0.5 mrad to 0.66 mrad were 0.67 mm/mrad, 0.80 mm/mrad, and 0.69 mm/mrad, respectively. According to the results of stability experiment, the spot position sensitivity of the PSD was approximately 0.2 μm; therefore, the angle measurement sensitivity in the above three angular ranges were 0.29 μrad, 0.25 μrad, and 0.28 μrad, respectively. The point-to-point deviations varied from −1.3 μrad to 1.8 μrad, and the standard deviation was about 0.82 μrad. As shown in Figure 8b, the sensitivity was approximately 0.095 μrad in the range of 0.21 mrad. The point-to-point deviations varied from −0.38 μrad to 1.0 μrad, and the standard deviation was approximately 0.34 μrad. As shown in Figure 8c, the sensitivity was about 0.034 μrad in the range of 0.084 mrad. The point-to-point deviations varied from −0.39 μrad to 0.63 μrad, and the standard deviation was approximately 0.23 μrad.

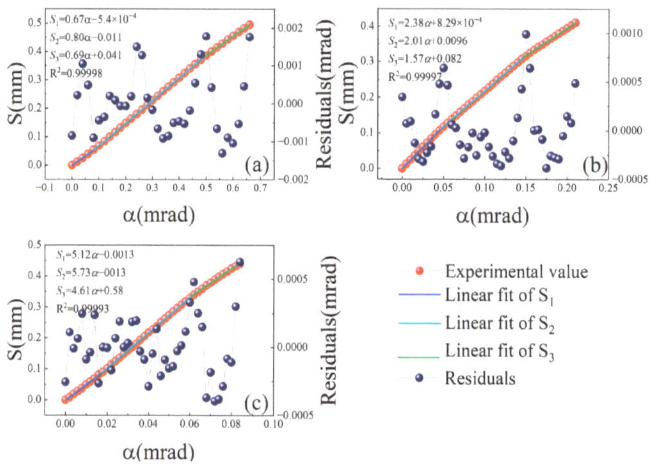

Figure 8. Calibration experiments of the proposed method with different values of L_1: (**a**) L_1 = 3.5 mm; (**b**) L_1 = 1.5 mm; (**c**) L_1 = 0.5 mm.

Corresponding to the three different values of L_1, the sensitivities calculated by Equation (9) are 0.26 μrad, 0.10 μrad, and 0.032 μrad, respectively; the experimental results are consistent with the theory. It can be observed from the numerical variations in the sensitivity, residuals, and standard deviation in Figure 8a–c that as L_1 decreases, the sensitivity of the system increases, and residuals and standard deviations decrease correspondingly, but the decreasing trend of residuals and standard deviation gradually slows down, and the ratio of residuals and standard deviation to sensitivity increases gradually, which means that the relative stability of the system decreases. The residuals are mainly affected by the electronic noise and environmental disturbances, such as light drift and so on. According to the measuring principle of the proposed method, with a decrease in L_1, the influence of a factor such as light drift increases in same proportion to the sensitivity of the system, whereas the influence of other factors such as electronic noise is basically unchanged. So, the signal-to-noise ratio increases with the decrease in L_1, and residuals and standard deviations correspondingly decrease. The experimental results and analysis show that the sensitivity of the proposed method can reach or even exceed 0.032 μrad, but at the same time, the standard deviation is close to seven times the sensitivity. The sensitivity determined by the results of the stability experiments, shown in Figure 7, corresponds to 1/10,000 of the relative position of the spot on the PSD. If the standard deviation is at least 10 times higher than the sensitivity, which means that the uncertainty bit of the measured value moves forward by 1 bit, the actual sensitivity of the system should correspond to 1/1000 of the relative position of the spot on the PSD. When the influence of light drift is dominant relative to the electronic noise, the further improvement of sensitivity is of limited help to the improvement of measurement accuracy. Proper compensation of light drift will be an important way to further improve the measurement accuracy of the system. Moreover, the smaller L_1 is, the smaller the size of the light spot on KE is, so the measuring range is limited and there is a trade-off between large measuring range and high sensitivity.

We also performed a calibration experiment on the conventional autocollimation measurement method shown in Figure 1. The parameters of each device in the experimental system were the same as those in the system shown in Figure 6 above.

The experimental results are shown in Figure 9. In the range of 3.2 mrad, the sensitivity was approximately 2 μrad, the point-to-point deviations varied from −8.0 μrad to 9.5 μrad, and the standard deviation was approximately 5.3 μrad.

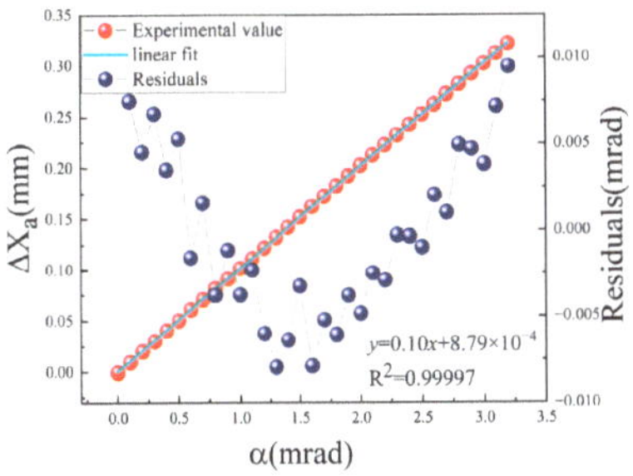

Figure 9. Calibration experiment of the autocollimation method based on the PSD in normal mode.

Clearly, the method based on the conventional working mode of the PSD has a larger linear measurement range; however, the detection sensitivity of the method based on the defect spot mode of the PSD was improved by approximately 57 times, which could not be realized by conventional methods.

3.3. Comparison Experiments

We conducted an angle measurement comparison experiment on the system corresponding to Figure 8b with a PZT in the rotation stage (Core Morrow Technology Co., Ltd., S21.R7S) with a resolution of 0.2 μrad as the standard meter. The PZT in the rotation stage rotates at a random angle, 12 times in total, and the corresponding measurement was performed by the proposed system. The experimental results are shown in Figure 10.

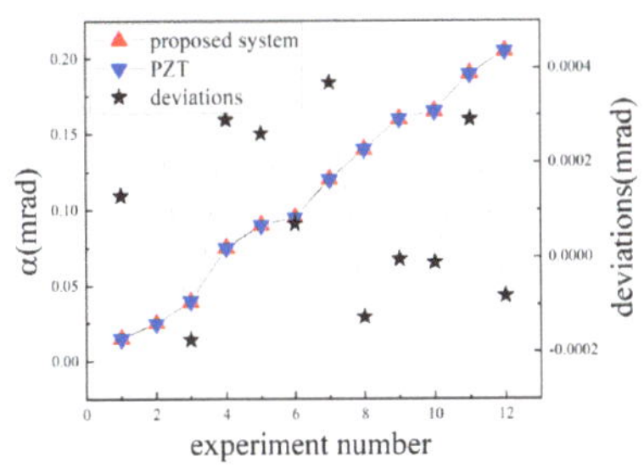

Figure 10. Comparison experiment results.

As shown in Figure 10, the deviations between the proposed system and the PZT were approximately −0.18 μrad to 0.37 μrad. Considering that the resolution of the PZT was approximately 0.2 μra, and that the standard deviation of the corresponding experiment was approximately 0.19 μrad in Figure 8b, these experimental data are quite consistent with the expectation, which confirms the feasibility and reliability of the proposed method.

3.4. Error Analysis

The main error sources of the proposed system include electronic noise, light drift, and the instability of the mechanical structure. Combining with Figure 4 and Equations (2)–(9), mechanical structural instability, such as the translation, rotation, or vibration of the KE, affects the numerical stability of L_1 and d. Therefore, it is necessary to analyze the angular measurement errors caused by the fluctuation of L_1 and d.

According to Equations (2)–(4), the fluctuation of d will cause the fluctuation of S.

$$\delta S_d = X'(D)(L_2/L_1)\delta \tag{10}$$

From Equation (9), the angular measurement error caused by the fluctuation δd can be expressed as follows:

$$\delta \alpha_d = X'(D)(L_2/L_1)\delta d/\left[2X'(D)(L_2/L_1)(f+L_1) - 2(f+L_2)\right]. \tag{11}$$

According to Equation (9), the angular measurement error caused by the fluctuation δL_1 can be expressed as follows:

$$\delta \alpha_{L_1} = SX'^{(D)L_2} f \delta L_1/\left[2X'^{(D)L_2} f + 2\left(X'^{(D)L_2} - f - L_2\right)L_1\right]^2. \tag{12}$$

The values of the structural parameters of the system corresponding to Figure 8b are substituted into the Equations (11) and (12), and the corresponding errors can be simulated by using MATLAB, as shown in Figure 11. The simulation results show that the angular measurement errors caused by fluctuations of 0.1 μm in d and 1 μm in L_1 were 1.1 μrad and 0.038 μrad, respectively. The structural stability of KE is very important.

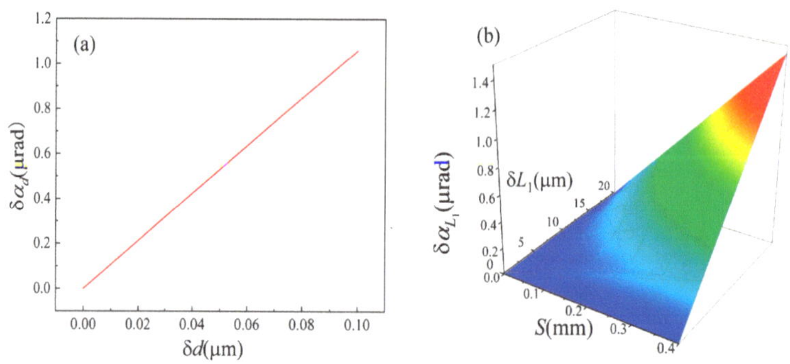

Figure 11. The angular measurement error caused by δd and δL_1. (**a**) Error introduced by δd; (**b**) Error introduced by δL_1.

Moreover, the method proposed in this paper is based on the detection of the energy center of the light spot on the PSD; the diffraction caused by the KE needs to be considered. The straight edge diffraction of the focused Gaussian beam and the point source uniform spherical wave are analyzed in references [25,26], respectively. Although the analysis processes and methods are different, the results of the two studies are similar. The closer the KE is to the focus point, the wider the diffraction fringes on the receiving plane, especially when the distance approaches 0 (the waist region); as a result, the receiving region contains only the first fringes, which leads to a symmetrical distribution of the light field, and the light field distribution based on the geometric projection relation used in proposed method is no longer applicable. So, the distance between the KE and the focus is the key to the feasibility of the proposed method.

The light field distribution of the incident light wave passing through the KE can be seen as the interference of two superimposing waves: the geometrical wave from the primary source of light and the boundary diffraction wave from the secondary source (KE) [25].

In our case, a slightly divergent Gaussian beam is incident on KE, as shown in Figure 12. The field of Gaussian beam can be expressed as follows:

$$U^{(g)} = exp\left[-\left(x^2+y^2\right)/\omega^2(z)\right]exp\left\{-j\left[k\left(z+\left(x^2+y^2\right)/2R\right)-arctan(z/F)\right]\right\}/\omega(z), \quad (13)$$

where $\omega(z)$ is the radius of the Gauss's spot, $k = 2\pi/\lambda$ denotes the wave vector, R is the radius of curvature of the equal phase plane of the Gaussian beam whose propagation axis intersects at the observation point, and F denotes the confocal parameter of the Gaussian beam.

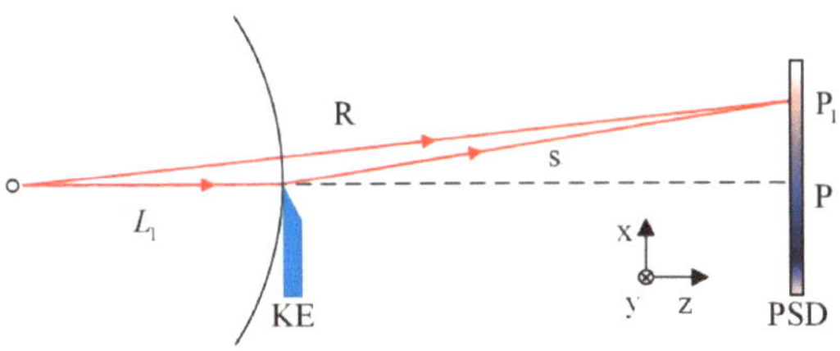

Figure 12. Optical structure of knife-edge diffraction.

The boundary diffraction wave can be expressed as follows:

$$U^{(d)} = \int_{\Sigma} exp\left[-l^2/\omega^2(z)\right]exp\left\{-j\left[k\left(z+s+l^2/2R\right)-arctan(z/F)\right]\right\}dl/[\omega(z)s], \quad (14)$$

where Σ denotes the boundary of the illuminated part of the KE, dl is an infinitesimal element situated on Σ, l denotes the distance from dl to the center of the Gaussian beam profile, and s denotes the distance from dl to the observation point.

The corresponding irradiance at the observation plane can be express as follows:

$$I = \left|U^{(g)}\right|^2 + \left|U^{(d)}\right|^2 - 2U^{(g)}U^{(d)}cos(\varphi), \quad (15)$$

where φ denotes the phase difference between two beams.

In Equation (15), the first term represents the projection of the Gaussian beam, and the other two terms represent the diffraction term and interference term, respectively. Therefore, the light intensity distribution on the PSD can be regarded as the superposition of the direct projection spot and the disturbance spot caused by the diffraction and interference effect. In order to understand the influence of the disturbance spot on the direct projection spot, combining the system parameters with Equations (13)–(15), the intensity of direct projection light and diffraction light near the center (0.5 mm, 0) and the edge (1.5 mm, 0) of the photosensitive surface of the PSD were simulated using MATLAB. The intensity ratios of distribution light and direct projection light at the two test points, using the system parameters in Figure 8a (L_1 = 3.5 mm), were approximately 0.3% and 1.3%, respectively, and those corresponding to Figure 8c (L_1 = 0.5 mm) were approximately 0.9% and 3.5%, respectively. The simulation results show that the influence of diffraction increases with the decrease in L_1 and the increase in x, but the influence is weak under the given system parameters, which is consistent with the experimental results in Figure 3. Based on existing research [25,26] and the above analysis, by approximating the diffraction limit spot size to the minimum waist of the Gauss beam and combining the focused beam projection relationship with the system parameters, the minimum distance (the beam waist region) between the KE and the theoretical focus of the lens in the proposed system can be estimated to be about 0.15 mm. The experimental and simulation results show that the proposed method is feasible when the KE is not in the beam waist region.

Other factors that can change the position of the energy center of the spot can be considered an interference spot directly superimposed on the spot of the direct projection light. Because the KE and the structure of the system are invariable, the effects of the interference spot are almost invariable and can be eliminated via system calibration as system errors. The influence of random errors such as electronic noise can be partially reduced by means of the mean algorithm.

4. Conclusions

This paper proposed a simple and high-precision small-angle measurement method. An innovative defect spot working mode of the PSD is used for angle measurement for the first time. Owing to the output characteristics of the PSD in defect spot mode and the size transformation properties of the focused beam, the small-angle measurement sensitivity was amplified using the same multiple as the spot size. Under the set parameters, compared with the current PSD-based autocollimation method, the sensitivity of the proposed method based on the PSD in defect mode is increased by 57 times. The main purpose of the experiments in this paper is to verify the feasibility and reliability of the proposed method, and the detection sensitivity of 0.034 μrad, given by the experiments, is not the best result of the proposed method. If the PSD with higher sensitivity [27] and the lens with larger focal length are selected and, considering the diffraction limit, the collimation system with the larger numerical aperture is selected, the sensitivity will be further improved, and it is expected to reach the level of digital collimators based on CCD and image processing algorithms, while the proposed method has the advantages of being a simple system with good real-time performance. In addition, the expansion of the measurement range is also important [13,28]. If the proposed angle measurement system is used as a "Null Detector", and the deviation (i.e., deflection angle) of the "Null Detector" is used to modulate the position of the KE via a feedback compensation system based on a PZT, the measurement method is expected to improve through the expansion of the measurement range. This study used the most common circular Gaussian as an example. The other types of light spots have different energy distributions, and accordingly, the linear interval and slope of the PSD output characteristic curve shown in Figure 3 are different. How to obtain higher angular sensitivity gain or a larger linear measurement range via beam/edge shaping is an interesting and practical question to explore in future work. In any case, the key property derived from the defect mode of the PSD remained unchanged; that is, the magnification of the asymmetrical defect spot area on the PSD could be converted into the magnification of detection sensitivity. Due to its comprehensive advantages in sensitivity, system complexity, and real-time responsiveness, the proposed method has broad application prospects in high-precision motion error detection, as well as in shape and position measurement.

Author Contributions: Conceptualization, Y.Z. (Yusheng Zhai); Methodology, Y.Z. (Yusheng Zhai); Software, R.W.; Validation, G.W.; Formal analysis, G.W. and Y.Z. (Yiheng Zhao); Investigation, G.W. and Y.Z. (Yiheng Zhao); Resources, Z.Z. and R.Z.; Data curation, Y.Z. (Yiheng Zhao) and R.W.; Writing—original draft, Y.Z. (Yusheng Zhai) and Y.Z. (Yiheng Zhao); Writing—review & editing, G.W.; Supervision, Z.Z.; Project administration, Y.Z. (Yusheng Zhai); Funding acquisition, Y.Z. (Yusheng Zhai), L.Z., Z.S., Z.Z. and P.Y. All authors have read and agreed to the published version of the manuscript.

Funding: This work is supported by National Natural Science Foundation of China (U1304507); Henan Major Science and Technology Project (221100230100); Key Research and Development Special Projects in Henan Province (241111230300, 231111222800); Key Research Projects of Higher Education Institutions in Henan Province (23B416001).

Institutional Review Board Statement: Not applicable.

Informed Consent Statement: Not applicable.

Data Availability Statement: Data are contained within the article.

Conflicts of Interest: Author Lin Zhang was employed by the company White Pigeon Abrasive Tools Co., Ltd. Author Zhan Su was employed by the company Henan Academy of Sciences Institute of Applied Physics Co., Ltd. The remaining authors declare that the research was conducted in the absence of any commercial or financial relationships that could be construed as a potential conflict of interest.

References

1. Ren, C.; Tan, Y.; Zhang, S. Angle measurement based on orthogonally polarized lasers. *Opt. Tech.* **2010**, *36*, 193–199.
2. Chen, X.; Liao, J.; Gu, H.; Zhang, C.; Jiang, H.; Liu, S. Remote Absolute Roll-Angle Measurement in Range of 180° Based on Polarization Modulation. *Nanomanufacturing Metrol.* **2020**, *3*, 228–235. [CrossRef]
3. Liu, Z.; Wang, L.; Li, K.; Ban, J.; Wang, M. A calibration method for the errors of ring laser gyro in rate-biased mode. *Sensors* **2019**, *19*, 4754. [CrossRef] [PubMed]
4. Ravaille, A.; Feugnet, G.; Debord, B.; Gérôme, F.; Benabid, F.; Bretenaker, F. Rotation measurements using a resonant fiber optic gyroscope based on Kagome fiber. *Appl. Opt.* **2019**, *58*, 2198–2204. [CrossRef]
5. Yuan, Y.; Xu, D.; Cheng, R.; Gao, Y.; Hoenders, B.J.; Cai, Y. 0–360 Degrees angular measurements using spatial displacement. *Appl. Phys. Lett.* **2023**, *122*, 111106. [CrossRef]
6. Huang, P.S.; Kiyono, S.; Kamada, O. Angle measurement based on the internal-reflection effect: A new method. *Appl. Opt.* **1992**, *31*, 6047–6055. [CrossRef]
7. Liu, Y.; Kuang, C.; Ku, Y. Small angle measurement method based on the total internal multi-reflection. *Opt. Laser Technol.* **2012**, *44*, 1346–1350. [CrossRef]
8. Yang, B.; Wang, C.; Zhang, J.; Ni, J. A small-angle self-mixing measurement system with improved detection resolution based on a rotatable pentagonal prism. *Opt. Commun.* **2018**, *429*, 29–34. [CrossRef]
9. Wang, X.; Su, J.; Yang, J.; Miao, L.; Huang, T. Modified homodyne laser interferometer based on phase modulation for simultaneously measuring displacement and angle. *Appl. Opt.* **2021**, *60*, 4647–4653. [CrossRef]
10. Zhu, F.; Tan, J.; Cui, J. Common-path design criteria for laser datum based on measurement of small angle deviations and laser autocollimation method in compliance with the criteria with high accuracy and stability. *Opt. Express* **2013**, *21*, 11391–11403. [CrossRef]
11. Guo, Y.; Cheng, H.; Wen, Y.; Feng, Y. Three-degree-of-freedom autocollimator based on a combined target reflector. *Appl. Opt.* **2020**, *59*, 2262–2269. [CrossRef]
12. Shimizu, Y.; Tan, S.L.; Murata, D.; Maruyama, T.; Ito, S.; Chen, Y.; Gao, W. Ultra-sensitive angle sensor based on laser auto collimation for measurement of stage tilt motions. *Opt. Express* **2016**, *24*, 2788–2805. [CrossRef]
13. Gao, W.; Ohnuma, T.; Satoh, H.; Shimizu, H.; Kiyono, S. A Precision Angle Sensor Using a Multi-cell Photodiode Array. *CIRP Ann.-Manuf. Technol.* **2004**, *53*, 425–428. [CrossRef]
14. Kuang, C.; Hong, E.; Feng, Q. High-accuracy method for measuring two-dimensional angles of a linear guideway. *Opt. Eng.* **2007**, *46*, 051016.
15. Li, R.; Xie, L.; Zhen, Y.; Xiao, H.; Wang, W.; Guo, J.; Konyakhin, I.; Nikitin, M.; Yu, X. Roll angle autocollimator measurement method based on a cylindrical cube-corner reflector with a high resolution and large range. *Opt. Express* **2022**, *30*, 7147–7161. [CrossRef]
16. Korolev, A.N.; Gartsuev, A.I.; Polishchuk, G.S.; Tregub, V.P. A digital autocollimator. *J. Opt. Technol.* **2009**, *76*, 624–628. [CrossRef]
17. Gao, M.; Dong, Z.; Bian, Z.; Ye, Q.; Fang, Z.; Qu, R. Robust CCD photoelectric autocollimator for outdoor use. *Chin. Opt. Lett.* **2011**, *9*, 091201.
18. Yuan, J.; Long, X. CCD-area-based autocollimator for precision small-angle measurement. *Rev. Sci. Instrum.* **2003**, *74*, 1362–1365. [CrossRef]
19. Cowsik, R.; Srinivasan, R.; Kasturirengan, S.; Kumar Senthil, A. Design and performance of a sub-nanoradian resolution autocollimating optical lever. *Rev. Sci. Instrum.* **2007**, *78*, 035105. [CrossRef]
20. Chen, Y.L.; Shimizu, Y.; Tamada, J.; Nakamura, K.; Matsukuma, H.; Chen, X.; Gao, W. Laser autocollimation based on an optical frequency comb for absolute angular position measurement. *Precis. Eng.* **2018**, *54*, 284–293. [CrossRef]
21. Li, K.; Kuang, C.; Liu, X. Small angular displacement measurement based on an autocollimator and a common-path compensation principle. *Sci. Instrum.* **2013**, *84*, 015108. [CrossRef] [PubMed]
22. St. John, W.D. Cylinder gauge measurement using a position sensitive detector. *Appl. Opt.* **2007**, *46*, 7469–7474. [CrossRef] [PubMed]
23. Zhai, Y.; Zhao, Y.; Yan, S.; Zhang, Z.; Geng, L.; Zhang, R.; Yang, K.; Kuang, C. High-precision detection method for an object edge based on a position-sensitive detector. *Appl. Opt.* **2023**, *62*, 1815–1821. [CrossRef] [PubMed]
24. Feng, Q.; Zhang, B.; Cui, C.; Kuang, C.; Zhai, Y.; You, F. Development of a simple system for simultaneously measuring 6DOF geometric motion errors of a linear guide. *Opt. Express* **2013**, *21*, 25805–25819.
25. Kumar, R.; Kaura, S.K.; Sharma, A.K.; Chhachhia, D.P.; Aggarwal, A.K. Knife-edge diffraction pattern as an interference phenomenon: An experimental reality. *Opt. Laser Technol.* **2007**, *39*, 256–261. [CrossRef]
26. Bibikova, E.A.; Al-wassiti, N.; Kundikova, N.D. Diffraction of a Gaussian beam near the beam waist. *J. Eur. Opt. Soc.-Rapid Publ.* **2019**, *15*, 17. [CrossRef]

27. Ren, W.; Cui, J.; Tan, J. Precision roll angle measurement system based on autocollimation. *Appl. Opt.* **2022**, *61*, 3811–3818. [CrossRef]
28. Li, R.; Xiao, H.; Xie, L.; Feng, T.; Ma, Y.; Guo, J.; Zhou, M.; Nikitin, M.; Konyakhin, I. Autocollimation angle-measurement method with a large range based on spot deformation. *Opt. Express* **2022**, *30*, 38727–38744. [CrossRef]

sensors MDPI

Article

Welding Defect Monitoring Based on Multi-Scale Feature Fusion of Molten Pool Videos

Chenbo Shi [1], Lei Wang [1], Changsheng Zhu [1], Tengyue Han [2], Xiangyu Zhang [1], Delin Wang [1] and Chun Zhang [1,2,*]

[1] College of Intelligent Equipment, Shandong University of Science and Technology, Taian 271019, China; skd996523@sdust.edu.cn (C.S.); 202283230040@sdust.edu.cn (L.W.); zcs@sdust.edu.cn (C.Z.); 202383230063@sdust.edu.cn (X.Z.); 202383230017@sdust.edu.cn (D.W.)

[2] Beijing Botsing Technology Co., Ltd., Beijing 100176, China; hty@botsing.com

* Correspondence: 13811989049@163.com; Tel.: +86-178-1077-0984

Abstract: Real-time quality monitoring through molten pool images is a critical focus in researching high-quality, intelligent automated welding. However, challenges such as the dynamic nature of the molten pool, changes in camera perspective, and variations in pool shape make defect detection using single-frame images difficult. We propose a multi-scale fusion method for defect monitoring based on molten pool videos to address these issues. This method analyzes the temporal changes in light spots on the molten pool surface, transferring features between frames to capture dynamic behavior. Our approach employs multi-scale feature fusion using row and column convolutions along with a gated fusion module to accommodate variations in pool size and position, enabling the detection of light spot changes of different sizes and directions from coarse to fine. Additionally, incorporating mixed attention with row and column features enables the model to capture the characteristics of the molten pool more efficiently. Our method achieves an accuracy of 97.416% on a molten pool video dataset, with a processing time of 16 ms per sample. Experimental results on the UCF101-24 and JHMDB datasets also demonstrate the method's generalization capability.

Keywords: molten pool; welding defect monitoring; dynamic characteristics; multi-scale feature fusion; light spots

1. Introduction

GMAW (Gas Metal Arc Welding) is widely used in various modern manufacturing industries, such as shipbuilding and storage tank construction, due to its advantages in automation and mechanization [1]. Despite the significant advancements in welding technology, the advanced manufacturing industry continues to demand higher welding production efficiency, intelligent automation, and superior welding quality. However, various welding defects remain unavoidable in actual welding processes. Among these, porosity is a critical issue affecting the quality of welded structures. The presence of porosity reduces the cross-sectional area at the welded joints, leads to uneven stress distribution, and severely impacts the quality of the weld. Therefore, monitoring porosity defects during welding is an urgent problem that needs to be addressed.

Welding is a dynamic, interactive, and nonlinear process. Experienced welders can improve weld quality and reduce defects by observing the molten pool and making real-time adjustments during the welding process [2]. However, extended observation of the molten pool can lead to welder fatigue, making it difficult to detect defects promptly and adjust the process accordingly. Additionally, welding produces irritating gases that pose health risks to welders. As a result, automated monitoring of the welding process using molten pool images to detect defects has become a research focus among scholars worldwide, aiming to replace the manual observation of the molten pool with automated methods.

When using features from molten pool images to identify welding defects, the critical challenge is linking the image features to the welding defects and establishing a mapping model from molten pool image features to welding defects. Research based on molten pool images can be broadly categorized into two types: one focusing on welding defect detection using single-frame molten pool images, and the other utilizing sequences of molten pool images (i.e., molten pool videos). For single-frame molten pool image analysis, one approach involves performing a multi-level statistical analysis of the geometric features of the molten pool, such as area, shape, and aspect ratio, to determine the state of the molten pool when defects occur [3–8]. This method, based on geometric feature information, is highly interpretable. However, this approach requires extensive statistical analysis, making it time-consuming. Moreover, due to variations in welding techniques and types, molten pool images may differ significantly, making it challenging to extend these geometric feature-based methods to other images. With the continuous advancement of deep learning technology, data-driven approaches like deep learning have been widely applied in image classification, object detection, speech recognition, and natural language processing [9–12]. In the context of molten pool image analysis, deep learning allows for the direct, end-to-end adaptive learning and extraction of molten pool features, replacing manual feature extraction. This approach not only enhances efficiency but also achieves excellent results in defect recognition [13,14]. To address the issue of CNNs (Convolutional Neural Networks) often being perceived as black boxes and the lack of large datasets of welding defects, Di Wu et al. [15] proposed a method that combines deep learning-extracted features with manually designed geometric features for prediction. This approach improves model accuracy and enhances the interpretability of the network. In studies involving molten pool image sequences, some works [1,16] utilize LSTM (Long Short-Term Memory) networks [17] to capture the differential features of the molten pool before and after changes during the welding process. These studies infer future frames of molten pool images and identify welding defects in those future frames, achieving early quality warnings.

In [2], Tianyuan Liu et al. proposed a CNN-LSTM model for online defect recognition in CO_2 welding. This model stretches the extracted features into two-dimensional representations, treating each row as a time series. By leveraging the strengths of LSTM in processing sequential data, the model effectively selects features in the spatial dimension, enabling accurate welding defect recognition. Although this approach utilizes LSTM, it only processes single-frame molten pool images and does not take advantage of the dynamic information inherent in the welding process. In contrast, Jun Lu et al. [1] developed a MPOM (Molten Pool Online Monitoring) model for monitoring the welding process, incorporating prediction and classification networks. The prediction model uses LSTM to capture the differences in molten pool states caused by temperature variations, allowing for predicting future molten pool shapes up to 10 time intervals in advance. These predicted future frames are then used for welding defect classification, highlighting the importance of molten pool features during the welding process for defect identification. However, research focusing on using dynamic features from the welding process for defect detection remains scarce.

Traditional RNNs (Recurrent Neural Networks) were introduced to handle sequential tasks, but they still need to be improved regarding gradient explosion and long-term dependency issues [18]. LSTM [17] addresses the gradient vanishing and explosion problems often occurring in long-sequence processing. However, LSTM models have a large number of parameters and face challenges when dealing with even more extended sequences. The C3D (3D Convolutional Network) [19] was introduced to handle three-dimensional spatial features in data, effectively capturing spatiotemporal characteristics, but it requires significantly more computation and resources compared to 2D convolution. Therefore, we employ the TSM (Temporal Shift Module) [20] based on 2D CNNs in this study. TSM shifts a portion of the feature channels from previous and subsequent frames along the temporal dimension, facilitating temporal information exchange without adding additional computational burden.

Due to variations in the camera's focal length, different installation angles, and welding scenarios, the position and size of the molten pool within the images are not fixed. Additionally, the redundant background information in the molten pool images poses challenges for welding defect detection. To address this, we segment the molten pool region in the images during preprocessing and crop the images based on the segmentation results. However, as shown in Figure 1, the size and position of the molten pool in the cropped images remain inconsistent, which complicates defect recognition. Inspired by the spatial pyramid structures in [21–23] and the feature fusion modules in [24,25], we added a MFF (Multi-Scale Feature Fusion) module at the head of the network. This module increases the receptive field of the network, enabling it to capture multi-scale features of the molten pool region. Additionally, we found that the attention mechanism further aids in weighting the features, enhancing the propagation of molten pool region features.

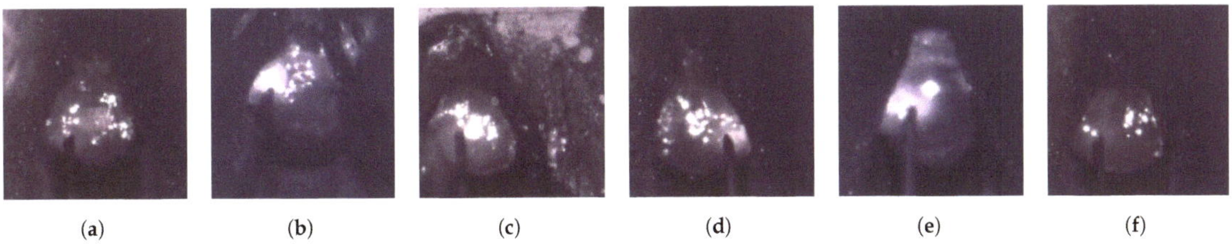

Figure 1. Images of the melt pool in different scenes. (**a**,**b**) are differences in the shape of the melt pool; (**c**,**d**) are differences in the camera viewpoint; and (**e**,**f**) are differences in size.

In the Gas Metal Arc Welding (GMAW) process, porosity defects often arise due to insufficient shielding gas flow, high humidity in the air, unclean weld seams, or moisture contamination on the welding plates. When dense porosity occurs internally, the shape of the molten pool and the surface light spots exhibit erratic and unstable changes. In contrast, a normal molten pool and its surface light spots show stable variations. Figure 2 illustrates the differences in molten pool behavior between porosity defects and normal conditions. This study focuses on porosity defects, aiming to extract critical information from the spatial characteristics of surface light spots and the dynamic features of molten pool image sequences to enhance the identification of welding defects. Our main contributions are as follows:

(1) We propose a lightweight multi-scale feature fusion module that improves feature propagation and fusion, capturing features from different scales and directions in molten pool images. The module enhances the model's expressive capability and its adaptability to molten pool size variations.
(2) We introduce an attention module that combines features from different directions and attention mechanisms to improve the model's ability to recognize both large objects and fine details, facilitating better propagation of features in the molten pool region.
(3) We establish a mapping model that links the temporal dependencies in molten pool image sequences to welding defects, leveraging the dynamic characteristics of the molten pool during the welding process to achieve efficient defect recognition.

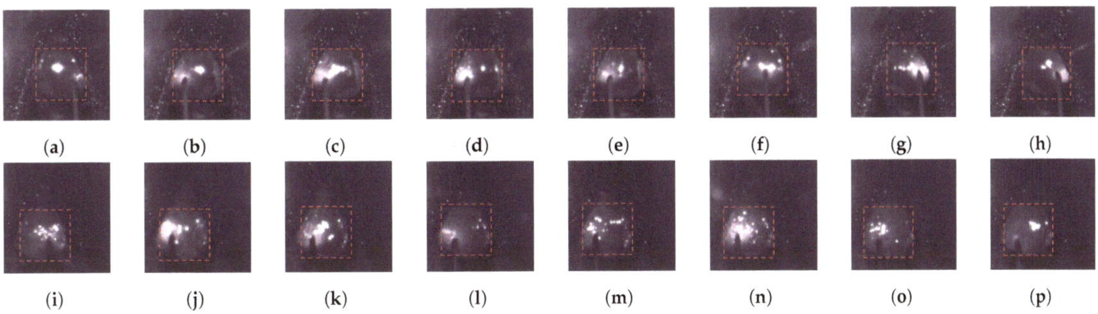

Figure 2. Eight time intervals of molten pool surface changes. (**a**–**h**) is the normal molten pool image of molten pool surface spot changes; (**i**–**p**) is the molten pool image of molten pool surface spot changes in the case of porosity defects.

2. Methods

Figure 3 illustrates the process of welding defect monitoring and identification. Images captured by the molten pool vision system are first preprocessed, and then a sequence of 8 consecutive frames is fed into the defect monitoring model. The model identifies the type of defect and issues a warning if necessary.

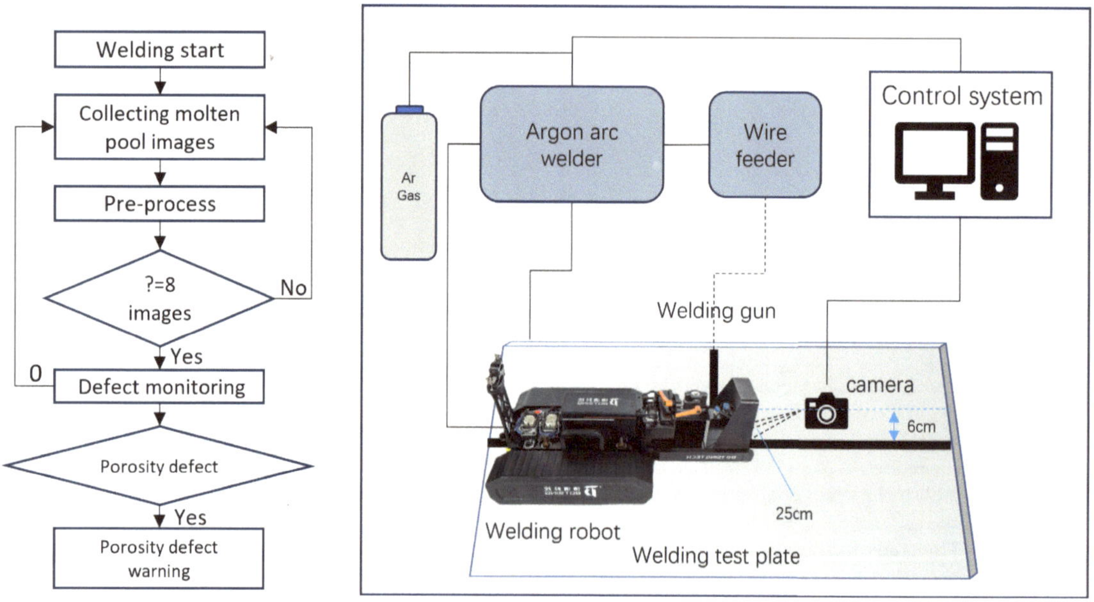

(a) (b)

Figure 3. Welding defect recognition process and melt pool vision system schematics: (**a**) welding defects monitoring process; (**b**) schematic diagram of molten pool vision system.

2.1. Molten Pool Visual System

This study collected data and conducted related experiments using the trackless crawling welding robot and the molten pool camera vision system developed by Beijing BOTSING Technology Co., Ltd., Beijing 100176, China. Figure 3 shows the molten pool vision system equipment used in this study, including the molten pool camera, the trackless crawling welding robot, and the industrial control computer.

The molten pool camera captures videos with a 640 × 512 pixels resoulution at a frame rate of up to 30 FPS. The camera is controlled by an industrial control computer, allowing for flexible recording. The experiments were conducted within a GMAW welding system. During GMAW welding, arc light can obscure many details in the weld pool image, decreaseing image quality. This study utilized the company's second-generation molten pool camera, which can filter out most of the arc light, thereby revealing more details in the molten pool images and reducing the negative impact of arc light on image quality.

2.2. Network Architecture

In the field of welding, there are stringent requirements for weld quality. Additionally, detecting welding defects in real-time and making adjustments promptly can significantly improve weld quality. To address this, we propose a lightweight welding defect detection model that processes video sequences as input. These sequences are represented as $A \in R^{NTCHW}$, where N is the batch size, T is the temporal dimension, C is the number of channels, and H and W are the spatial dimensions. In our model, a CNN extracts enhanced and effective features from each frame of the molten pool images, and the TSM module is used to capture the temporal dependencies of these features, enabling accurate identification of welding defects.

As shown in Figure 4, the Multi-Scale Feature Fusion Network based on Molten Pool Video (MFVNet) consists of a backbone and a head. The backbone processes the original molten pool image sequence to extract feature information. The head then further processes these features and performs the final classification. The head comprises three main components: a multi-scale feature fusion (MFF) module, an attention module (AM), and a fully connected layer. The MFF module has three branches and extracts and fuse feature information from the feature maps. Each layer stacks convolutions with different kernel sizes to further process the extracted molten pool features, combining branches with different receptive fields. The MFF module allows the model to detect targets at various scales. Additionally, we incorporate an attention module to enhance the model's ability to recognize both large objects and fine details. This is achieved by applying convolutional block attention module (CBAM) [26] attention and stacked row-column convolutions to the channel-shuffled features, weighting them for more effective feature processing. To further improve the accuracy of welding defect detection, we add Temporal Shift Modules (TSM) [20] after each layer of the backbone and after the MFF and AM modules. These TSM modules capture dynamic feature information from the video sequence, which is crucial for recognizing welding defects. Given the real-time requirements of welding defect detection, we use a unidirectional TSM module, as shown in Figure 5, to shift features from the previous frame to the current frame.

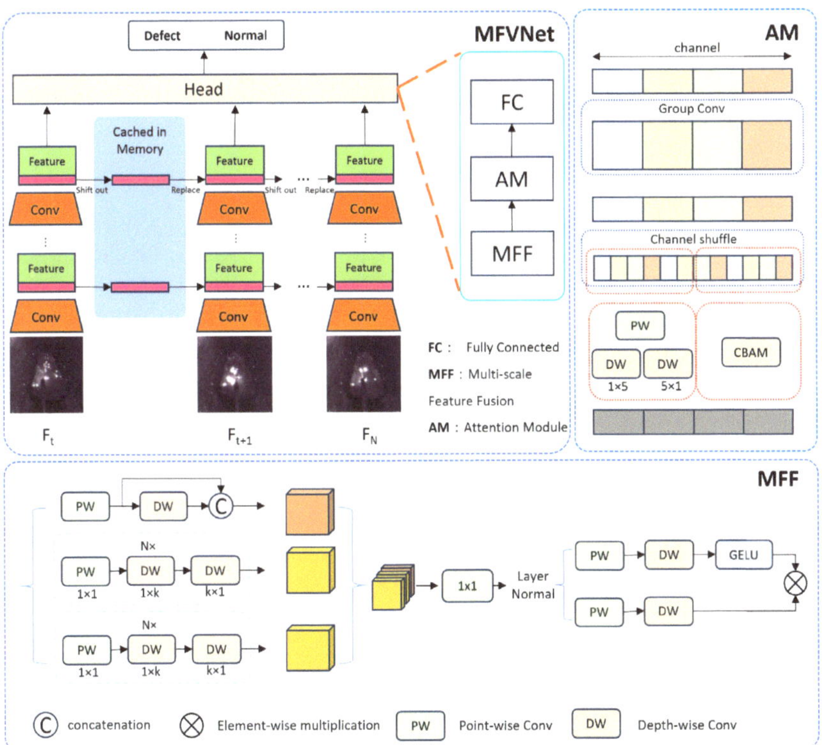

Figure 4. Structure of the proposed model MFVNet. the MFVNet consists of a backbone network and a header, with multi-scale feature fusion module and attention module embedded in the header and temporal shift module (temporal shift) in each layer.

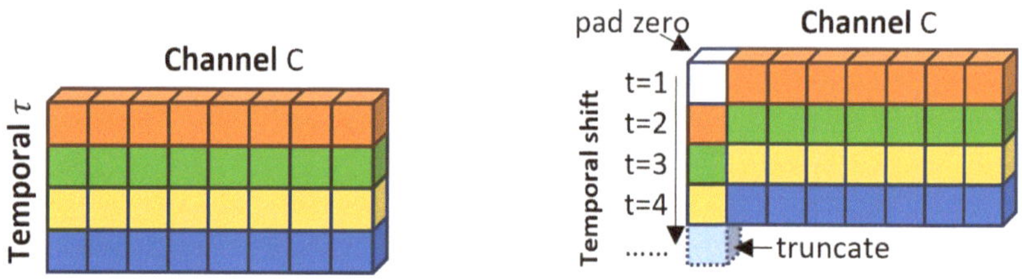

(a) (b)

Figure 5. The Temporal Shift Module (TSM) performs efficient temporal modeling by shifting the feature map along the time dimension. The unidirectional TSM mixes the past frame with the current one. (**a**) The original tensor without shift. (**b**) Temporal shift (uni-direction).

In the backbone section, we designed a relatively lightweight backbone network due to the high real-time requirements of industrial applications for welding defect detection. We chose the MobileNetV2 [9] as the backbone, which utilizes depthwise separable convolutions and inverted residual structures. This design maintains a lightweight architecture while achieving excellent performance, extracting features from molten pool images to enhance welding defect recognition. Similar to the approach in [20], we integrated TSM modules into each backbone layer, as illustrated in Figure 6. The TSM modules use residual shifts to fuse temporal information within the residual branches, further improving the model's ability to detect welding defects.

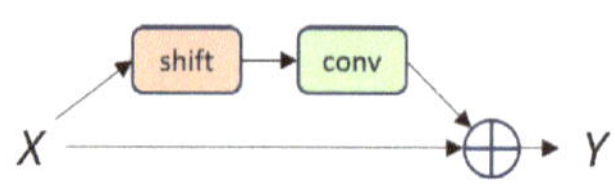

Figure 6. Residual TSM.

In this study, we propose a lightweight Multi-Scale Feature Fusion (MFF) module, which integrates concepts from feature pyramids to enhance the model's ability to handle multi-scale features. The MFF module consists of three branches and a feature fusion component, each branch using convolutions with different kernel sizes to improve the model's ability to extract multi-scale features. The first branch comprises a pointwise convolution layer (PW) and a depthwise convolution layer (DW). After the pointwise convolution, the number of feature channels is reduced to half of the input. Then, the original number of channels is restored through a residual structure and group convolution at the feature layer. This operation significantly reduces the number of parameters and accelerates the model's inference speed. Inspired by the feature pyramids in [21–23], the second and third branches also aim to capture features at different scales by increasing the receptive field. However, unlike those works, we do not use dilated convolutions to achieve varying receptive field sizes. Instead, we employ pointwise convolutions combined with row and column group convolutions of different kernel sizes, capturing features at different scales in various directions. In each branch, we stack these operations to enable the model to capture both detailed and global information at the same level, thereby enhancing the model's ability to handle complex backgrounds and intricate details.

Before feature fusion, we set a relatively small batch size due to the equipment's limitations. To mitigate this impact on model performance, we introduced a Layer Normalization (LN) layer. The LN layer normalizes all features within each sample, ensuring that features from different sources have similar distribution ranges, thereby eliminating the influence of batch size on the model's performance. In the feature fusion module, we combined depthwise convolution layers with a 3 × 3 kernel size and pointwise convolution layers with a 1 × 1 kernel size. This setup allows for efficient information fusion and encoding. We also implemented a gating mechanism, adding an extra path after the GELU activation function as a gate. This mechanism facilitates the effective propagation and fusion of features, enabling the model to focus on finer details.

Attention mechanisms are widely used in deep learning to recognize large objects and distinguishing between foreground and background. To enhance the model's performance, we designed a custom attention module. As shown in the Figure 7, we divided the feature channels into different groups using grouped convolutions, followed by a channel shuffle operation to increase interaction between different groups in the grouped convolution. In one branch, we employed the Convolutional Block Attention Module (CBAM), while in another branch, we used pointwise convolution combined with row and column group convolutions. CBAM is a lightweight attention mechanism that combines channel attention [27] and spatial attention [28] and can be flexibly integrated into CNN networks [26]. CBAM assigns higher weights to important targets, allowing the model to focus more on them while suppressing irrelevant features like background noise by assigning them lower weights. This helps the model effectively filter out background interference and focus on the critical features of the molten pool. Combining mixed attention with features from different directions improves the model's capability to detect fine details, enhancing its ability to detect subtle changes, such as size and shape.

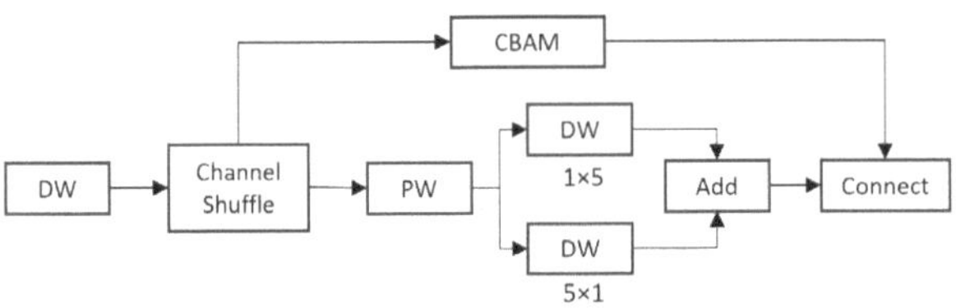

Figure 7. Attention mechanism flowchart.

2.3. Loss Function

Cross-entropy loss is a commonly used loss function in classification problems, particularly in deep learning models like Convolutional Neural Networks (CNNs). It measures the difference between the predicted probability distribution and the actual probability distribution. In binary classification problems, the model's output is typically a probability value representing the likelihood that a sample belongs to the positive class. The cross-entropy loss can be expressed as

$$Loss = \frac{1}{N}\sum_{i}^{N} -[y_i \log(p_i) + (1 - y_i)\log(1 - p_i)]. \tag{1}$$

where y_i denotes the truth value of the sample i , with 1 for the positive class and 0 for the negative class. p_i denotes the probability that the sample i is predicted to be in the positive class.

3. Experiments Design

3.1. Dataset

This study used GMAW welding, two types of welding wire (solid wire and flux-cored wire), and 980 high-strength steel with welds having different gaps and groove angles, and a molten pool vision system was employed to collect regular and porosity defect molten pool videos from horizontal and vertical welding on 30 welding plates. The data collected from four plates in both the normal and porosity defect samples were used as the test set. The captured videos had a 640 × 512 resolution at 30 frames per second. The videos were segmented into 1 s samples, with one sample taken every 4 s. Eight frames were evenly selected from each sample, with approximately 600 normal molten pool samples and 600 porosity defect samples. Since the molten pool images include parts of the weld seam, base metal, welding torch, and molten pool, with the molten pool region being the primary focus for detecting welding defects, this region occupies only a small portion of the image. High resolution and unnecessary background details can negatively impact the performance and speed of the neural network. Therefore, the molten pool region was segmented in each of the eight frames per sample. Based on the segmentation results, the molten pool region was cropped from the original image with a 1:1 aspect ratio, and the size was then resized to 224 × 224 pixels without altering the original aspect ratio of the molten pool region. This resizing preserves the image details and minimizes the impact on accuracy while improving processing speed by four times. The resulting dataset was named WELDPOOL.

3.2. Test Environment

In the experiment, we used a desktop computer with a GeForce RTX 3060 GPU and a 12th Gen Intel(R) Core(TM) i9-12900H CPU running Windows 11 operating system. The WELDPOOL dataset was utilized, with the dataset divided into training, validation, and test sets in a ratio of 6:1:3. The training parameters were set as follows: an initial learning rate of 0.0001, a weight decay coefficient of 0.00005, and, since we used pre-trained weights from MobileNetV2, we followed the settings in [20] and set the number of training epochs to 50. Parameters were updated using the gradient descent method. The batch size for each training iteration was set to 4 due to GPU limitations.

3.3. Performance Metrics

To evaluate the effectiveness of the proposed model in identifying porosity defects objectively, this study assesses the model from both performance and real-time capability perspectives. We use accuracy (Top-1 and Top-5), recall, and F1-score as the key metrics for performance evaluation. We consider the number of parameters and computational complexity for real-time capability as evaluation criteria. Following the settings in [20], we also include latency and throughput as real-time evaluation metrics. Latency refers to the time taken to process a single sample during inference, which in this study is the delay in processing a sequence sample consisting of 8 frames. Latency directly reflects the model's real-time capability. Conversely, throughput indicates the amount of data that can be processed per second, serving as an essential metric for evaluating the model's efficiency.

3.4. Experiments

The experiments were conducted in the environment specified in Section 3.2. First, we conducted benchmark tests to validate the proposed approach and the effectiveness of the model. To further assess the effectiveness of the proposed modules, we performed ablation experiments using the TSM model with MobileNetV2 as the backbone. Specifically, the Multi-Scale Feature Fusion (MFF) module and the Attention Module (AM) were tested as validation components, and their impact was further illustrated using confusion matrices. Next, to verify that the dynamic features of the molten pool can enhance the model's recognition capability and to assess the impact of sample frame count on the model, we compared the model's result under different sample frame counts. Additionally, to explore the effect of different backbones on the model's performance and real-time capability, we conducted comparative experiments by replacing the backbone with several lightweight alternatives and comparing them with the proposed model. Finally, to demonstrate the superiority of our algorithm, we compared it with several existing video classification algorithms. To further verify the generalizability of the proposed algorithm, we also conducted classification tests on the UCF101-24 and JHMDB subsets of the action recognition datasets UCF101 [29] and HMDB51 [30] to evaluate the performance of the proposed method.

4. Experiments Results

4.1. Benchmark Testing

This section primarily discusses the comparison between the proposed model and the baseline model. We enhance defect recognition by utilizing the dynamic features of the molten pool. To improve the model's adaptability to different molten pool scales, we apply multi-scale feature fusion. Additionally, we use attention mechanisms to strengthen the model's ability to identify the molten pool region. Based on this approach, we set up a baseline model that includes TSM, a spatial pyramid module, and CBAM.

As shown in Table 1, the proposed model has a lower parameter count and computational complexity compared to the baseline model. It achieves higher accuracy in Top-1 classification, although slightly lower in Top-5 accuracy. This indicates that the effective stacking of row and column group convolutions at different scales, combined with efficient feature fusion, is superior to using dilated convolutions with various dilation factors and pooling layers. This approach allows the proposed model to extract multi-scale features from different directions accurately and effectively while maintaining lower parameters and computational load. Moreover, incorporating the attention mechanism within the channel shuffle branch proves to be more effective than using the CBAM attention mechanism alone. Additionally, the proposed model demonstrates a significant advantage in processing speed.

Furthermore, we evaluated the proposed model using seven-fold cross-validation. The model achieved an average accuracy of 97.16%, with a standard deviation of 0.24%. This result indicates high accuracy and remarkable stability across different folds, demonstrating the model's generalization ability.

Table 1. Results for different benchmark configurations on the WELDPOOL dataset.

Moudle	#Test-Top1	#Test-Top5	#Param	#FLops	#Thrput
Baseline	96.833%	**99.911%**	13.108 M	6.912 G	22 ms
Baseline-AM	96.997%	98.95%	13.108 M	6.914 G	22 ms
Baseline-MFF	97.322%	98.962%	4.839 M	**3.669 G**	**15 ms**
MFVNet	**97.416%**	99.728%	4.840 M	3.671 G	16 ms

4.2. Ablation Study

The results of validating the Multi-Scale Feature Fusion module (MFF) and the Attention Module (AM) on the TSM baseline model with MobileNetV2 as the backbone are shown in Table 2. The experimental results indicate that when the proposed Attention Module (AM) is integrated directly into the baseline model, the Top-1 accuracy improves by 0.258%, and the recall rate increases by 0.2%. The results also demonstrate that the AM module helps the model leverage more effective features, significantly enhancing overall performance. The model incorporating the Multi-Scale Feature Fusion module (MFF) outperforms the baseline model on our dataset due to its enhanced capability for multi-scale feature extraction. The Top-1 accuracy improved by 1.292%, and the recall rate and F1 score of the baseline model and the model with only the AM module were also lower than those of the TSM model with the MFF module. Consequently, we integrated both modules into the TSM model to improve performance. Compared to other model combinations, MFVNet demonstrated superior results across the board. Therefore, from these experiments, we can conclude that the proposed Attention Module (AM) and Multi-Scale Feature Fusion module (MFF) effectively capture molten pool features and are well-suited for real-time monitoring and identification of welding defects in our welding scenarios.

Table 2. Results of different model configurations on the WELDPOOL dataset.

Moudle	#Test-Top1	#Test-Top5	#Recall	#F1	#Δacc
TSM	95.866%	98.325%	95.9%	95.9%	0%
TSM-AM	96.124%	98.85%	96.1%	96.1%	0.258%
TSM-MFF	97.158%	98.9%	97.2%	97.2%	1.292%
MFVNet	**97.416%**	**99.728%**	**97.4%**	**97.4%**	**1.55%**

To further demonstrate the effectiveness and impact of each module, we also calculated the confusion matrix for the inference results under each model configuration, as shown in Figure 8. In the confusion matrix, each column represents the predicted class, with the total number in each column indicating the count of data predicted as that class. Each row represents the actual class, with the total number in each row showing the number of instances of that class. The experimental results indicate that, regardless of the model configuration, a minimal number of normal samples are misclassified as porosity defects, possibly due to other unstable factors in the welding process causing instability in the molten pool surface light spots. However, after integrating the proposed modules, the model's misclassification rate decreases, demonstrating that the MFF and AM modules enhance the model's ability to identify welding defects.

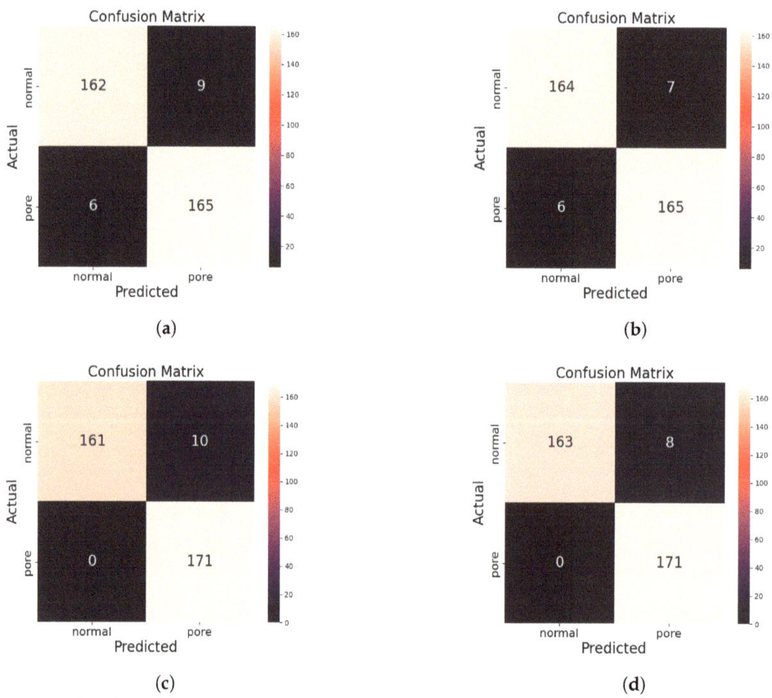

Figure 8. Confusion matrix results for different configuration models on the WELDPOOL test set. (**a**) TSM. (**b**) TSM-AM. (**c**) TSM-MFF. (**d**) MFVNet.

To further investigate the causes of model misclassification, we conducted tests on normal molten pool samples with different conditions: the presence or absence of spatter, variations in shape, and differences in size. Case1 represents no spatter, while Case2 represents the presence of spatter. The results are shown in the Table 3. It can be observed that variations in shape and size have minimal impact on the model's performance, while the presence of spatter has a more significant effect. This is likely due to the bright spots caused by metal spatter, which interfere with the model's ability to accurately recognize the true state of the molten pool. Therefore, improving the model's robustness to spatter interference remains a critical direction for further optimization.

Table 3. Results of the proposed model on normal molten pool samples with different conditions.

Conditions	#Case1	#Case2
splashes	99.617%	94.839%
shapes	99.774%	99.735%
size	99.769%	99.803%

Figure 9 presents the class activation map (CAM) visualizations for each model configuration. As shown in Figure 9b,c, the proposed MFF and AM modules significantly enhance the model's ability to identify the molten pool spots. Furthermore, Figure 9d demonstrates that combining the MFF and AM modules enables the model to focus more effectively on critical regions, thereby improving detection accuracy.

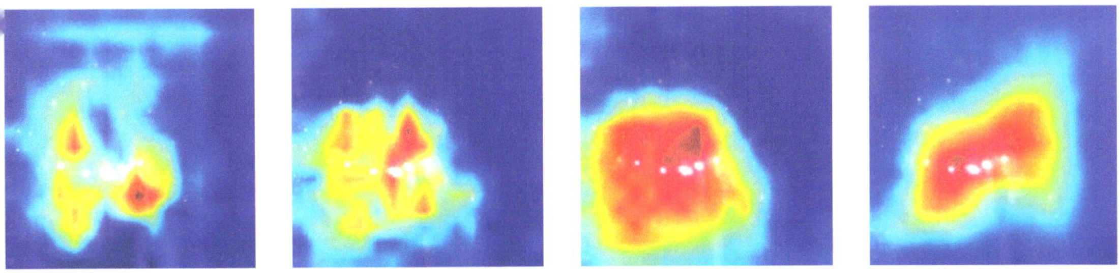

(a) (b) (c) (d)

Figure 9. In Class Activation Map (CAM), the importance of features is visualized using different colors, where the heatmap colors range from blue, yellow, orange to red, representing the lowest to highest feature importance, respectively. Class Activation Map (CAM) results for different configuration models on the WELDPOOL test set: (**a**) TSM; (**b**) TSM-AM; (**c**) TSM-MFF; (**d**) MFVNet.

4.3. *Impact of Sample Frames Number on the Model*

This section discusses the impact of different sample frame counts on the model's real-time performance and accuracy. When the sample frame count is 1, only a single frame is in the time series, and no feature shifting is applied. Unidirectional feature shifting is used for sample frame counts ranging from 2 to 16. To ensure a fair comparison, we use the inference time of a single frame as a metric for real-time performance in this experiment.

As shown in Figure 10a, when the sample frame count is between 2 and 16, the model's performance improves compared to when the frame count is 1, indicating that dynamic features in the molten pool images enhance the model's ability to identify welding defects. Furthermore, when the sample frame count reaches 12 to 16 frames, the model's ability to extract dynamic features from the molten pool video stabilizes. From a real-time perspective, as illustrated in Figure 10b, the inference time per image decreases as the sample frame count increases. However, after the sample frame count reaches 8, the improvement in inference speed becomes negligible. Therefore, the experimental results suggest that with a sample frame count of 8, the model's accuracy and real-time performance meet the requirements for real-time welding defect monitoring in industrial applications.

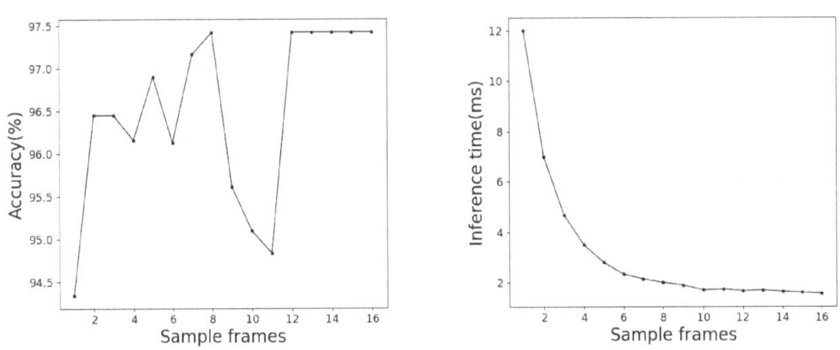

(a) (b)

Figure 10. Impact of different sample frame sizes on the real-time and performance of the model. (**a**) Accuracy of models with different sample frame sizes. (**b**) Time to reason about single-frame images for different sample frame number models.

4.4. Comparison with Other Backbone

The comparison results with two other popular lightweight backbone networks, ShuffleNetV2 [31] and GhostNetV2 [32], are shown in Table 4. The model using MobileNetV2 as the backbone performs the best across several evaluation metrics, including Top-1 accuracy, recall, F1 score, and inference latency. While GhostNetV2 and MobileNetV2 achieve the same accuracy, GhostNetV2 has a significantly slower inference latency of 27ms compared to MobileNetV2's 16ms, likely due to the higher number of parameters in GhostNetV2. Although ShuffleNetV2 matches MobileNetV2 in inference latency, it falls short in Top-1 accuracy. We speculate that the extensive channel shuffle operations in ShuffleNetV2, while enhancing information flow, may have slightly compromised its feature extraction capability on our dataset. The model with MobileNetV2 as the backbone achieves an accuracy of 97.416% with an inference latency of 16ms, meeting the accuracy and real-time requirements necessary for monitoring welding defects during the welding process.

Table 4. Results of the proposed model using different backbone on the WELDPOOL dataset.

Backbone	#Test-Top1	#Test-Top5	#FLOPs	#Param	#Latency	#Thrput
MFVNet (ShuffleNetV2)	96.641%	99.128%	**2.435 G**	**4.260 M**	**16 ms**	**62.5 V/s**
MFVNet (GhostNetV2)	**97.416%**	**99.728%**	3.619 G	9.048 M	27 ms	37 V/s
MFVNet (MobileNetV2)	**97.416%**	**99.728%**	3.671 G	4.840 M	**16 ms**	**62.5 V/s**

4.5. Comparison with Other Methods and Dataset

In this section, we analyzed the work presented in this paper in comparison to the latest research on molten pool analysis. For a fair comparison, we made adjustments to AMSegNet (additive manufacturing–SegNet) [33]. Specifically, we added Temporal Shift Modules to the convolutional layers after each downsampling and upsampling step to better handle time-series data. Additionally, we modified the final output to categorical. As shown in the Table 5, although AMSegNet demonstrates excellent performance in terms of top-1 accuracy, recall, and F1 score, its computational complexity is significantly higher, reaching 64.003 G FLOPs, with 17.064 M parameters. This results in a higher latency of 71 ms and a lower throughput of only 14.1 V/s. Although AMSegNet uses lightweight CNN modules, modifying its input to image sequences considerably increased the number of parameters and computational load. However, with profit from the attention mechanism, AMSegNet still achieved outstanding performance in detecting molten pool defects.

Table 5. The latest research on the WELDPOOL dataset.

Model	#Top1	#recall	#F1	#FLOPs	#Param	#Latency	#Thrput
AMSegNet	97.158%	97.2%	97.2%	64.003 G	17.064 M	71 ms	14.1 V/s
MFVNet	**97.416%**	**97.4%**	**97.4%**	**3.671 G**	**4.840 M**	**16 ms**	**62.5 V/s**

The experimental results of CNN+LSTM, C3D [19], Video Swin (Tiny) [34], and our proposed MFVNet across different datasets are shown in Table 6. The performance trends of the models are consistent across all three datasets: they perform well on our custom WELDPOOL dataset and the UCF101-24 dataset but show weaker performance on the JHMDB dataset, possibly due to the weaker temporal relationships in JHMDB. Our model has been optimized for the WELDPOOL dataset, primarily by capturing features from molten pool images at different scales and improving defect recognition through temporal modeling. This optimization also leads to strong performance on the action recognition dataset UCF101-24, with similar improvements observed in the CNN+LSTM model when these modules are incorporated. On our custom dataset, the proposed modules enable our model to achieve the best performance, with significantly lower inference latency compared to other models. The Video Swin model's performance is moderate, possibly be due to the reduced number of layers in the tiny version. Although our model demonstrates a clear advantage in inference latency on the UCF101-24 dataset, its accuracy is only slightly better than that of the CNN+LSTM model. As noted in the literature [20], methods that integrate temporal fusion across all layers generally outperform those like CNN+LSTM, which only apply temporal fusion in later feature extraction stages. On the JHMDB dataset, our model's accuracy is 0.989% lower than that of Video Swin, but it still outperforms the other two models and achieves the best inference latency.

Table 6. Experimental results of different models on different datasets.

Model	WELDPOOL	UCF101-24	JHMDB
CNN-LSTM	96.425% (50.9 ms)	96.596% (52.6 ms)	69.681% (52.2 ms)
C3D	95.85% (123.89 ms)	98.234% (119.53 ms)	68.989% (126.48 ms)
video swin	96.37% (62.26 ms)	**98.338%** (55.49 ms)	**71.372%** (66.14 ms)
MFVNet(our)	**97.416%** **(16 ms)**	97.819% **(13 ms)**	70.383% **(17.5 ms)**

In summary, our proposed MFVNet demonstrates excellent performance across multiple datasets, with a significant advantage in inference speed. While its accuracy on certain datasets, such as UCF101-24, is slightly lower than the top-performing model (e.g., Video Swin), MFVNet's substantial advantage in inference latency, combined with superior accuracy on our custom welding defect dataset, makes real-time welding defect monitoring feasible during the welding process.

5. Discussion

This study presents MFVNet, a video-based model for welding defect classification. The model uses the lightweight MobileNetV2 as the backbone for feature extraction and incorporates lightweight temporal shift modules in each layer, making the model more efficient. To address the characteristics of our molten pool images, we added a Multi-Scale Feature Fusion (MFF) module, which captures features from different scales and directions of the molten pool and integrates them across feature channels. The structure allows the model to better capture temporal dependencies within the sequence. Additionally, we introduced an attention mechanism module (AM) that combines attention with features from different directions, enabling the model to focus on the most important features. The dataset used in this study was created using a trackless crawling welding robot from Beijing BOTSING Technology, with molten pool videos captured by a molten pool camera, consisting of 1200 samples, with eight frames uniformly sampled from each video for training and validation. Experimental results demonstrate that the proposed model achieves low latency (16ms per sample) and high throughput (62.5 video samples per second), meeting the real-time requirements of practical applications. The model also achieved a welding defect recognition accuracy of 97.416%, laying a solid foundation for molten pool video-based welding defect detection. Furthermore, experiments on the UCF101-24 and JHMDB datasets indicate that our model is generalizable.

While our model has shown high performance and good real-time capabilities on our custom dataset, some limitations remain. The current algorithm focuses on identifying potential porosity defects, but further exploration is needed to extend it to identify a wider range of welding defects. Although we built a dataset with 1,200 samples, deep learning models typically benefit from larger datasets for improved generalization and robustness. Future research should delve deeper into the model, welding defects, and molten pool characteristics, and collect more diverse and larger-scale welding defect video data to enhance the model's performance.

Author Contributions: Conceptualization, C.S., L.W., C.Z. (Changsheng Zhu), T.H. and C.Z. (Chun Zhang); Data curation, L.W., X.Z. and D.W.; Formal analysis, L.W.; Funding acquisition, C.S.; Investigation, L.W.; Methodology, L.W.; Project administration, T.H. and C.Z. (Chun Zhang); Resources, L.W.; Software, L.W.; Supervision, C.S., C.Z. (Changsheng Zhu), T.H. and C.Z. (Chun Zhang); Validation, L.W., X.Z. and D.W.; Visualization, L.W.; Writing—original draft, L.W.; Writing—review and editing, C.S., L.W., C.Z. (Changsheng Zhu), T.H., X.Z., D.W. and C.Z. (Chun Zhang). All authors have read and agreed to the published version of the manuscript.

Funding: This work was supported by the Shandong Province Science and Technology SMEs Innovation CapabilityEnhancement Project (No. 2023TSGC0576 and No. 2023TSGC0605).

Institutional Review Board Statement: Not applicable.

Informed Consent Statement: Not applicable.

Data Availability Statement: The data presented in this study are available on request from the corresponding author.

Acknowledgments: Thanks to Beijing Botsing Technology Co., Ltd. for providing the welding robot experimental equipment and welding tests.

Conflicts of Interest: Author Tengyue Han was employed by the company Beijing Botsing Technology Co., Ltd. The remaining authors declare that the research was conducted in the absence of any commercial or financial relationships that could be construed as a potential conflict of interest.

References

1. Lu, J.; Xie, H.; Chen, X.; Han, J.; Bai, L.; Zhao, Z. Online welding quality diagnosis based on molten pool behavior prediction. *Opt Laser Technol.* **2020**, *126*, 106126. [CrossRef]
2. Liu, T.; Bao, J.; Wang, J.; Zhang, Y. A Hybrid CNN–LSTM Algorithm for Online Defect Recognition of CO_2 Welding. *Sensors* **2018**, *18*, 4369. [CrossRef] [PubMed]
3. García-Moreno, A.I. A fast method for monitoring molten pool in infrared image streams using gravitational superpixels. *J. Intell. Manuf.* **2022**, *33*, 1779–1794. [CrossRef]
4. Hong, Y.; Yang, M.; Chang, B.; Du, D. Filter-PCA-Based Process Monitoring and Defect Identification During Climbing Helium Arc Welding Process Using DE-SVM. *IEEE Trans. Ind. Electron.* **2023**, *70*, 7353–7362. [CrossRef]
5. Meng, X.; Qin, G.; Zou, Z. Investigation of humping defect in high speed gas tungsten arc welding by numerical modelling. *Mater. Des.* **2016**, *94*, 69–78. [CrossRef]
6. Hong, Y.; Zhang, C.; Lu, J.; Bai, L.; Zhao, Z.; Han, J. In-process monitoring of lack of fusion in ultra-thin sheets edge welding using machine vision. *Sensors* **2018**, *18*, 2411. [CrossRef]
7. Ai, Y. ; Jiang, P.; Wang, C.; Mi, G.; Geng, S.; Liu, W.; Han, C. Investigation of the humping formation in the high power and high speed laser welding. *Opt. Lasers Eng.* **2018**, *107*, 102–111. [CrossRef]
8. Luo, L. ; Qian, E.; Lu, T.; Pan, J.; Liu, M.; Liu, C.; Guo, Y.; Bi, L. Vision-Based Estimation of Force Balance of Near-Suspended Melt Pool for Drooping and Collapsing Prediction. *Sensors* **2024**, *24*, 3270. [CrossRef]
9. Sandler, M.; Howard, A.; Zhu, M.; Zhmoginov, A.; Chen, L.C. MobileNetV2: Inverted Residuals and Linear Bottlenecks. In Proceedings of the 2018 IEEE/CVF Conference on Computer Vision and Pattern Recognition, Salt Lake City, UT, USA, 18–23 June 2018; pp. 4510–4520. [CrossRef]
10. Ge, Z.; Liu, S.; Wang, F.; Li, Z.; Sun, J. YOLOX: Exceeding YOLO Series in 2021. *arXiv* **2021**, arXiv:2107.08430.
11. Han, W.; Zhang, Z.; Zhang, Y.; Yu, J.; Chiu, C.C.; Qin, J.; Gulati, A.;Pang, R.; Wu, Y. ContextNet: Improving Convolutional Neural Networks for Automatic Speech Recognition with Global Context. *arXiv* **2020**, arXiv:2005.03191.
12. Devlin, J.; Chang, M.W.; Lee, K.; Toutanova, K. BERT: Pre-training of Deep Bidirectional Transformers for Language Understanding. *arXiv* **2018**, arXiv:1810.04805.
13. Liu, T.; Zheng, P.; Chen, H.; Zhang, L. An attention-based bilinear feature extraction mechanism for fine-grained laser welding molten pool/keyhole defect recognition. *J. Manuf. Process.* **2023**, *87*, 150–159. [CrossRef]
14. Zhang, Y.; You, D.; Gao, X.; Zhang, N.; Gao, P.P. Welding defects detection based on deep learning with multiple optical sensors during disk laser welding of thick plates. *J. Manuf. Syst.* **2019**, *51*, 87–94. [CrossRef]
15. Wu, D.; Hu, M.; Huang, Y.; Zhang, P.; Yu, Z. In situ monitoring and penetration prediction of plasma arc welding based on welder intelligence-enhanced deep random forest fusion. *J. Manuf. Process.* **2021**, *66*, 153–165. [CrossRef]
16. Wang, Y.; Zhang, C.; Lu, J.; Bai, L.; Zhao, Z.; Han, J. Weld Reinforcement Analysis Based on Long-Term Prediction of Molten Pool Image in Additive Manufacturing. *IEEE Access* **2020**, *8*, 69908–69918. [CrossRef]
17. Hochreiter, S.; Schmidhuber, J. Long Short-Term Memory. *Neural Comput.* **1997**, *9*, 1735–1780. [CrossRef]
18. Bengio, Y.; Simard, P.; Frasconi, P. Learning long-term dependencies with gradient descent is difficult. *IEEE Trans. Neural Netw.* **1994**, *5*, 157–166. [CrossRef] [PubMed]

19. Tran, D.; Bourdev, L.; Fergus, R.; Torresani, L.; Paluri, M. Learning spatiotemporal features with 3d convolutional networks. In Proceedings of the IEEE International Conference on Computer Vision (ICCV), Santiago, Chile, 7–13 December 2015; pp. 4489–4497. [CrossRef]
20. Lin, J.; Gan, C.; Han, S. TSM: Temporal Shift Module for Efficient Video Understanding. *arXiv* **2018**, arXiv:1811.08383.
21. Chen, L.C.; Papandreou, G.; Kokkinos, I.; Murphy, K.; Yuille, A.L. DeepLab: Semantic Image Segmentation with Deep Convolutional Nets, Atrous Convolution, and Fully Connected CRFs. *arXiv* **2016**, arXiv:1606.00915. [CrossRef]
22. Xie, G.W.; Yin, F.; Zhang, X.Y.; Liu, C.L. Dewarping Document Image by Displacement Flow Estimation with Fully Convolutional Network. In *Proceedings of the Document Analysis Systems;* Springer International Publishing: Cham, Switzerland, 2020; pp. 131–144._1. [CrossRef]
23. Verhoeven, F.; Magne, T.; Sorkine-Hornung, O. Neural document unwarping using coupled grids. *arXiv* **2023**, arXiv:2302.02887.
24. Ruan, L.; Bemana, M.; Seidel, H.P.; Myszkowski, K.; Chen, B. Revisiting image deblurring with an efficient ConvNet. *arXiv* **2023**, arXiv:2302.02234.
25. Zamir, S.W.; Arora, A.; Khan, S.; Hayat, M.; Khan, F.S.; Yang, M.H. Restormer: Efficient transformer for high-resolution image restoration. In Proceedings of the IEEE/CVF Conference on Computer Vision and Pattern Recognition, New Orleans, LA, USA, 18–24 June 2022; pp. 5728–5739. [CrossRef]
26. Woo, S.; Park, J.; Lee, J.Y.; Kweon, I.S. CBAM: Convolutional Block Attention Module. In *Proceedings of the European Conference on Computer Vision;* Springer: Cham, Switzerland, 2018; pp. 3–19. [CrossRef]
27. Jin, X.; Xie, Y.; Wei, X.S.; Zhao, B.R.; Chen, Z.M.; Tan, X. Delving deep into spatial pooling for squeeze-and-excitation networks. *Pattern Recognit.* **2022**, *121*, 108159. [CrossRef]
28. Jaderberg, M.; Simonyan, K.; Zisserman, A. Spatial transformer networks. *Adv. Neural Inf. Process. Syst.* **2015**, *28*. [CrossRef]
29. Soomro, K.; Zamir, A.R.; Shah, M. UCF101: A Dataset of 101 Human Actions Classes from Videos in the Wild. *arXiv* **2012**, arXiv:1212.0402.
30. Kuehne, H.; Jhuang, H.; Garrote, E.; Poggio, T.; Serre, T. HMDB: A large video database for human motion recognition. In Proceedings of the IEEE International Conference on Computer Vision (ICCV), Barcelona, Spain, 6–13 November 2011; pp. 2556–2563. [CrossRef]
31. Ma, N.; Zhang, X.; Zheng, H.T.; Sun, J. ShuffleNet V2: Practical Guidelines for Efficient CNN Architecture Design. In *Proceedings of the European Conference on Computer Vision;* Springer: Cham, Switzerland, 2018; pp. 122–138._8. [CrossRef]
32. Tang, Y.; Han, K.; Guo, J.; Xu, C.; Xu, C.; Wang, Y. GhostNetV2: Enhance Cheap Operation with Long-Range Attention. In *Proceedings of the 36th International Conference on Neural Information Processing Systems (NIPS'24);* Curran Associates Inc.: Red Hook, NY, USA, 2024; pp. 9969–9982.
33. Li, W.; Lambert-Garcia, R.; Getley, A. C. M.; Kim, K.; Bhagavath, S.; Majkut, M.; Leung, C. L. A. AM-SegNet for additive manufacturing in situ X-ray image segmentation and feature quantification. *Virtual Phys. Prototyp.* **2024**, *19*, 1. [CrossRef]
34. Liu, Z.; Ning, J.; Cao, Y.; Wei, Y.; Zhang, Z.; Lin, S.; Hu, H. Video Swin Transformer. In Proceedings of the IEEE/CVF Conference on Computer Vision and Pattern Recognition, New Orleans, LA, USA, 18–24 June 2022; pp. 3202–3211.

sensors

MDPI

Article

Intelligent Sensing of Thermal Error of CNC Machine Tool Spindle Based on Multi-Source Information Fusion

Zeqing Yang [1,2], Beibei Liu [1], Yanrui Zhang [1,*], Yingshu Chen [1], Hongwei Zhao [3,4], Guofeng Zhang [3], Wei Yi [4] and Zonghua Zhang [1]

1 School of Mechanical Engineering, Hebei University of Technology, Tianjin 300130, China; yangzeqing@hebut.edu.cn (Z.Y.); 15031244682@163.com (B.L.); chenyingshu@hebut.edu.cn (Y.C.); zhangzonghua@hebut.edu.cn (Z.Z.)
2 Key Laboratory of Hebei Province on Scale-Span Intelligent Equipment Technology, Hebei University of Technology, Tianjin 300130, China
3 State Key Laboratory for Manufacturing Systems Engineering, Xi'an Jiaotong University, Xi'an 710049, China; 17622351275@163.com (H.Z.); cw211512@foxmail.com (G.Z.)
4 State Key Laboratory of Strength and Structural Integrity, Aircraft Strength Research Institute of China, Xi'an 710065, China; wei853193@gmail.com
* Correspondence: zhangyanrui@hebut.edu.cn

Abstract: Aiming at the shortcomings of single-sensor sensing information characterization ability, which is easily interfered with by external environmental factors, a method of intelligent perception is proposed in this paper. This method integrates multi-source and multi-level information, including spindle temperature field, spindle thermal deformation, operating parameters, and motor current. Firstly, the internal and external thermal-error-related signals of the spindle system are collected by sensors, and the feature parameters are extracted; then, the radial basis function (RBF) neural network is utilized to realize the preliminary integration of the feature parameters because of the advantages of the RBF neural network, which offers strong multi-dimensional solid nonlinear mapping ability and generalization ability. Thermal-error decision values are then generated by a weighted fusion of different pieces of evidence by considering uncertain information from multiple sources. The spindle thermal-error sensing experiment was based on the spindle system of the VMC850 (Yunnan Machine Tool Group Co., LTD, Yunnan, China) vertical machining center of the Yunnan Machine Tool Factory. Experiments were designed for thermal-error sensing of the spindle under constant speed (2000 r/min and 4000 r/min), standard variable speed, and stepped variable speed conditions. The experiment's results show that the prediction accuracy of the intelligent-sensing model with multi-source information fusion can reach 98.1%, 99.3%, 98.6%, and 98.8% under the above working conditions, respectively. The intelligent-perception model proposed in this paper has higher accuracy and lower residual error than the traditional BP neural network perception and wavelet neural network models. The research in this paper provides a theoretical basis for the operation, maintenance management, and performance optimization of machine tool spindle systems.

Keywords: spindle; thermal-error modeling; multi-source information fusion; intelligent perception; machine tool

1. Introduction

The spindle system plays a crucial role in machine tools, and its stability and accuracy directly affect the machining quality. Under complex working conditions, the uneven distribution of the temperature field of the spindle system may lead to unpredictable thermal deformation, which, in turn, affects the machining accuracy. Therefore, studying intelligent-sensing methods for spindle thermal errors is crucial for improving the performance and intelligence of CNC machine tools. The ability of the spindle system to independently perceive its own state and environmental conditions can allow the system to realize real-time monitoring and adjustment of the machining process, thereby improving machining accuracy and stability. The research on this intelligent-perception method not only helps to improve the machining quality of the machine tool but also enhances the intelligent level of operation and maintenance management of the machine tool, making it more adaptable to the complex and changing machining environment and its needs [1,2].

The research on intelligent sensing of thermal error in machine tool spindles mainly has two aspects: thermal-error signal analysis and thermal-error modeling. Thermal-error signal analysis mainly collects the relevant signals in the machine tool and the surrounding machining environment through temperature sensors, displacement sensors, etc. It performs signal analysis and feature extraction on them. Brecher, C. et al. [3] used the unscented Kalman filter (UKF) to estimate machine tool kinematic error model parameters. The kinematic error model of a machine tool contains the time-varying errors, both geometric and thermal. The researchers used an unscented Kalman filter to fuse three-dimensional probe data with a low sampling rate, three-dimensional probe data with a high sampling rate, and comprehensive deformation sensor data with a high sampling rate for the real-time calibration of thermal-error models. This reduces the impact of modeling errors caused by nonlinearity and measurement noise and improves the machining accuracy and stability of the machine. Guo et al. [4] proposed a static thermal deformation modelling method (ST-CLSTM) for machine tools based on a spatiotemporal correlation hybrid CNN-LSTM. They used a convolutional neural network (CNN) to extract temperature features and construct the dataset and a long short-term memory (LSTM) network to capture the temperature change features, considering the sequential nature of the temperature data. The experiment verifies that the model has higher prediction accuracy than the traditional model and solves the problem of temperature-sensitive point selection in thermal-error modeling. Jia et al. [5] constructed a thermal-error prediction model using a one-dimensional convolutional neural network-gated recurrent unit (1DCNN-GRU-Attention). The convolution module is used to replace the traditional temperature-sensitive point selection method. The experiment's results show that the prediction accuracy of the proposed model is 81.53% under multi-coupled factors. The root-mean-square error (RMSE) is 40% lower than that of the traditional method.

Regarding thermal-error modeling, there are presently mainly thermal-error modeling methods based on heat transfer theory, and polynomial fitting or neural network modeling methods based on experimental data [6]. The thermal-error modeling method based on heat transfer theory is mainly based on the energy conservation equation of heat conduction–convection–thermal radiation used to solve the temperature field and the corresponding displacement field of the machine tool spindle or key components. The classical methods are the centralized mass method and the finite element method [7]. The centralized mass method simplifies the geometry and material distribution of the analyzed object, and by reasonably selecting the location and mass value of the concentrated mass points, connecting each of them with each other using thermal resistance, and establishing the energy conservation equation, a thermal-error model can be obtained in order to predict the characteristics and response of the structure [8].

Kim et al. [9] used the centralized mass method to model the thermal error of the ball screw feed drive system to calculate the temperature distribution and thermal deformation of the ball screw feed drive system. Huang et al. [10], who used the centralized mass method to model the thermal errors of tension rods and bending beams, investigated the relationships between thermal deformation and temperature and heat. In this case, the thermal-error model of the spindle was established by using the thermoelastic mechanics theory and the lumped heat capacity method, and the average fitting accuracy of the model reached 91.3%. The finite element analysis method uses finite element software to model the structure and material properties of the machine tool. Then, it analyzes the thermal error of the machine tool under different operating conditions [11]. Wu et al. [12] used the finite element method to analyze the thermal characteristics of a ball screw feed drive system under long-term operating conditions. By estimating the heat source intensity, based on the temperature profile, through inverse analysis, the primitive domain of the ball screw is divided into multiple units so that the discrete system is equivalently replaced by a continuous system. The temperature distribution is then converted into transient heat transfer in a non-deforming medium, and the thermal expansion of the ball screw is simulated based on the calculated heat flux. Yang et al. [13] have numerically simulated the thermal expansion process of a high-speed motorized spindle under normal operating conditions using a transient thermal–structural coupled finite element analysis method. The finite element prediction results were also compared with the measured temperatures and deformations, and it was found that the thermal model can be used to predict the transient thermal characteristics under various operating conditions. Ma et al. [14] developed a three-dimensional finite element model, considering thermal contact resistance and bearing stiffness, for transient thermal–structural coupling analysis of a high-speed electrical spindle. They verified the validity of the model through thermal balance experiments.

Thermal-error modeling methods based on heat transfer theory generally involve complex mathematical and physical equations, which may require significant computational resources and time, to solve complex structures. The centralized mass method and the finite element method usually require some approximations to reduce the computational complexity, but these approximations, to a certain extent, affect the accuracy of the analysis results, which need to be evaluated and verified according to the actual situation in the specific application.

A polynomial fitting or neural network modeling method based on experimental data is used to construct a correlation model using internal and external information from the machine, such as temperature rise at sensitive points and machine speed obtained from thermal-error experiments, as inputs, and thermal drift as output. Among these methods, the polynomial fitting method usually uses a polynomial function to fit the thermal-error data of a machine tool in order to model and predict it. In contrast, the neural network method trains a neural network model with a large amount of data, thus realizing the modeling and prediction of the thermal error of the machine tool. These classical methods have been widely used and studied in practical engineering.

Gowda Chethana, R. et al. [15] used multiple linear regression methods to develop a prediction model for thermal errors in CNC machine tools. They used the experimentally measured diameter deviation as the dependent variable and the temperature data as the independent variable. They obtained the regression coefficients in multiple linear regressions using the least-squares method to determine the deviation between the tool point and the workpiece. The experiment's results show that the method effectively predicts the radial deviation of CNC machine tools. Zhao et al. [16] proposed a three-dimensional thermal-error analysis method based on rotation error vector and translation error vector, obtained six vectors of thermal error in the three-dimensional space of the spindle through the testing experiments, and used the thermal-error compensation technique of space coordinate transformation parameters to verify the machining of S samples before and after the compensation for the thermal error. The machining accuracy of the parts was improved by 34.1%, which laid a theoretical foundation for the detection of and compensation for thermal error in asymmetric spindles in the same kind of high-torque CNC machine tools. Li [17] and others established a least-squares support vector machine (LSSVM) prediction model optimized by Aquila Optimizer (AO), and the experiment's results showed that the prediction accuracy of the AO-LSSVM prediction model for the thermal error of the electrical spindle can reach 94%, and it has a good stability and generalization ability. Huang et al. [18] introduced a genetic algorithm to optimize the initial weights and thresholds of the traditional back-propagation neural network. They used the combination of genetic algorithm and neural network in the thermal-error prediction of high-speed spindles, which showed advantages in solving the global minimum search problem quickly, compared with the traditional back-propagation neural network model. Lee et al. [19] applied fuzzy logic decision to thermal-error modeling, and many other scholars [20–22] have applied a gray model and an artificial neural network to spindle thermal-error prediction, an approach which has fewer learning samples and avoids the loss of information in a single modeling approach. All of these methods reflect the thermal error by modeling the correlation between the temperature-sensitive points and the thermal deformation, so selecting the temperature-sensitive points is particularly important.

The methods mainly used for selecting temperature-sensitive points are the empirical correlation coefficient and the cluster analysis [23]. The empirical analysis method empirically analyses the components in which thermal deformation occurs and uses the temperature-sensitive points associated with them as model inputs. Among these efforts, Wu et al. [24] used three temperature measuring points, associated with spindle speed, spindle movement, and coolant system, as temperature-sensitive points; Yang et al. [25] used measuring points related to the spindle base, X-axis screw, and spindle column as the temperature-sensitive points, according to empirical statistics; the correlation coefficient method involves screening the temperature measuring points by describing the correlation coefficient between the temperature field and the thermal error to determine its temperature-sensitive point. Using this, Guo et al. [26] used the correlation coefficient to classify the 12 groups of temperature measuring point data. A group of data was selected in each category to reflect the temperature information in the group. Three groups of temperature-sensitive-point data were obtained as inputs for the model after screening. Liu et al. [27] used the correlation coefficient method to evaluate the correlation between temperature-sensitive points and thermal errors. They found that this method has advantages in selecting temperature-sensitive points that remain stably correlated with thermal errors over time. The cluster analysis method of screening temperature-sensitive points generally uses a clustering algorithm to group temperature data, calculate the distance between data points and the clustering centers, and group data points with similar temperatures into one category. Fu et al. [28] used correlation analysis and K-means clustering to select combinations of global temperature sensitivities for machine tools. Liu et al. [29] used a fuzzy clustering algorithm combined with average influence values to optimize the temperature collection points, which ensured the robustness of the model by classifying the variables and prompting the selection of typical variables to reduce the inputs. Hu [30] used fuzzy C-mean (FCM) clustering and correlation analysis to select temperature-sensitive points in the thermal-error modeling and introduced the Dunn index to determine the optimal number of clusters, a tactic which can effectively suppress the multicollinearity problem between temperature measuring points. Li et al. [31] used the fuzzy C-mean clustering algorithm to screen the temperature-sensitive points and then used the Pearson correlation coefficient to improve the covariance and correlation between the temperature variables; the covariance and correlation problems between the temperature variables were effectively weakened.

In summary, most existing studies only consider the spindle temperature field's influence on the spindle's thermal error. In contrast, the spindle speed, spindle load, cooling system, spindle structure, ambient temperature, lubrication, and the thermal conductivity of processing materials, and so on, will affect the thermal error of the spindle system in the actual working process, so it is necessary to analyze the internal and external information from the machine tool in a comprehensive perception of the tool and the use of multi-source information fusion can help to obtain a more accurate thermal-error model.

The rest of this paper is structured as follows: Section 2 firstly briefly analyzes the generation mechanism of the thermal error of the CNC machine tool spindle and then systematically introduces the proposed intelligent-sensing method and model of spindle thermal error based on multi-source information fusion and describes in detail the multi-source information feature extraction and information fusion algorithm in the model. Section 3 describes the experimental design testing thermal-error sensing for spindles. Then, the experiment's results are analyzed and discussed in Section 4, and the proposed intelligent thermal-error sensing method based on multi-source information fusion is compared and analyzed with the existing algorithms. Finally, conclusions are drawn in Section 5.

2. Intelligent-Sensing Method of Thermal Error

2.1. Mechanism Analysis of Thermal Error of CNC Machine Tool Spindle

During the machining process, CNC machine tools are affected by internal and external heat sources such as cutting heat, friction heat, and the surrounding environment. The heat generated {Q} is transferred to the machine components by radiation, convection, and conduction. This can result in an uneven distribution of the temperature field {ϕ} and the thermal deformation {u} of the machine tool due to the incomplete symmetry of the structure of the machine tool, the different materials of the internal components, and the differences in the degree of heat dissipation on the surface of the machine tool. This causes the CNC machine to produce a change in the relative positions of the components compared to the standard steady-state condition, which ultimately leads to a relative displacement {δ} between the workpiece and the tool, affecting the machining accuracy, as shown in Figure 1.

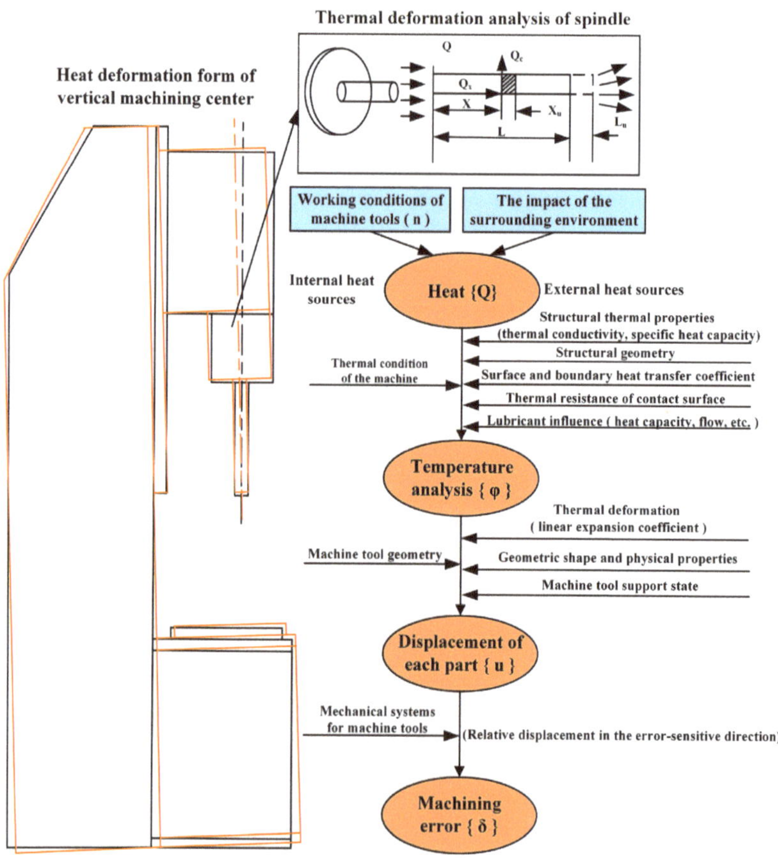

Figure 1. Thermal deformation mechanism of the vertical machining center.

As a critical core component of CNC machine tools in high-speed operation, the spindle will be subject to friction, cutting heat, and other factors, resulting in a rise in the spindle's local temperature and causing thermal deformation. Meanwhile, the spindle is affected by high temperatures during the working process, and the local area will be affected by thermal stress, which will change the shape of the spindle and cause thermal error. In addition, with the complexity and variability of the machining environment and working conditions, the spindle in the operation process or machining process produces unpredictable thermal errors, so it is vital to carry out intelligent sensing of the spindle thermal error of the machine tool under complex working conditions. In the actual operation of the machine tool, machine settings, staff operation, maintenance, calibration, and other aspects of these factors may have impacts on the thermal error. However, the impact on the thermal error is relatively small, so in order to simplify the model analysis, these factors can be ignored.

Existing thermal-error sensing models generally model the correlation between temperature-sensitive points and thermal deformation. However, the selection of the sensitive points, given that the thermal-error model has been established using a single sensor, has certain limitations. These points cannot accurately reflect the correlation of temperature field and thermal deformation under different working conditions. In order to improve the robustness of the thermal-error sensing model, the factors affecting the thermal error of the spindle system are analyzed, and it is found that the main signals related to the thermal error of the spindle are the spindle temperature field, the spindle working-condition parameters, and the motor current signal. In addition, elements of working-condition information such as the spindle reach, thermal balance time, spindle speed, and size of the cutting force directly affect the thermal deformation of the spindle; when the spindle is at a high speed, the friction and cutting heat will increase, resulting in rises in the spindle's temperature, increasing the possibility of thermal error. The spindle motor current signal reflects the spindle load condition and working status, and the increase in cutting load will increase the spindle force, generate more heat, and aggravate the thermal error of the spindle system.

Therefore, the work of utilizing sensors to collect multi-source thermal information associated with both the machine tool spindle itself and the machining environment, and analyzing, processing, and fusing it in real-time to construct a thermal-error model is indeed critical to improving the performance and accuracy of the machine tool. Such an intelligent-sensing system allows the machine tool to know, in real time, its own thermal state and the influence of the surrounding environment, allowing it to make timely adjustments and optimizations during the machining process to improve the quality and efficiency of its machining.

2.2. CNC Machine Tool Spindle Thermal-Error Sensing Methods

Thermal-error sensing of machine tool spindles generally requires thermal performance tests under different operating conditions to obtain temperature rise data and thermal deformation data at measurement points. Then, using the test data, with the corresponding method for the optimization of the selection of each temperature measurement point, the experimental thermal deformation data and optimized measurement point data are used to establish the spindle system's temperature-sensitive points and determine the thermal deformation using the correlation model. This employs a large number of samples to obtain the parameters of the correlation model, and then relies on the measured temperature rise data for thermal-error sensing prediction.

Referring to the structural characteristics of the vertical machining center spindle and the thermal deformation mechanism, this paper proposes an intelligent-sensing method architecture for spindle thermal error based on multi-source information fusion, as shown in Figure 2. This consists of three main layers: the perception layer, the analysis layer, and the reasoning decision-making layer. The perception layer senses the internal and external signals of the machine tool spindle system by arranging a number of sensors, including an infrared thermal imager (Japan Avionics Co., Ltd. Yokohama, Japan) and eddy current displacement sensors, and inputs them into the PC to complete the acquisition of signals. The analysis layer prepares the multi-source information by filtering, denoising, and utilizing other pre-processing and feature extraction techniques to achieve the screening of information, in order to obtain the spindle temperature, motor current, and spindle operating conditions, as well as other characteristics and parameters, enabling high-precision intelligent sensing, to provide more accurate information to the decision-making layer. Data preprocessing reduces the complexity of the data by considering only the extracted or selected data for modeling, thus improving the performance of the model [32]. The decision-making layer fuses the multi-source information, takes the multi-physical-domain fusion information as the input to the intelligent-sensing model of the thermal error of the machine tool spindle, adopts the corresponding intelligent algorithm to process to determine the estimated result of the thermal error, and analyzes the decision to detetermine the optimal strategy by considering the specific constraints [33].

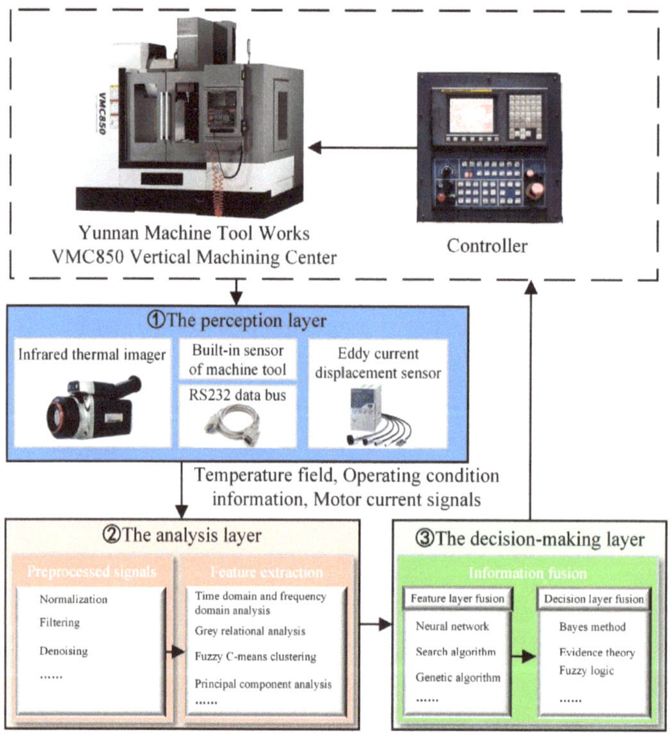

Figure 2. Thermal error intelligent-perception architecture.

2.3. *Intelligent-Perception Model of Spindle Thermal Error Based on Multi-Source Information Fusion*

By analyzing the causes of thermal error in the spindle systems of CNC machine tools, it can be seen that the primary sources of information related to thermal error are the spindle temperature field, operating condition information, and motor current signals. According to the thermal error intelligent-sensing architecture, the thermal-error sensing model of the spindle system is constructed as shown in Figure 3, including key steps such as signal acquisition, signal preprocessing, feature extraction, feature fusion, and decision fusion. In order to obtain comprehensive information about the temperature field and to avoid arranging a large number of contact temperature sensors that would interfere with normal processing, a non-contact infrared thermometer is used to obtain the temperature value of the measurement point from the thermal image. Since the spindle speed and motor current have large influences on the temperature field distribution and thermal deformation of the spindle system, the built-in speed sensor and current sensor of the machine tool are used here to obtain the spindle speed signal and motor current signal, respectively. Signal preprocessing requires filtering and denoising of multi-source signals to improve the efficiency and accuracy of the system's signal processing. The spindle temperature field, operating condition information, and motor current associated with thermal error are extracted to construct a collection of evidence bodies in the feature space. In view of the advantages of the RBF neural network, with strong multi-dimensional nonlinear mapping ability, generalization ability, and clustering analysis ability, the RBF neural network is used to perform feature-layer fusion on the collection of evidence bodies in the feature space. At the same time, considering the uncertain information from multiple sources, the improved D-S evidence theory is further used to fuse the fusion results of the feature layer at the decision-making layer to solve the problem of accurate sensing of thermal errors of CNC machine tool spindles under complex machining environments. By weighting the fusion of different evidence, the degrees of contribution of different information sources can be more accurately reflected, making the final fusion result closer to the real situation. The specific steps are as follows:

Step 1 Analyze the main causes of thermal error, identify multiple sources of information associated with thermal error, and specify the type of sensor.

Step 2 Obtain the signal from the sensors, perform signal preprocessing, extract the feature parameters associated with the thermal error, and construct a collection of evidence bodies in the feature space to complete the training of the temperature rise–thermal error neural network model for temperature-sensitive points.

Step 3 The RBF neural network is used to diagnose the body of evidence in each feature space separately, and the set of preliminary diagnostic results is obtained.

Step 4 Calculate the basic credibility of each preliminary diagnostic result set.

Step 5 The reliability interval of each evidence body in the recognition framework under the action of a single evidence body is calculated according to the basic credibility assignment of the evidence body in each feature space.

Step 6 According to the weighted evidence fusion algorithm, the spindle thermal-error prediction value is calculated, and the thermal-error sensing result is obtained.

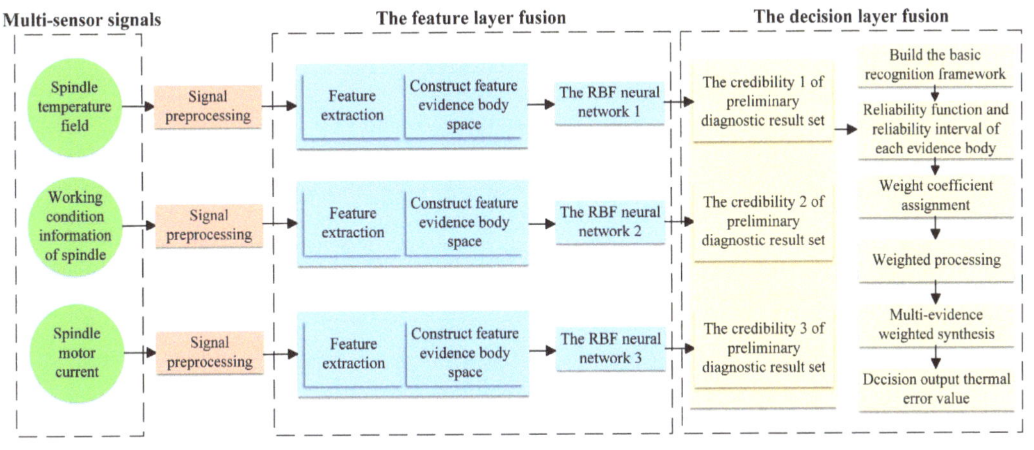

Figure 3. Thermal-error perception model of machine tool based on multi-source information fusion.

The image of the temperature field of the spindle system obtained by the infrared thermal imager contains a large number of temperature measuring points. In order to reduce the amount of computation associated with the sensing model and to ensure the accuracy of the sensing model, it is necessary to optimize the selection of the temperature measuring points. A combination of gray correlation and principal component analysis is used to screen the temperature sensitivities of the temperature field. Firstly, the temperature measuring points highly correlated with thermal error are roughly screened by gray correlation analysis. Then its n uncorrelated principal components are extracted by principal component analysis. The principal components are used instead of the original temperature data as inputs for the thermal-error sensing model; the steps are shown in Figure 4.

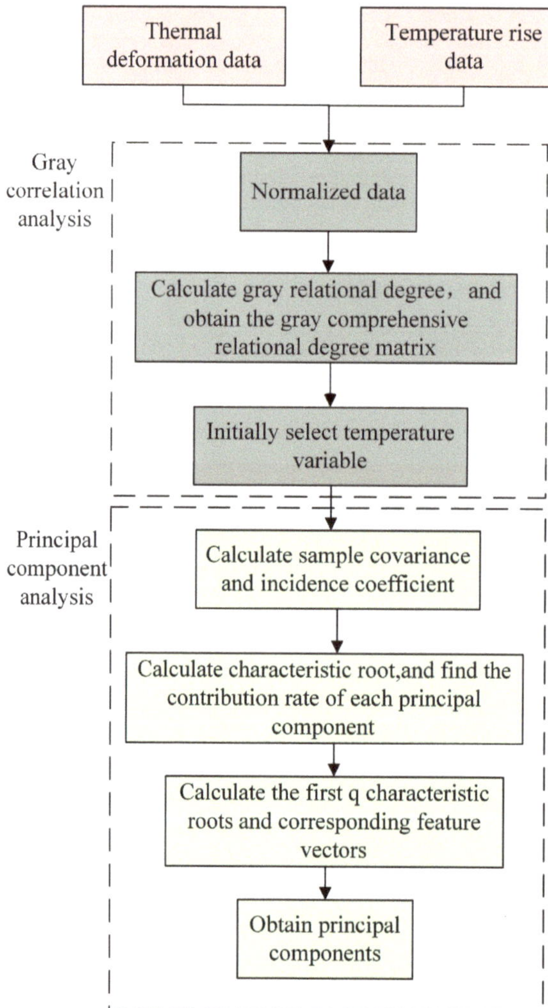

Figure 4. Temperature-sensitive point optimization process.

Firstly, the reference sequence in the gray correlation analysis is determined to be the thermal deformation data, and the comparison sequence is the temperature rise data of the temperature measuring points; the two sequences of data are standardized. Then, the gray comprehensive relational grade between each comparison sequence and the reference sequence is calculated, and the comparison sequence corresponding to the gray comprehensive relational grade is selected as the primary temperature rise variable, that is, the sample sequence for principal component analysis. The covariance between different factors of the sample series is calculated. Specifically, the temperature rise data of different temperature measuring points are ascertained; the contribution rates of the principal components corresponding to each characteristic root are found by calculating the characteristic root, and then selecting the first q characteristic roots and the corresponding characteristic vectors according to the contribution rate to obtain the principal components. Finally, based on the characteristic vectors, the contribution rates of the temperature measuring points as to each principal component are obtained, and the optimal temperature-sensitive points are determined.

The feature extraction of the spindle condition information needs to correspond to the sampling frequency of the temperature field to determine the speed of the spindle at the current moment, as does the signal processing of the rotational speed. Data dimension correspondence is first needed to ensure that the spindle speed sampling time is aligned with the temperature field sampling time. The data are then normalized, and the dimensionless numbers obtained from the normalization are used as features of the speed signal.

The current signal collected through RS232 is a continuous sinusoidal quantity, so the spindle motor current feature extraction must first carry out a Fourier-transform (FFT) on the continuous current flow. Spectral analysis of the current is performed to find the statistical characterization of the current signal in the frequency domain, that is, the center of gravity frequency; the calculation is shown in Equation (1). The center of gravity frequency and the effective value of the motor current are used as current signal characteristics.

$$f_c = \frac{\sum_{i=1}^{n} f_i p_i}{\sum_{i=1}^{n} p_i} \quad (1)$$

where q_i is the frequency domain signal of the current, the center of gravity frequency is f_c, and n is the number of points of spectral data obtained after FFT.

In order to improve the performance of the thermal-error sensing model, the temperature features, operating condition features, and current features from different data sources are fused in the feature layer to construct a collection of evidence bodies in the feature space. Then, the RBF neural network is used to initially fuse the body of evidence in each feature space separately; the specific steps are shown in Figure 5. The number of hidden nodes is first estimated based on an empirical formula, and the number of data centers is determined using the trial-and-error method. The K-means clustering algorithm determines the primary function's data centers and expansion constants. Then, the pseudo-inverse method is utilized to calculate the output layer weights, the model error is calculated, and the training is completed after reaching the standard. When the trained RBF neural network is modeled and saved, the preliminary fusion results of a thermal error are recorded, as well as the accuracy of each neural network for decision-making layer fusion.

The RBF neural network obtains the preliminary fusion result set after fusing the feature layers of the body of evidence in each feature space, which reduces the dimensionality of the interest and extracts the effective features in it. In order to improve the accuracy of the model and make the final fusion result closer to the actual situation, this paper introduces evidence fusion at the decision-making layer; the degrees of contribution of different sources of information can be more accurately reflected through evidence fusion of different evidence.

Since the spindle mechanism speed, motor current, and temperature field of a CNC machine tool are interrelated with each other, the signals related to each thermal error cannot be independent of each other. Classical D-S evidence theory is sensitive to the basic probability assignment function and lacks robustness [34]. It suffers from fusion failure in the face of conflicting evidence. This paper employs a weighted evidence fusion theory at the decision level to address this issue. By introducing weight parameters to weight different evidence, the evidence with higher credibility or more importance has more influence in the fusion process, thus effectively solving the problem of conflicting evidence. The weighted evidence fusion is calculated as follows.

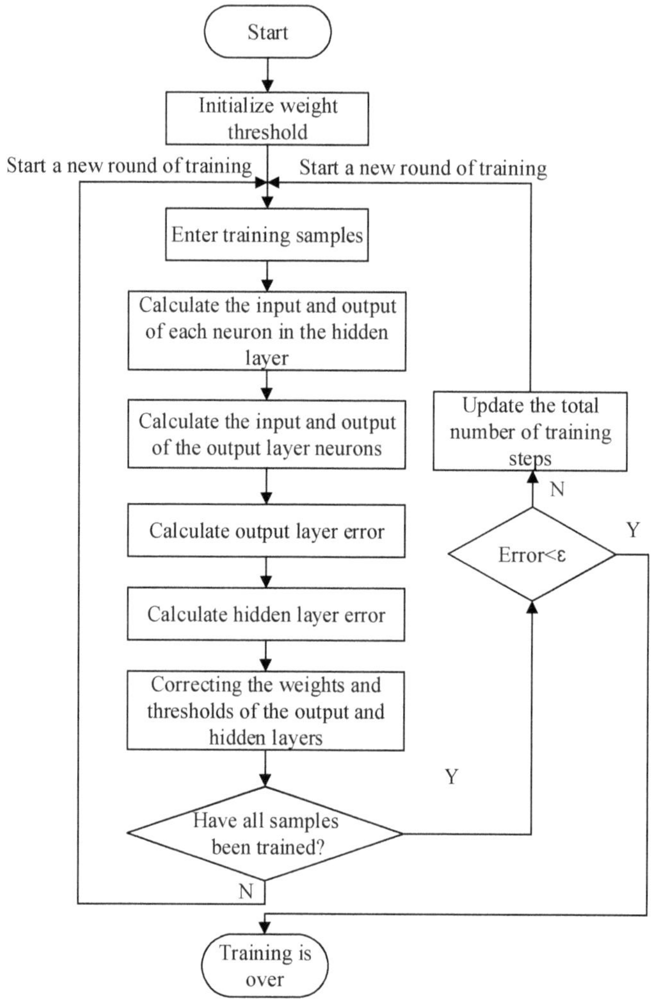

Figure 5. Training process of the generalized RBF neural network model.

It is assumed that the set of independent possible conclusions of a problem is the identification framework $\Theta = \{A_1, A_2, \ldots A_n\}$, A_i is the basic-element of Θ, and 2^Θ is the power set of Θ. If the set function mapping $m : 2^\Theta \to [0,1]$ satisfies

$$\begin{cases} m(\varnothing) = 0 \\ \sum\limits_{A\subseteq\Theta} m(A) = 1, A \neq \varnothing \end{cases} \tag{2}$$

then the mapping $m : 2^\Theta \to [0,1]$ is called the basic probability distribution function on the identification framework Θ. $\forall A \subseteq \Theta$, $m(A)$ is called the basic probability assignment of A. The identification framework Θ is defined by

$$\begin{cases} Bel(A) = \sum\limits_{B\subseteq A} m(B) \\ Pl(A) = 1 - Bel(A) = \sum\limits_{B\cap A=\varnothing} m(B) \end{cases} \tag{3}$$

The mapping $Bel : 2^{\Theta} \rightarrow [0,1]$ is the belief function on the identification framework Θ. Mapping $Pl : 2^{\Theta} \rightarrow [0,1]$ is the plausibility function of *Bel*. For $\forall A \subseteq \Theta$, $[Bel(A), Pl(A)]$ is called the belief interval of A. The belief interval describes the upper and lower bounds on the degree of confidence that the current evidence holds in proposition A.

When two pieces of evidence are combined, m_1 and m_2 are the basic probability distribution functions on the same identification framework Θ. The basic-elements are $E_1, E_2, \ldots E_k$ and $F_1, F_2, \ldots F_n$, respectively. If $\forall A \subseteq \Theta$ and

$$\mathrm{N} = \sum_{E \cap F \neq \varnothing} m_1(E) \cdot m_2(F) > 0 \tag{4}$$

then, the synthesized basic probability distribution function $m : 2^{\Theta} \rightarrow [0,1]$ is

$$\begin{cases} m(\varnothing) = 0 \\ m(A) = \frac{1}{N} \sum\limits_{E \cap F = A} m_1(E) \cdot m_2(F), A \neq \varnothing \end{cases} \tag{5}$$

In the above formula, N is a normal number. The function is to assign the lost reliability on the empty set to the non-empty set, in proportion, to meet the requirements of the probability assignment. The N value can reflect the degree of evidence conflict; the greater the evidence conflict, the smaller the N value. The above formula is called the orthogonal sum, denoted by $m_1 \oplus m_2$. Moreover, the combination of evidence is independent of the order of operation. Therefore, the calculation of multiple evidence combinations can be recursively derived from the calculation of two evidence combinations to obtain

$$m = \{[(m_1 \oplus m_2) \oplus m_3] \oplus \ldots\} \oplus m_n \tag{6}$$

In the weighted evidence theory, the evidence weight factor is determined by the degree to which that evidence conflicts with other evidence. For the same identification framework Θ, let the reliability of each body of evidence for identifying n propositions in the identification framework be $R(A) \rightarrow [0,1], \forall A \subset \Theta$, then by

$$W(A) = \frac{n \cdot R^k(A)}{\sum_{A \subset \Theta} R^k(A)} \tag{7}$$

The mapping $W(\cdot)$ is a weight coefficient assignment function on the identification framework Θ. $\forall A \subseteq \Theta$, *W(A)* is referred to as the weight coefficient assignment of the body of evidence to *A*; in the formula, and when the data on the reliability of the evidence for the identification of each proposition is more reliable, k takes a larger value. The weight coefficient *W(A)* of the evidence reflects that the evidence has different degrees of reliability in identifying the propositions in the identification framework.

The basic probability assignment function *m*:2^Θ→[0,1] is weighted to take full account of the weight of each piece of evidence for each proposition when the evidence is combined. $\forall A \subseteq \Theta$, and then there are

$$\begin{cases} W_m(A) = \frac{W(A) \cdot m(A)}{m(\Theta) + \sum\limits_{A \subset \Theta} W(A) \cdot m(A)} \\ W_m(\Theta) = 1 - \sum\limits_{A \subset \Theta} W_m(A) \end{cases} \tag{8}$$

The mapping $Wm : 2^{\Theta} \rightarrow [0,1]$ is a weighted probability assignment function on the identification framework Θ. $\forall A \subseteq \Theta$, $Wm(A)$ is called the weighted probability assignment of A.

Therefore, the weighted synthesis rule for multiple evidence is

$$Wm = \{[(Wm_1 \oplus Wm_2) \oplus Wm_3] \oplus \ldots\} \oplus Wm_n \tag{9}$$

As a result of the weighting of the basic probability assignment function, rational evidence is strengthened, irrational evidence is weakened, and conflicts between evidence are significantly reduced. Therefore, the improved weighted evidence theory can more widely meet various practical applications. The improved weighted evidence theory fully utilizes the information from the evidence sources and eliminates the incompleteness and uncertainty of the information contained in a single data source.

3. Thermal-Error Perception Experiment Setup

3.1. CNC Machine Tool Spindle Thermal-Error Perception Experiment

A large number of scholars' experimental research has found that the spindle, in the X and Y directions, does not produce significant thermal error [35]. In contrast, the Z-direction thermal error is very obvious. So the spindle thermal error is generally established as a mapping relationship between the spindle temperature measuring points and the Z-direction thermal error. This paper's intelligent-sensing model of thermal error mainly establishes the relationship between the temperature field at the corresponding moment, the working conditions, the spindle motor current and thermal deformation measurement point data. Therefore, the data to be measured in the thermal-error experiment are the temperature field, spindle speed, spindle current, and Z-direction thermal deformation.

The experiment takes the Yunnan Machine Tool Factory VMC850 vertical machining center as the perceived object. This paper's experiments select an eddy current displacement sensor to measure the thermal error of the spindle system. The contact temperature sensor has many installation elements and cumbersome wiring is employed in the actual measurement, leading to inaccurate measurement results. So, in this paper, we select an infrared thermal imager to collect the temperature data from the temperature measurement points of the spindle system of the vertical machining centers. Then, the data is transmitted to the computer through USB transmission. Speed and motor current can be monitored and collected in real time by connecting the spindle's built-in sensor to the PC through the RS232 transmission bus. The eddy current displacement sensor can realize the real-time measurement of thermal deformation of the spindle under the idle state of the machine tool through non-contact measurement. The experimental platform built for each device is shown in Figure 6.

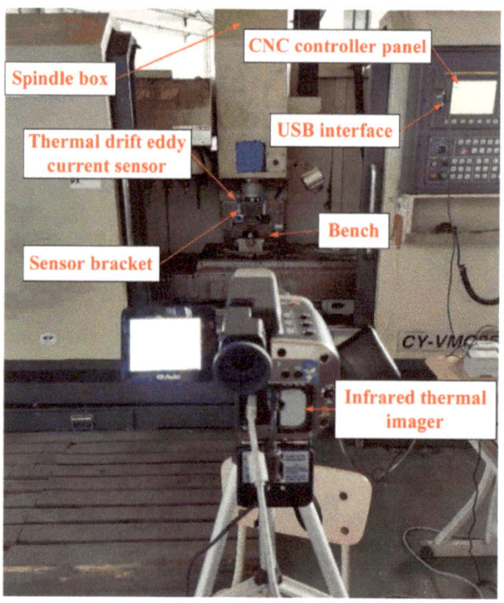

Figure 6. Experimental platform of thermal error intelligent-sensing system.

Considering the huge structure of the machine tool and the large number of sensors required, this paper adopts an infrared imager to take pictures of the spindle system. The thermal image and its temperature measuring points are arranged as shown in Figure 7. Points a, b, c, g, h, and i are the spindle box shell temperatures; points d, e, and f are the points for the spindle motor shell temperature; points j and k are for the temperature of spindle bearing; l is the temperature of the spindle claw disc; and the m and n points will not produce noticeable temperature changes with the extension of the spindle working time, so they are used to represent the workbench and the ambient temperature.

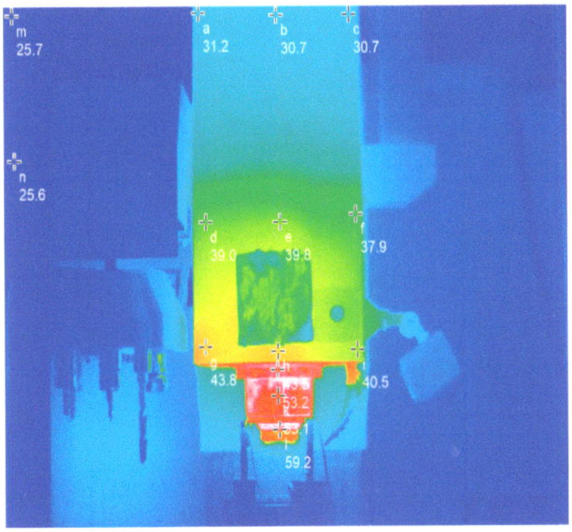

Figure 7. Layout of temperature measuring points at a certain time during machine operation.

In order to simulate the change of spindle speed during the actual machining process, the experiment was carried out in no-load mode. We edited the working procedure of the vertical machining center according to the set working conditions, collected data every 1 min, and produced four groups of data in total. Constant speed and ISO-variable-speed experimental data were used for modelling, and stepped variable speed data were used to validate the model's predictive accuracy.

3.2. Experimental Conditions Design

In order to simulate the thermal drift of the spindle under actual processing conditions through no-load experiments, this project carried out thermal-error test experiments. ISO [36] testing standards identified three no-load experiments at different speeds. The temperature, spindle motor current, spindle speed, and thermal-error data were collected under different speed conditions to provide basic data for intelligent-perception modelling and model verification of thermal error. The three groups of no-load test conditions are composed of one group at constant speed conditions, and two groups at variable speed conditions. The constant speed idling was 2000 r/min for 2 h and 4000 r/min for 2 h, respectively. The two sets of variable speed conditions comprised ISO-standard variable speed conditions and stepped variable speed conditions, as shown in Figures 8 and 9.

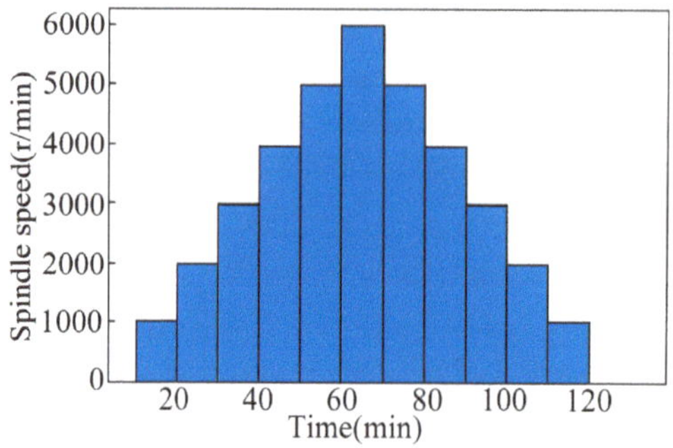

Figure 8. ISO-standard variable speed operation: speed–time diagram.

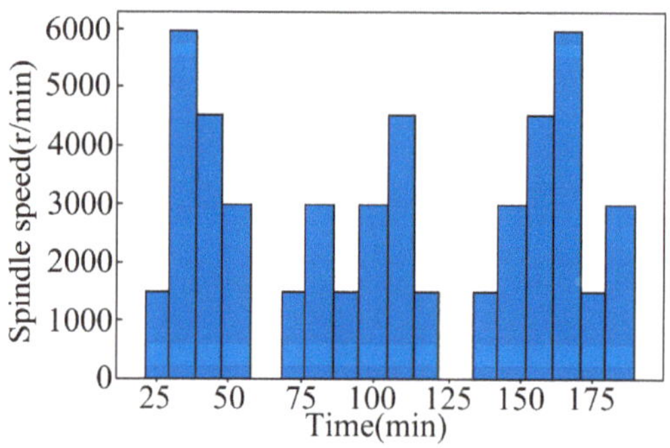

Figure 9. Stepped variable speed operation: speed–time diagram.

The two working conditions of constant speed and ISO-standard variable speed were used for model training, and the stepped variable speed condition was used for model testing.

4. Results of the Experiment and Discussion

4.1. Results of the Experiment and Data Analysis

The thermal characteristic experiment was carried out at the two sets of off-line speeds described in Section 3.2; the temperature rises for the measured points in the temperature field of the spindle at a constant speed of 2000 r/min, a constant speed of 4000 r/min, and a variable speed of ISO are shown in Figure 10.

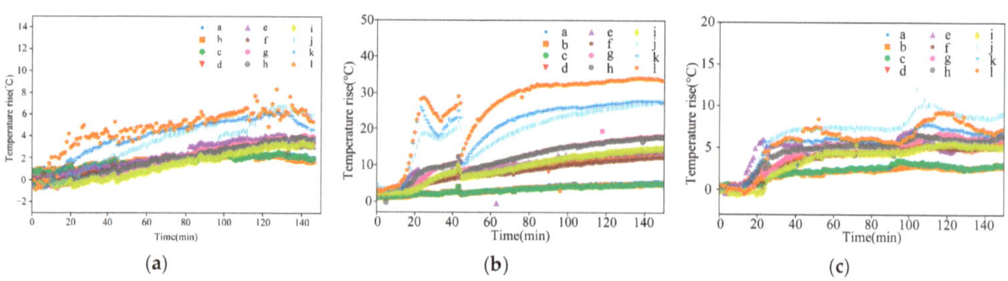

Figure 10. Temperature rises at temperature measuring points at 2000 r/min, 4000 r/min constant speed, and ISO-standard variable speed: (**a**) 2000 r/min constant speed; (**b**) 4000 r/min constant speed; and (**c**) ISO-standard variable speed.

The figure shows that under a constant speed of 2000 r/min, the temperature field no longer produces apparent changes after the spindle is run for 95 min. Point l (temperature measuring point of spindle claw disc) has the most drastic temperature change, with an eventual stable temperature rise of about 7.8 °C. Under the conditions of 4000 r/min constant speed, the spindle bearing and claw disc at first produced a significant temperature rise; it then dropped back to normal temperature as the cooling air conditioning started working. The spindle bearing and claw disc reached a stable temperature rise after 80 min of operation, within which the temperature rise at point l was the most obvious, rising to about 34.8 °C. After 100 min, the other temperature measuring points no longer produced significant temperature rises, and reached a stable state. Under ISO variable speed conditions, the spindle temperature field produced different temperature rises with speed, and the stable thermal equilibrium time is no longer apparent. The temperature change at point g (the temperature measuring point of the spindle bearing) is most drastic under variable speed conditions. This proves that the temperature field variation of the spindle is closely related to the speed. Between 0 and 20 min, the rotation speed of the spindle is small, and the temperatures of the relevant measuring points also change little in this period. Between 20 and 40 min, the speed of the spindle is larger, and the temperature rise of the measuring point is also larger. Between 40 and 100 min, the speed of the spindle decreases, and the temperature rise of the measuring point does not change significantly. Between 100 and 120 min, the spindle speed increases, so the measuring point has a significant temperature rise. After 120 min, the speed of the spindle decreases, and the temperature of the measuring point slowly falls back and tends to be stable.

In the off-line experiment of spindle thermal characteristics, the Z-direction thermal deformations produced by the spindle at 2000 r/min constant speed, 4000 r/min constant speed, and ISO variable speed are shown in Figure 11. As can be seen from the figure, the maximum change in thermal error when the spindle is running at constant speed occurs between 10 min and 20 min; the main reason is that the temperature of the machine tool is low after the tool is switched on, and the heat only begins to gather. The machine tool then works for a period of time, producing temperature field changes, and thus causing thermal deformation of the spindle. After 80 min, the machine reaches thermal equilibrium, and the thermal error reaches a stable value. Moreover, according to the graph for the variable speed, it can be seen that due to its lower speed, the change of thermal error in the initial state is much smaller than that in the constant speed state. It can be seen that the thermal error of the spindle is positively correlated with the speed of the spindle, and the thermal error is generated with a certain lag.

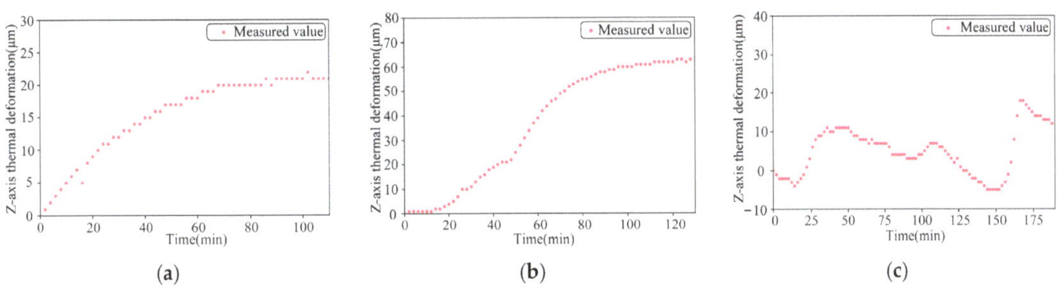

Figure 11. Z-axis thermal deformation at 2000 r/min constant speed, 4000 r/min constant speed, and ISO variable speed: (**a**) 2000 r/min constant speed; (**b**) 4000 r/min constant speed; and (**c**) ISO-standard variable speed.

4.2. Comparison and Analyses of Prediction Performance of Thermal Error Intelligent Perception Models

The BP neural network is the most widely used network in thermal-error modeling, one which has good results for thermal-error prediction and is universal across different machine tools [37]. Moreover, the combination of wavelet analysis and BP neural networks can solve artificial neural networks' slow-convergence problems and transition fitting. The wavelet neural network optimized with the genetic algorithm has higher accuracy and faster convergence. Therefore, this paper uses two thermal-error modeling methods, the traditional BP neural network and the wavelet neural network optimized by genetic algorithm, for the thermal-error comparison model.

The performance of the three models is compared for thermal-error prediction under three different operating conditions, namely, 2000 r/min constant speed, 4000 r/min constant speed, and ISO variable speed, respectively, as shown in Figures 12–14. In order to verify the robustness of the model proposed in this paper, the experimental data measured under the working conditions of a stepped variable speed are input into the thermal-error prediction model as test data and compared with the measured Z-axis thermal-error, as shown in Figure 15.

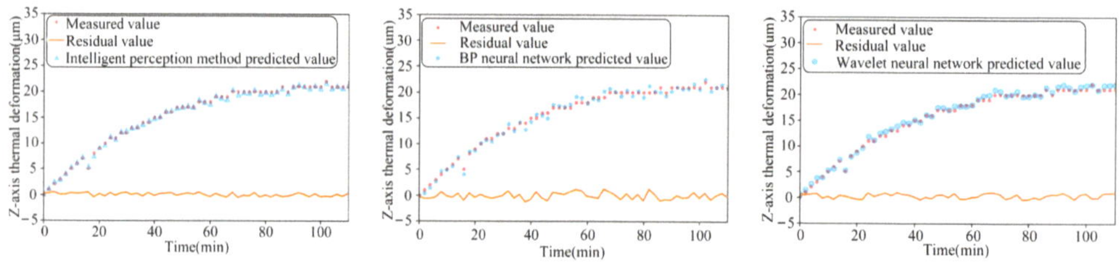

(a) (b) (c)

Figure 12. The comparison between the measured value of 2000 r/min idling and the predicted value of the three models: (**a**) Intelligent-perception method of multi-source information fusion; (**b**) BP neural network speed; and (**c**) Wavelet neural network.

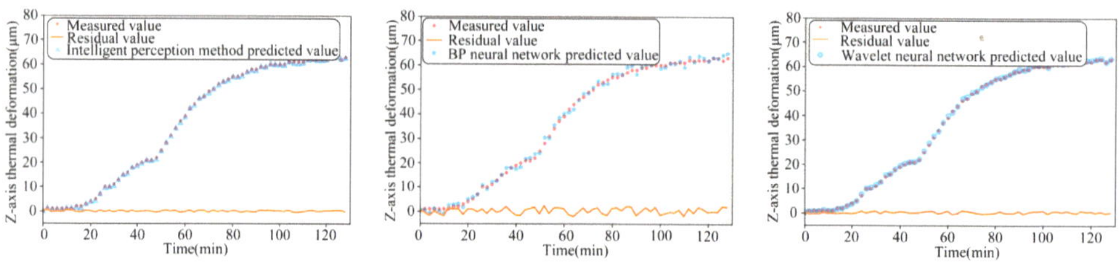

(a) (b) (c)

Figure 13. The comparison between the measured value of 4000 r/min idling and the predicted value of the three models: (**a**) Intelligent-perception method of multi-source information fusion; (**b**) BP neural network speed; and (**c**) Wavelet neural network.

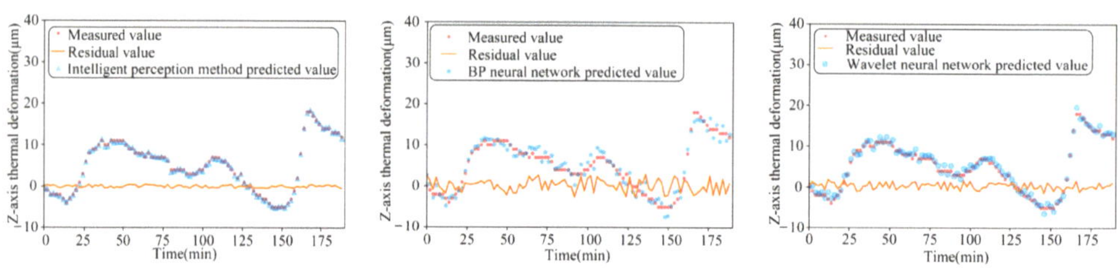

(a) (b) (c)

Figure 14. The comparison between the measured value of ISO-standard variable speed idling and the predicted value of the three models: (**a**) Intelligent-perception method of multi-source information fusion; (**b**) BP neural network speed; and (**c**) Wavelet neural network.

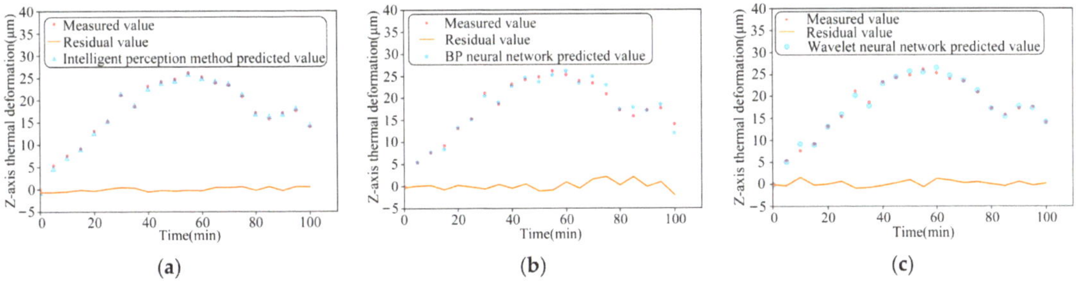

Figure 15. The comparison between the measured value of idling with step change speed and the predicted value of the three models: (**a**) Intelligent-perception method of multi-source information fusion; (**b**) BP Neural Network speed; (**c**) Wavelet neural network.

As can be seen from the figure, the maximum residuals of the thermal error intelligent-sensing method based on multi-source information fusion, BP neural network and wavelet neural network optimized by genetic algorithm are 0.49 μm, 1.26 μm, and 0.96 μm, respectively, under the idling at a constant rotational speed of 2000 r/min. The maximum residuals of the thermal error intelligent-sensing method based on multi-source information fusion, BP neural network, and wavelet neural network optimized by genetic algorithm are 0.47 μm, 2.41 μm, and 0.99 μm, respectively, under idling at a constant rotational speed of 4000 r/min. The maximum residuals of thermal errors of the main shaft based on the intelligent-sensing method of thermal error with multi-source information fusion, BP neural network, and wavelet neural network optimized by genetic algorithm, under online prediction of step-variable rotational speed conditions, are 0.57 μm, 2.09 μm, and 1.52 μm, respectively. The model's superiority under complex working conditions is further verified. The intelligent-sensing model performs well under a variety of operating conditions, proving that it is robust and able to adapt to different operating conditions and environmental changes, which reduces the need for model retraining and parameter tuning.

The results of the comparison of these three models are shown in Table 1, which com-pares and analyses the effectiveness of thermal-error prediction of the spindle under constant speed, standard variable speed, and stepped variable speed conditions. We found that the prediction accuracy of the intelligent-perception model based on multi-source in-formation fusion reaches 98.8%, which is 6.6% higher than the traditional BP neural network perception model. The traditional BP neural network has low prediction accuracy under high and variable speed conditions, is easily affected by noise and non-linearity, and has poor robustness. The intelligent-perception model based on the fusion of multi-source information improved by 4.3% over the wavelet neural network optimized by genetic algorithm. Although the optimization effect of the wavelet neural network optimized by genetic algorithm is significant, there is still room for improvement under dynamic and complex working conditions. The maximum residual errors of the thermal error intelligent-sensing model based on multi-source information fusion proposed in this paper are generally smaller than those of the traditional thermal-error prediction model. The stepped variable speed as test data in the intelligent-perception model obtained good prediction results with small maximum residual errors and high accuracy.

Table 1. Comparison of the prediction-effectiveness of the three models.

Model	BP Neural Network Model		Wavelet Neural Network Model		Intelligent-Perception Model	
	Maximum Residual Error (μm)	Fitting Accuracy	Maximum Residual Error (μm)	Fitting Accuracy	Maximum Residual Error (μm)	Fitting Accuracy
2000 r/min off-line detection	1.26	94.7%	0.96	96.3%	0.49	98.1%
4000 r/min off-line detection	2.41	96.2%	0.99	98.5%	0.47	99.3%
ISO variable speed off-line detection	2.97	85.2%	1.79	91.4%	0.48	98.6%
Stepped variable speed online prediction	2.09	92.6%	1.52	94.5%	0.57	98.8%

5. Conclusions

The spindle system is an essential core component of CNC machine tools, and its operational status plays a vital role in machining quality. The main idea of realizing the intelligent sensing of thermal error in the spindle system is to utilize multiple sensors to monitor the relevant information of spindle thermal errors. Through suitable mathematical models, these pieces of information are comprehensively analyzed, modeled, and processed in order to make decisions, resulting in an intelligent perception of thermal error. The decision given is the result of intelligent sensing of thermal error. The realization of intelligent spindle autonomous perception plays a vital role in its subsequent active control and predictive reasoning. This paper introduces the evidence theory, based on a neural network, establishes the intelligent-sensing model of spindle thermal error, and adopts the feature layer–decision layer approach to realize the fusion of multi-source information at different levels, which further improves the accuracy and robustness of the thermal-error sensing.

(1) Aiming at the shortcomings of single-sensor-based information, namely, characterization ability, high contingency, and susceptibility to being interfered with by external environmental factors, this paper proposes a multi-source and multi-level information fusion intelligent-sensing method. A multi-source and multi-layer thermal error intelligent-sensing model is established by feature extraction and fusion of multiple related signals affecting the thermal error of the spindle. In the model construction, the RBF neural network is used for the initial fusion of feature layers to improve the model's generalizing ability. At the same time, the weighted evidence fusion theory is introduced, which can more accurately reflect the contribution degree of different information sources by weighted fusion of varying evidence, making the final fusion results closer to the actual value. Thus, the intelligent-perception model has a higher prediction accuracy and lays a specific theoretical foundation for developing intelligent spindles.

(2) This project conducted thermal-error perception experiments on the spindle system of the VMC850 vertical machining center manufactured by Yunnan Machine Tool Factory, under the conditions of constant speed, standard variable speed, and stepped variable speed. Then we compared the effectiveness of the traditional thermal-error sensing model and the intelligent-sensing model with multi-source information fusion. The experiment's results show that the prediction accuracy of the multi-source and multi-level information fusion intelligent-sensing model proposed in this paper can reach 98.8%, which is significantly better than the traditional model. This shows that the method proposed in this paper has significant advantages and application potential in solving the thermal-error sensing problem of the CNC machine tool spindle.

(3) This paper summarizes and analyzes the signals related to thermal error inside and outside the machine tool and finally selects three signals for multi-source information fusion to obtain the thermal-error model. The amount of data is increased compared with the traditional thermal-error model. However, because the thermal-error influencing factors of the machine tool under complex working conditions are multifarious,

the number and types of training samples can be gradually increased in the next study to further train and optimize the model's parameters.

(4) In addition, in order to obtain a more comprehensive thermal-error model under the various working conditions of the machine tool, it is necessary to measure the thermal error changes under different working conditions, especially under the cutting working condition. However, due to the limitations of the experimental conditions, this paper simulates the thermal error changes of the spindle under different working conditions only through the spindle's different rotational speeds. It does not take into account the effect of the cutting heat on the thermal error. Therefore, the experiments, as to the working-condition information, can be further enriched, and a more complete thermal-error model can be obtained.

Author Contributions: Conceptualization, Z.Y.; Methodology, software, validation, and resources, Z.Y.; Data curation, Z.Y.; Writing—original draft, B.L.; Writing—review and editing, Z.Y., Y.Z., Y.C., G.Z., W.Y., Z.Z. and H.Z.; Funding acquisition, Z.Y. All authors have read and agreed to the published version of the manuscript.

Funding: This work is partially supported by National Natural Science Foundation of China (Grant number 52175461, U1864208, and U2341275); Intelligent Manufacturing Project of Tianjin (Grant number 20201199).

Institutional Review Board Statement: Not applicable.

Informed Consent Statement: Not applicable.

Data Availability Statement: Data are contained within the article.

Conflicts of Interest: The authors declare no conflicts of interest.

References

1. Josef, M.; Jerzy, J.; Eckart, U. Thermal issues in machine tools. *CIRP Ann.* **2012**, *61*, 771–791.
2. Abele, E.; Altintas, Y.; Brecher, C. Machine tool spindle units. *CIRP Ann. Manuf. Technol.* **2010**, *59*, 781–802. [CrossRef]
3. Christian, B.; Matthias, B.; Michel, K.; Tae, H.; Filippos, T. Application of an unscented Kalman filter for modeling multiple types of machine tool errors. *Procedia CIRP* **2017**, *63*, 449–459.
4. Guo, J.H.; Xiong, Q.Y.; Chen, J.; Miao, E.M.; Wu, C.; Zhu, Q.W.; Yang, Z.Y.; Chen, J. Study of static thermal deformation modeling based on a hybrid CNN-LSTM model with spatiotemporal correlation. *Int. J. Adv. Manuf. Technol.* **2022**, *119*, 2601–2613. [CrossRef]
5. Jia, G.G.; Zhang, X.; Wang, X.Z.; Zhang, X.P.; Huang, N.D. A spindle thermal error modeling based on 1DCNN-GRU-Attention architecture under controlled ambient temperature and active cooling. *Int. J. Adv. Manuf. Technol.* **2023**, *127*, 1525–1539. [CrossRef]
6. Wei, G.; Soichi, I.; Alkan, D.; Daisuke, K.; Mayer, J.R.R.; Chen, Y.L.; Karoly, S.; Andreas, A.; Jean-Marc, L.; Norikazu, S. Machine tool calibration: Measurement, modeling, and compen-sation of machine tool errors. *Int. J. Mach. Tool. Manuf.* **2023**, *187*, 104017.
7. Jakstas, A.; Kausinis, S.; Barauskas, R.; Kasparaitis, A.; Barakauskas, A. Thermal error analysis in precision length measurements Measurement. 2014, 51, 133–146. *51*.
8. Liu, T.; Gao, W.G.; Zhang, D.W.; Chang, W.F.; Liang, C.M.; Tian, Y.L. Analytical modeling for thermal errors of motorized spindle unit. *Int. J. Mach. Tools Manuf.* **2017**, *112*, 53–70. [CrossRef]
9. Kim, S.K.; Cho, D.W. Real-time estimation of temperature distribution in a ball-screw system. *Int. J. Mach Tools Manuf.* **1997**, *37*, 451–464. [CrossRef]
10. Huang, S.G.; Feng, P.F.; Xu, C.; Ma, Y.; Ye, J.; Zhou, K. Utilization of heat quantity to model thermal errors of machine tool spindle. *Int. J. Adv. Manuf. Technol.* **2018**, *97*, 1733–1743. [CrossRef]
11. Li, Y.; Zhao, W.H.; Lan, S.H.; Ni, J.; Wu, W.W.; Lu, B.H. A review on spindle thermal error compensation in machine tools. *Int. J. Mach. Tools Manuf.* **2015**, *95*, 20–38. [CrossRef]
12. Wu, C.H.; Kung, Y.T. Thermal analysis for the feed drive system of a CNC machine center. *Int. J. Mach. Tools Manuf.* **2003**, *43*, 1521–1528. [CrossRef]
13. Yang, A.S.; Yu, X.H.; Zhuang, J.R.; Lee, C.Y.; Hsieh, W.H. DOE-FEM based design improvement to minimize thermal errors of a high speed spindle system. *Therm. Sci. Eng. Prog.* **2018**, *8*, 525–536. [CrossRef]
14. Ma, C.; Yang, J.; Zhao, L.; Mei, X.S.; Shi, H. Simulation and experimental study on the thermally induced deformations of high-speed spindle system. *Appl. Therm. Eng.* **2015**, *86*, 251–268. [CrossRef]
15. Chethana, R.G.; Rahul, D.; Aslam, T.P.; Ravi, L. Use of multiple linear regression to compensate for diametrical deviations in machined components due to thermal errors. *Mater. Today* **2022**, *56*, 2630–2639.
16. Zhao, C.Y.; Xia, Y.M.; Chen, X.Z.; Jiang, Y.F.; He, Y.; Pan, S.L.; Fei, Y. Thermal error detection and compensation technology for spindle of horizontal CNC machine tool with large torque. *Int. J. Adv. Manuf. Technol.* **2020**, *107*, 85–96. [CrossRef]

17. Li, Z.L.; Wang, Q.H.; Zhu, B.; Wang, B.D.; Zhu, W.M. Thermal error modeling of high-speed electric spindle based on Aquila Optimizer optimized least squares support vector machine. *Case Stud. Therm. Eng.* **2022**, *39*, 102432. [CrossRef]
18. Huang, Y.Q.; Zhang, J.; Li, X.; Tian, L.J. Thermal error modeling by integrating GA and BP algorithms for the high-speed spindle. *Int. J. Adv. Manuf. Technol.* **2014**, *71*, 1669–1675. [CrossRef]
19. Lee, J.H.; Lee, J.H.; Yang, S.H. Thermal Error Modeling of a Horizontal Machining Center Using Fuzzy Logic Strategy. *J. Manuf Process.* **2012**, *3*, 120–127. [CrossRef]
20. Guo, Q.J.; Fan, S.; Cheng, X.; Zhao, G.Y.; Yang, J.G. Thermal Error Optimization Modeling of a Five-axis Machine Tool. *Chin. J Mech. Eng.* **2017**, *30*, 746–753. [CrossRef]
21. Li, T.J.; Sun, T.Y.; Zhang, Y.M.; Zhao, C.Y. Prediction of thermal error for feed system of machine tools based on random radial basis function neural network. *Int. J. Adv. Manuf. Technol.* **2021**, *114*, 1545–1553. [CrossRef]
22. Ma, C.; Zhao, L.; Mei, X.S.; Shi, H.; Yang, J. Thermal error compensation of high-speed spindle system based on a modified BP neural network. *Int. J. Adv. Manuf. Technol.* **2017**, *89*, 3071–3085. [CrossRef]
23. Li, Y.; Yu, M.L.; Bai, Y.M.; Hou, Z.Y.; Wu, W.W. A review of t.hermal error modeling methods for machine tools. *Appl. Sci.* **2021**, *11*, 5216. [CrossRef]
24. Wu, C.H.; Kung, Y.T. Thermal analysis and compensation of a double-column machining center. *Proc. Inst. Mech. Eng. Part B J Eng. Manuf.* **2006**, *220*, 109–117. [CrossRef]
25. Yang, J.G.; Yuan, J.X.; Ni, J. Thermal error mode analysis and robust modeling for error co mpensation on a CNC turning center. *Int. J. Mach. Tools Manuf.* **1999**, *39*, 1367–1381. [CrossRef]
26. Guo, Q.; Yang, J.; Wu, H. Application of ACO-BPN to thermal error modeling of NC machine tool. *Int. J. Adv. Manuf. Technol.* **2010**, *50*, 667–675. [CrossRef]
27. Liu, H.; Miao, E.M.; Zhang, L.Y.; Tang, D.F.; Hou, Y.L. Correlation Stability Problem in Selecting Temperature measuring point of CNC Machine Tools. *Machines.* **2022**, *10*, 132. [CrossRef]
28. Fu, G.Q.; Tao, C.; Xie, Y.P.; Lu, C.J.; Gao, H.L. Temperature-sensitive point selection for thermal error modeling of machine tool spindle by considering heat source regions. *Int. J. Adv. Manuf. Technol.* **2021**, *112*, 2447–2460. [CrossRef]
29. Liu, Z.H.; Yang, B.; Ma, C.; Wang, S.L.; Yang, Y.F. Thermal error modeling of gear hobbing machine based on IGWO-GRNN. *Int. J. Adv. Manuf. Technol.* **2020**, *106*, 5001–5016. [CrossRef]
30. Shi, H.; Jiang, C.P.; Yan, Z.Z.; Tao, T.; Mei, X.S. Bayesian neural network-based thermal error modeling of feed drive system of CNC machine tool. *Int. J. Adv. Manuf. Technol.* **2020**, *108*, 3031–3044. [CrossRef]
31. Li, Z.L.; Wang, B.D.; Zhu, B.; Wang, Q.H.; Zhu, W.M. Thermal error modeling of electrical spindle based on optimized ELM with marine predator algorithm. *Case Stud. Therm. Eng.* **2022**, *38*, 102326. [CrossRef]
32. Marko, Q.; Nebojša, B.; Dragoljub, S.; Maja, Č.; Dragana, O.; Marko, P. From predictive to energy-based maintenance paradigm: Achieving cleaner production through functional-productiveness. *J. Clean. Prod.* **2023**, *408*, 137177.
33. Malgorzata, J.K.; Stanislaw, L.; Piotr, K. Maintenance 4.0 technologies—New opportunities for sustainability driven maintenance. *Manag. Prod. Eng. Rev.* **2020**, *11*, 74–87.
34. Chen, B.; Wang, J.F.; Chen, S.B. Prediction of pulsed GTAW penetration status based on BP neural network and D-S evidence theory information fusion. *Int. J. Adv. Manuf. Technol.* **2010**, *48*, 83–94. [CrossRef]
35. Cao, H.R.; Zhang, X.W.; Chen, X.F. The concept and progress of intelligent spindles: A review. *Int. J. Mach. Tools Manuf.* **2017**, *112*, 21–52. [CrossRef]
36. *ISO 230-3:2001*; Test Code for Machine Tools—Part 3: Determination of Thermal Effects. International Organization for Standardization: Geneva, Switzerland, 2001.
37. Dai, Y.; Yu, B.L.; Tao, X.S.; Wang, X.; He, S.; Wang, G. Thermal displacement prediction of variable preload motorized spindles based on speed reduction experiments and IABC-BP optimization models. *Case Stud. Therm. Eng.* **2024**, *53*, 103941. [CrossRef]

sensors MDPI

Article

Using Spectroradiometry to Measure Organic Carbon in Carbonate-Containing Soils

Piotr Bartmiński [1,*], Anna Siedliska [2] and Marcin Siłuch [1]

[1] Department of Geology, Soil Science and Geoinformation, Institute of Earth and Environmental Sciences, Maria Curie-Skłodowska University, al. Kraśnicka 2cd, 20-718 Lublin, Poland; marcin.siluch@mail.umcs.pl
[2] Institute of Agrophysics Polish Academy of Sciences, ul. Doświadczalna 4, 20-290 Lublin, Poland; a.siedliska@ipan.lublin.pl
* Correspondence: piotr.bartminski@mail.umcs.pl; Tel.: +48-81-5376889

Abstract: This study explores the feasibility of analyzing soil organic carbon (SOC) in carbonate-rich soils using visible near-infrared spectroscopy (VIS-NIR). Employing a combination of datasets, feature groups, variable selection methods, and regression models, 22 modeling pipelines were developed. Spectral data and spectral data combined with carbonate contents were used as datasets, while raw reflectance, first-derivative (FD) reflectance, and second-derivative (SD) reflectance constituted the feature groups. The variable selection methods included Spearman correlation, Variable Importance in Projection (VIP), and Random Frog (Rfrog), while Partial Least Squares Regression (PLSR), Random Forest Regression (RFR), and Support Vector Regression (SVR) were the regression models. The obtained results indicated that the FD preprocessing method combined with RF, results in the model that is sufficiently robust and stable to be applied to soils rich in calcium carbonate.

Keywords: VIS-NIR spectrometry; carbonate soils; SOC

1. Introduction

Visible near-infrared spectroscopy (VIS-NIR) has become an increasingly widely used research tool in recent years [1–3]. The technique is based on the phenomenon of reflectivity, and the material under examination is exposed to electromagnetic radiation of a certain range and intensity. Specific chemical bonds absorb the radiation, reaching the object in different ways; by estimating how much radiation has been reflected from the object in the electromagnetic spectrum range from 400 to 2500 nm, the content of selected compounds can be determined indirectly. The visualization of the acquired information—a characteristic reflectance curve within which there are substance-specific peaks and valleys—provides a spectral signature that, when properly processed, can provide valuable qualitative and quantitative data. The acquired data require the necessary processing—preprocessing—for which various tools are used, such as curve smoothing, moving averages, etc. [4]. In many cases, the curves need to be corrected in specific ranges at the interface between the measurement ranges of the sensors used in the instrument; in a broad spectrum, it is impossible to use a single sensor. The data, prepared in an appropriate manner, are analyzed statistically, with the analysis usually studying either whole spectral curves or selected continuous fragments of them, dedicated—usually on the basis of previous scientific studies—to specific chemical compounds [5].

VIS-NIR spectroscopy is used in many fields: for instance, in agriculture, including precision agriculture, it is related to the appropriate application of the right amounts of substances necessary for plant growth and the assessment of the condition of the plant cover [6,7]. It is also used in geological studies, providing important information on, among other things, the mineralogical composition or the broader genesis of bedrock [8]. It can also be used indirectly to predict erosion risk [9].

A key feature of VIS-NIR spectroscopy is its high throughput, which allows a significant number of samples to be analyzed in a relatively short time. This feature plays a key role in projects requiring extensive research or the continuous monitoring of a given object [10]. The real-time analysis provided by VIS-NIR spectroscopy is another important advantage, enabling qualitative and quantitative assessments on site or in the field. This real-time feature is invaluable in situations where rapid or immediate decisions are required [11].

The quantitative analysis capabilities of VIS-NIR spectroscopy contribute significantly to the understanding of soil properties. The method can provide information on a range of parameters, including organic matter content [12,13], moisture levels [14] and nutrient concentrations [15], soil contamination with different types of pollutants [16], and electric conductivity [17]. Moreover, VIS-NIR spectroscopy is multifunctional and offers a holistic view by simultaneously assessing different soil characteristics, such as texture and mineral composition [18]. This comprehensive approach enables a detailed understanding of the conditions in a soil environment [19,20].

VIS-NIR spectrometric measurements are usually performed under controlled laboratory conditions. Samples are suitably prepared, above all in terms of material homogenization. The predictive models for individual soil properties obtained in the laboratory have a high accuracy. Attempts are made to carry out tests under field conditions, but the data obtained deviate significantly from the values obtained under laboratory conditions, and the predictive models developed are much weaker in this case [3,21]. Importantly, however, it should always be considered whether it is more important to obtain a better model or obtain information in an easier and much faster way.

In order to precisely identify characteristics, so-called spectral libraries are created [19,22]. Reflectance curves are collected for specific soils with specific characteristics, and these characteristics should include both the parameter directly covered by the spectral survey (e.g., organic carbon) as well as other soil characteristics that may have a significant impact on the spectral response (e.g., grain size, carbonates, mineralogical composition, etc.). A predictive model based on a broad spectral library should, in principle, produce higher-quality results. However, due to the enormous variability in soils on a global scale, the use of such libraries may be effective on a local scale or for specific soil types, under the additional assumption of limiting the availability of these studies.

Mapping spatial variability, facilitated by the integration of VIS-NIR spectroscopy with geospatial technologies, represents a higher level of soil survey methodology [23,24]. Creating detailed maps showing differences in soil properties within a given area is invaluable for precision agriculture and land use optimization. By identifying spatial patterns, this approach allows for targeted interventions, optimizing resource use and improving overall land management practices [25].

Among the soil parameters that have been determined by researchers is grain size, in varying aspects, both in terms of individual granulometric fractions and of the individual finest clay fraction [26], organic carbon content [27], soil color [28], salinity [29], and calcium carbonate content, sometimes presented as calcium carbonate equivalent [30,31].

Good results in terms of prediction based on spectral response have so far been achieved precisely for the analysis of soil organic carbon [32,33]. VIS-NIR spectrometry data can also be successfully used to determine the spatial variability in soil organic carbon saturation at the field scale [10]. Nevertheless, researchers indicate that the evaluation of the parameter in question can be affected by various other additional factors, such as aggregate structure, moisture content, content of specific substances, etc. [34,35]. One of the substances that could interfere and significantly affect the quality of SOC prediction is calcium carbonate; a significant interrelation of the two components in the samples studied, in terms of spectral properties, was pointed out by Rasooli et al. [36], among others.

The purpose of this study is to determine the feasibility of analyzing soil organic carbon in carbonate-rich soils using VIS-NIR spectroscopy.

2. Materials and Methods

The materials for this analysis consisted of samples of soils, developed on the weathering of carbonate rocks—rendzinas—collected in Eastern Poland. The distinguishing feature of these soils is the high abundance of calcium carbonate and the associated high pH. All the soils at the time of collection were not in agricultural use and were perennial fallows, covered with vegetation of a natural-succession nature; nevertheless, no trees or shrubs were found. Soil samples were taken from the humus horizon of the soil, from the face wall of the excavated pit, from a depth of 5–15 cm, which ensured that the sample had not been disturbed due to processes on the ground surface, such as biological activity. The collected material was dried under room conditions and then sieved through a 2 mm sieve and placed in cardboard boxes. Chemical determinations were made on the material in the laboratory. Carbon was determined using a LECO TruSpec automatic analyzer, according to ISO 10694:1995 [37]. The analysis was performed in triplicate, in separate analytical samples. The carbonate (inorganic carbon) content was determined by volume, using a Scheibler apparatus, based on ISO 10693:2014 [38]. The proportion of organic carbon was at an average level, but the presence of carbonates affected the brightening of color, which, to some extent, masked the abundance of organic matter. The basic properties of the soils are summarized in Table 1.

Table 1. Descriptive analysis of soil organic matter and calcium carbonate equivalent in soil samples.

Sample Set	n	SOC			$CaCO_3$		
		Range [%]	Mean	Sd	Range [%]	Mean	Sd
All samples	68	0.1–5.1	1.5	1.2	0.1–86.1	44.1	25.4
Calibration set	51	0.1–5.1	1.5	1.2	0.1–86.1	43.6	26.4
Validation set	17	0.3–4.7	1.6	1.3	1.5–73.9	45.6	22.9

Spectroradiometric data acquisition was carried out in laboratory conditions, using a Spectral Evolution RS-3500 instrument. The instrument's operating range is between 350 and 2500 nm, and its resolution is 2.8nm (up to 700 nm), 8 nm (up to 1500 nm), and 6 nm (up to 2100 nm). The measurement was performed by contact, directly on samples homogenized immediately before the measurement, using a custom light source (Tungsten lamp). Due to the nature of the material tested (very light color), reflectance calibration was performed on a reference plate (Spectralon Reflectance Standard) before each measurement. At the end of the measurements, the results were digitally brought down to 1 nm resolution

2.1. Spectral Preprocessing

Prior to processing raw spectral data for the development of qualitative or quantitative analytical models in multivariate data analysis, mathematical transformations are applied. This essential preprocessing step is implemented to mitigate spectral variability and noise unrelated to the intended objectives of the models, while concurrently augmenting selectivity. In this study, two commonly used pretreatments for raw spectra, which included first (FD) and second (SD) using Savitzky–Golay smoothing filtering with 8 points and a polynomial order of 3, were selected. All the spectral pretreatments were performed in Python 3.8. To reduce the impact of low-intensity signals, two sections of the spectra spanning the complete wavelength spectrum were excluded: 350–400 nm. Thus, the spectral range from 400 to 2500 nm was retained for subsequent analysis.

2.2. Selection of Optimal Wavelengths

Due to the high dimensionality of hyperspectral data, selecting variables helps decrease the number of features to the most relevant ones. This process mitigates overfitting and enhances the predictive accuracy of regression models. For this investigation, Spearman correlation, Variable Importance in Projection (VIP), and Random Frog (Rfrog) were employed to select variables across three feature groups: raw reflectance, FD, and SD.

2.3. Spearman Correlation

A Spearman correlation analysis was conducted to evaluate the magnitude and direction of the monotonic relationship between the ranked response variable (the stem characteristics of each vine) and the ranked predictor variables (spectral data at various wavelengths). This analytical approach captures the tendency for paired variables to change in a synchronized manner, albeit not necessarily at a uniform rate, thereby facilitating the detection of nonlinear associations without imposing assumptions regarding the normal distribution of variables. The Spearman correlation coefficient, ranging from +1 to -1, signifies the strength of a monotonic relationship, with values closer to ± 1 indicating stronger associations. Spearman correlation coefficients were calculated to indicate the relationships among SOC, SIN, raw reflectance, FD, and SD spectra. Variables exhibiting coefficients surpassing the threshold of 0.6 were deemed significant for inclusion in this study. The correlation analysis was conducted using "spearmanr" from the scipy library in Python 3.8.

2.4. Variable Importance in Projection (VIP)

The assessment of *VIP* is pivotal in multivariate analysis, particularly in the context of Partial Least Squares (PLS) modeling. The *VIP* selection method utilizes coefficients derived from a fitted PLS model to evaluate the significance of individual wavelengths (variables) within the dataset. In this method, key matrices including the X-score matrix (T), the y-loading vector (q), and the normalized X-weight matrix (W) are instrumental. Here, N represents the number of samples, M denotes the number of features, and K signifies the number of latent variables. The VIPs are computed using the following equation:

$$VIP = \sqrt{M \frac{W^2 (q^2\ T^t\ T)^t}{\sum_k (q^2 T^t T)_k}} \tag{1}$$

This approach offers a comprehensive framework for assessing variable importance, aiding in feature selection and enhancing the interpretability of PLS models within scientific research and analysis [39]. Since the average of squared *VIP* scores equals 1, only influential wavelengths with a *VIP* score greater than 1 were kept in the calibration model.

2.5. Random Frog (Rfrog)

The Rfrog technique is an iterative selection method that commences with randomly chosen features, which are dynamically adjusted throughout the iteration process. During each iteration, a random subset or superset is generated and evaluated against the previously selected features through cross-validation. The Rfrog method maintains a counter for each feature, and the counters for all features within the "winning" set (i.e., achieving higher cross-validation scores) are incremented after each iteration. Following the completion of all iterations, the features with the highest selection frequencies are chosen for inclusion [40]. The number of iterations (N) was set to 50 in this study. In this study, VIP and Rfrog analyses were performed in Python 3.8 using the AUSWAHL (AUtomatic Selection of WAvelengtH Library) package.

2.6. Prediction Models

In this work, two types of datasets were proposed to achieve the best prediction accuracy for SOC estimation. The first type of dataset contained only spectral bands selected based on Spearman's rank correlation coefficients and the VIP and Rfrog methods. In the second type of dataset, hyperspectral data were combined with information about the SIC obtained from laboratory measurements. As a result, sixteen different dataset combinations were utilized as inputs for the SOC prediction models, employing RF and PLSR algorithms.

2.7. Random Forest

RF is an ensemble regression technique that employs multiple decision trees. It constructs m decision trees from the training dataset using bootstrap resampling on m samples. Each decision tree split is built using a random subset of the dataset to measure a random subset of features in each partition [41]. This introduces variability among individual trees, thereby reducing the risk of overfitting and enhancing the overall prediction performance [42]. During the prediction phase, the algorithm aggregates the results of all trees by averaging, fostering a collaborative decision-making process supported by multiple trees and their insights. This approach yields stable and precise results, making Random Forests a versatile and reliable tool for various regression tasks. Throughout the training of the RF algorithm, a range of values were explored for the tuning parameters. Ultimately, the following parameter settings were selected: n_estimators = 10; max_depth = 20; and random_state = 101.

2.8. Partial Least Squares Regression Method (PLSR)

The PLSR algorithm amalgamates aspects of multiple linear regression analysis, canonical correlation analysis, and principal component analysis, offering not just a fitting regression model but also a comprehensive expression of information. The method operates under the assumption that the dependent variable can be estimated through a linear combination of explanatory variables [43]. Particularly advantageous in scenarios where numerous variables exhibit multiple correlations and the sample sizes are small, PLSR provides a many-to-many linear regression modeling approach. Unlike traditional classical regression analysis, which can lead to overfitting due to correlations among independent variables, PLSR identifies new linearly independent variables to replace the original ones, maximizing the difference between them.

The performance of the Rf and PLSR models was evaluated utilizing the Scikit-Learn python machine learning library package on the Windows (Spyder) platform.

2.9. Model Evaluation

In all dataset variants, the soil samples were divided at a 75:25 ratio into a calibration set and a validation set. For merged datasets containing hyperspectral and SIC data, the preprocessing phase involved standardizing the data to ensure compatibility with the analysis algorithms. Data integration plays a vital role when handling diverse data sources, often requiring merging and integration to create a cohesive and comprehensive dataset. To standardize the data, the Scikit-Learn library's StandardScaler was utilized, providing a robust and efficient method for scaling features to a common mean and standard deviation.

The coefficient of determination (R^2), the root mean squared error (RMSE), and the mean squared error (MAE) were calculated as indexes to evaluate the performance of the used models. Generally, a well-performing model tends to achieve a high R^2 alongside low RMSE and MAE values, suggesting accurate predictions with minimal error.

3. Results and Discussion

The descriptive statistics of the soil organic matter and calcium carbonate equivalent in the soil samples are presented in Table 1 for both the calibration and validation sets. This includes calculations for the number of samples (N), the mean, the standard deviation (SDe), and the range. This finding indicated that the mean SOC for the calibration set and the validation set was 1.5% and 1.6%, respectively, whereas the average $CaCO_3$ content in the calibration and validation set was 43.6% and 45.6%, respectively. The dataset distribution closely resembled that of the entire dataset, suggesting a representative division. Clearly, the inorganic carbon content in the form of calcium carbonate exceeded the organic carbon content by an order of magnitude. This is the distinguishing feature of carbonate soils in this type of rendzina.

Figure 1 shows raw spectra and pretreated spectra with FD and SD. In the raw spectra, consistent shapes can be observed across all the samples. The course of the curves is typical for soil material, as reported by many authors (LIT). Notably, three distinct absorption peaks are discernible in the near-infrared region, attributed to the hydroxyl group of free water (at 1410 nm and 1900 nm) and the Al-OH group of clay minerals, at 2210 nm [13].

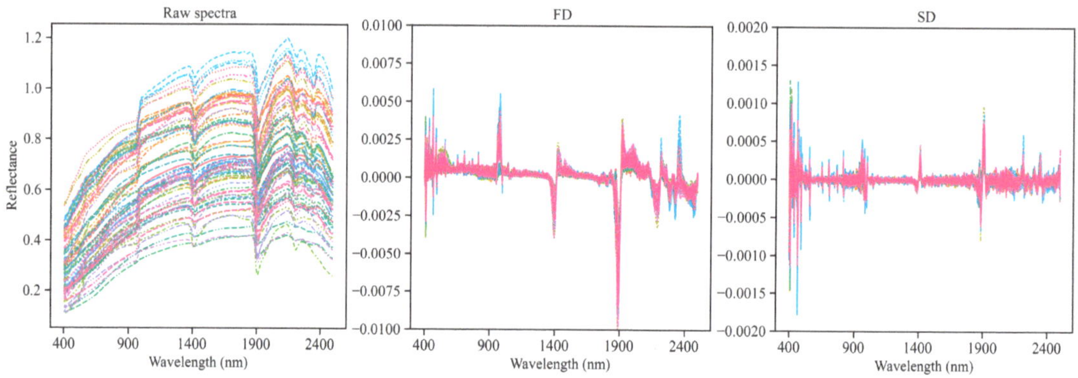

Figure 1. Raw and pretreated spectra. FD—first derivative; SD—second derivative.

Raw, unprocessed spectra showed reflectance in some sections exceeding the value of 1.0. This was due to the fact that the analyzed material had specific properties. First of all, the very bright color of the soil containing a high amount of carbonates resulted in a specific spectral response: the reflectance was locally higher than the reflection from the reference plate. In order to avoid error, calibration was performed before measuring each soil sample, as indicated in the Section 2. The highest recorded values, reaching 1.2 (reflectance at 120% relative to the reference), were recorded for soil samples containing up to 85% turbulent carbonates. This may be an important contribution to the discussion on the reference materials to be used for the heaviest soils (in terms of grain-size distribution) containing large amounts of inorganic carbon. In the literature, one does not encounter reflectance data exceeding 1. Nevertheless, the carbonate contents in the works analyzed are much lower (up to a maximum of 60%), and all the works showed very high correlations between reflectance and the amount of carbonate in the samples.

In Figure 2, Spearman's rank correlation coefficients are presented for both the raw spectra and the spectral bands after FD and SD preprocessing. Notably, a significant negative correlation, approximately −0.7, was detected between the SOC content and the raw spectral data across the entirety of the spectrum range. Moreover, the SOC and the FD and SD spectra exhibited significant negative correlations within specific wavelength ranges, including 400–550 nm, 1400–1500 nm, and 1900–2000 nm. Conversely, a statistically significant and notably strong positive correlation was observed within the spectral regions spanning from 1700 to 1900 nm and from 2200 to 2500 nm. The strong correlation of the FD spectrum at 2300 nm was influenced by the characteristic absorption peak of C–H. However, a distinct response pattern was observed for SIC. Unlike SOC, a positive correlation was observed between the raw spectral data and SIC across the entire spectral range. A strong negative correlation was evident around 2300 nm and 2500 nm, while a strong positive correlation (higher than 0.75) was observed near 2400 nm. This indicated a contrasting relationship between the spectral data and the SIC compared to the SOC.

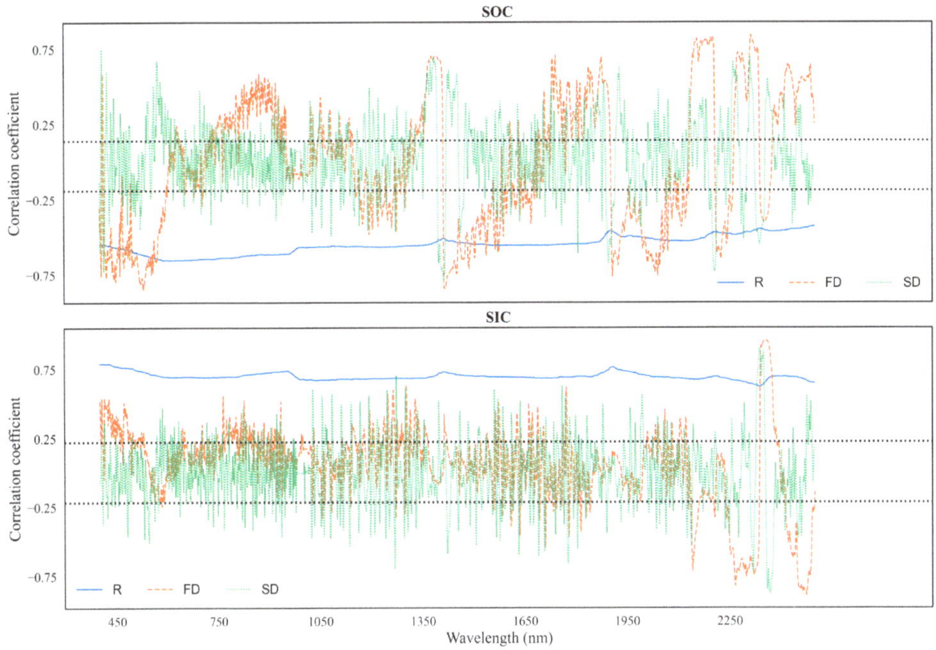

Figure 2. Spearman's correlation coefficient distribution between the SOC and the SIC content with different preprocessing methods of spectral reflectance soil spectral data. The black dotted lines indicate significance at $p < 0.01$.

After conducting the Spearman correlation analysis, 187 wavelengths were identified for the FD spectra, while 19 wavelengths were deemed relevant for the SD spectra, all with a Spearman correlation rank higher than 0.6. These selected wavelengths are illustrated in Figure 3. For the FD spectra, the chosen wavelengths are aggregated into four ranges: 515–538 nm, 1420–1433 nm, 2165–2207 nm, and 2310–2333 nm. Meanwhile, the majority of wavelengths in the SD spectra are concentrated within the range of 1406–1417 nm.

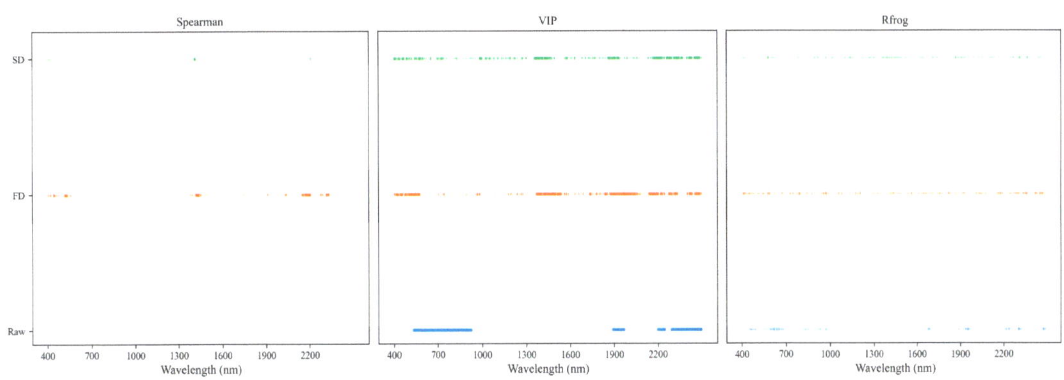

Figure 3. The distribution of spectral variables selected by a Spearman correlation analysis, VIP, and RFrog with different spectral pretreatments.

The VIP scores of the wavelengths obtained for the raw FD and SD spectra are depicted in Figure 4. In the case of the raw spectra (Figure 4A), three primary spectral zones were identified, spanning from 520 to 920 nm, around 1900 nm, and 2250–2500 nm. Additionally, five major spectral zones were distinguished as significant for the FD spectra, ranging from 450 to 570 nm, 1300–1520 nm, 1860–2600 nm, 2130–2200 nm, and 2270–2340 nm. Conversely, in the SD spectra, significance was observed across the entire spectrum. The implementation of the VIP method facilitated reduction, enabling the development of accurate and reliable models. The number of wavelengths decreased from 2100 to 765 for the raw spectra, resulting in a data size reduction of approximately 64%. Similarly, for the first- and second-order differentiations, the data size was reduced by approximately 57% and 61%, respectively. All the selected variables are illustrated in Figure 4. Four main spectral zones were observed for the raw spectra (525–927 nm, 1886–1974 nm, 2194–2250 nm, and 2287–2500 nm), while six spectral ranges were identified for the FD spectra pretreatments.

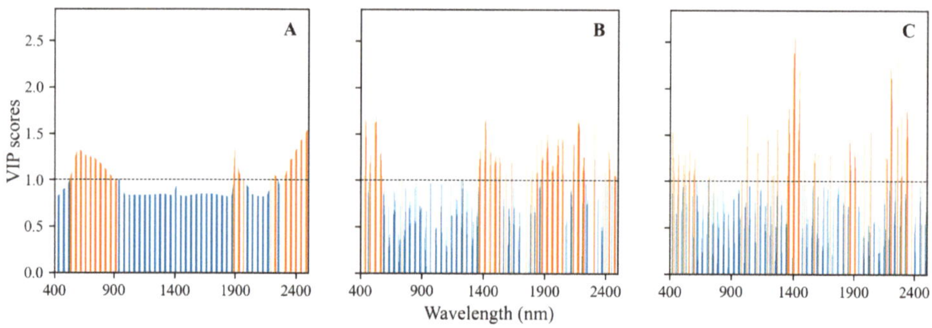

Figure 4. VIP scores of the wavelengths obtained for the raw (**A**), FD (**B**), and SD (**C**) spectra.

The feature wavelengths were selected by the Random Frog algorithm through the calculation of their selection probabilities within the spectrum. In Figure 5, the selection probabilities of each wavelength, determined by the Random Frog algorithm, are summarized for the raw and first- and second-order differentiations of the reflectance spectra. The threshold, established at 0.7 based on prior experience, was utilized to select important wavenumbers, with a selection probability surpassing this threshold as characteristic waves. Additionally, the number of model simulation iterations was set to 50 to ensure convergence. When employing a selection probability cutoff of 0.7, a total of 77, 169, and 93 significant wavelengths were identified for the raw, FD, and SD spectra, respectively (Figure 5).

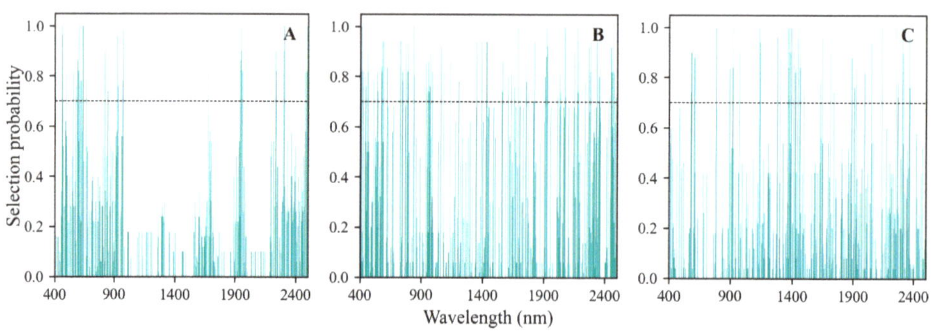

Figure 5. VIP scores of the wavelengths obtained for the raw (**A**), FD (**B**), and SD (**C**) spectra.

With the hyperspectral data as an independent variable, two methods, Random Forest (RF) and Partial Least Squares Regression (PLSR), were employed to predict the SOC. The performance metrics of the proposed models utilizing various feature variable extraction methods are presented in Table 2.

Table 2. The prediction results of SOC from the established RF and PLSR models, using Spearman, VIP, and Rfrog as data reduction methods on spectral input data.

Model	Feature Selection	Preprocessing	Calibration Set			Prediction Set		
			R^2	MAE	RMSE	R^2	MAE	RMSE
RF	–	-	0.88	0.28	0.42	0.51	0.59	0.89
		FD	0.96	0.18	0.25	0.79	0.43	0.59
		SD	0.91	0.27	0.37	0.82	0.42	0.54
	COR	FD	0.95	0.18	0.26	0.77	0.44	0.61
		SD	0.88	0.29	0.42	0.61	0.59	0.79
	VIP	-	0.91	0.26	0.36	0.38	0.69	0.99
		FD	0.90	0.28	0.38	0.74	0.51	0.65
		SD	0.92	0.25	0.34	0.59	0.60	0.81
	Rfrog	-	1	0.63	0.71	0.69	0.71	0.63
		FD	0.94	0.19	0.28	0.79	0.46	0.58
		SD	0.95	0.20	0.26	0.53	0.54	0.87
PLSR	–	-	0.27	0.74	1.02	0.41	0.69	0.98
		FD	0.55	0.61	0.8	0.65	0.61	0.76
		SD	0.57	0.57	0.79	0.62	0.79	0.57
	Spearman Cor	FD	0.64	0.72	0.53	0.73	0.50	0.66
		SD	0.51	0.56	0.74	0.66	0.56	0.75
	VIP	-	0.28	0.73	1.01	0.41	0.70	0.98
		FD	0.56	0.60	0.80	0.67	0.57	0.73
		SD	0.59	0.54	0.72	0.68	0.54	0.72
	RFrog	-	0.29	0.73	1.02	0.41	0.70	0.97
		FD	0.64	0.54	0.72	0.75	0.47	0.64
		SD	0.68	0.45	0.66	0.73	0.45	0.66

For all the studied variants, the prediction accuracies exceeded 65%. The models were constructed after band selection but still required fine-tuning to make better SOC predictions. The first derivative transformation of reflectance afforded the best predictions. The RF model attained the highest R^2 value of 0.79 when employing variable extraction by Rfrog and preprocessing using FD. The model constructed based on the 19 wavelengths selected through Spearman's correlation could predict the SOC with an R^2 value of 0.77.

The highest prediction accuracy was observed in the SOC prediction model based on the 169 wavelengths selected using the Rfrog method, with an R^2 value of 0.79 and an RMSEP of 0.58%. Remarkably, the linear PLSR model demonstrated an inferior performance compared to the nonlinear RF model. The prediction models constructed using the PLSR algorithm demonstrated an adjusted validation R^2 of between 0.41 and 0.75, with RMSE values of 0.98 and 0.64.

Scatter plots depicting the predicted versus the measured values of SOC, generated by these high-quality models, are illustrated in Figure 6.

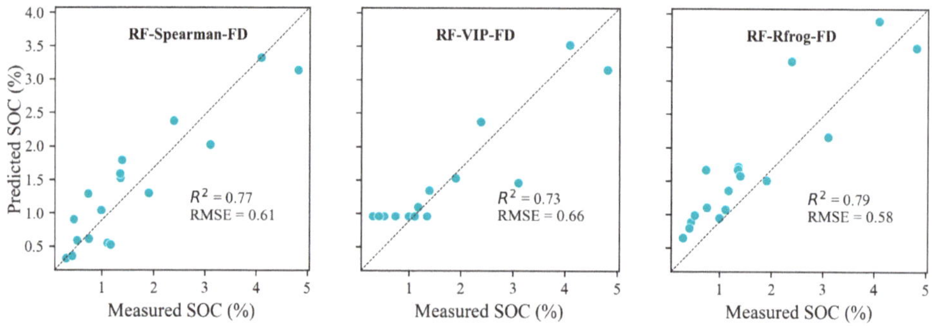

Figure 6. Scatter plots of the measured vs. predicted SOC values for different RF models based on spectral data. Black dotted lines represent the 1:1 lines.

To enhance the predictive capability of the models, the hyperspectral data were integrated with the measured values of the SIC (Table 3). Compared to the model constructed solely based on the spectral data, the fused models showed an increase in their prediction accuracies of up to 20% and 13% for the RF and PLSR models, respectively.

Table 3. The prediction results of SOC from the established RF and PLSR models, using Spearman, VIP, and Rfrog as data reduction methods on fused input data.

Model	Feature Selection	Preprocessing	Calibration Set			Prediction Set		
			R^2	MAE	RMSE	R^2	MAE	RMSE
RF	–	-	0.88	0.28	0.42	0.51	0.59	0.89
		FD	0.96	0.18	0.25	0.79	0.43	0.59
		SD	0.91	0.27	0.37	0.82	0.42	0.54
	COR	FD	0.95	0.27	0.18	0.88	0.32	0.45
		SD	0.93	0.32	0.23	0.77	0.46	0.61
	VIP	-	0.89	0.26	0.42	0.42	0.65	0.97
		FD	0.95	0.19	0.28	0.81	0.38	0.56
		SD	0.93	0.24	0.32	0.75	0.44	0.63
	Rfrog	-	0.87	0.28	0.44	0.54	0.61	0.86
		FD	0.95	0.20	0.28	0.86	0.32	0.47
		SD	0.93	0.24	0.32	0.58	0.62	0.82
PLSR	–	-	0.48	0.65	0.87	0.42	0.74	0.97
		FD	0.72	0.49	0.64	0.70	0.50	0.69
		SD	0.74	0.47	0.52	0.69	0.52	0.70
	Spearman Cor	FD	0.78	0.46	0.57	0.65	0.55	0.76
		SD	0.61	0.57	0.75	0.69	0.50	0.71
	VIP	-	0.53	0.65	0.83	0.51	0.60	0.89
		FD	0.78	0.45	0.56	0.77	0.45	0.61
		SD	0.80	0.42	0.54	0.74	0.51	0.65
	RFrog	-	0.56	0.63	0.79	0.55	0.57	0.85
		FD	0.81	0.40	0.52	0.81	0.40	0.55
		SD	0.84	0.37	0.48	0.68	0.61	0.72

From Table 3, it can be noticed that the prediction accuracy, as indicated by R^2, is satisfactory. However, there are notable discrepancies in the MAE and RMSE, attributed to the significant variability in the SOC samples. The best prediction model for the combined data was achieved with RF-Spearman-FD (R^2 = 0.88; RMSE = 0.45).

The obtained results indicated that the FD preprocessing method combined with RF, results in the model that is sufficiently robust and stable to be applied to soils rich in calcium carbonate. However, from Figures 2 and 3, some discrepancy between the predicted and the measured values of the SOC content can be observed. The prediction of soil organic carbon (SOC) content is influenced by a range of environmental and management factors. Key factors affecting SOC prediction include the mineral composition and soil texture and, indirectly, soil structure, biological activity, vegetation cover, and climate conditions. These factors interact in complex ways, leading to spatial and temporal variability in the SOC content. Thus, effective SOC prediction models need to consider these diverse factors to improve accuracy and reliability. Due to the complex influence of many factors on the quality of organic carbon prediction, it should be taken into account that the laboratory testing of standardized soil samples (of a homogeneous structure, with water removed, free of plant debris and plant fragments) eliminates many factors that affect the results in unpredictable ways. Hence, testing in a systematic way, according to a specific protocol, allows one to achieve more reliable test results that are, in addition, directly comparable with the results of other researchers [44].

Figures 6 and 7 illustrate the precision of the prediction model, observable through the dispersion of points along the Y-axis. A narrower spread of these points around the predicted values signifies a higher precision. However, the observed scatter indicates that the model has certain limitations in its precision. This dispersion may stem from various factors, including inherent model constraints, data variability, and potentially unaccounted-for variables. Comparing prediction precision and instrumental measurement precision is vital for the validation and reliability assessment of the SOC prediction model. The precision of instrumental measurements, such as those obtained through the laboratory analysis of soil samples, serves as a benchmark for the prediction model. When instruments demonstrate a high precision, the SOC prediction model should ideally achieve a comparable precision to be considered reliable. Comparing the variability in the model's predictions with the known precision of the instruments allows for a more detailed error analysis. If the model's predictions show greater variability than the instrument's measurements, this excess variability is likely due to the model's limitations rather than issues with the SOC data's quality.

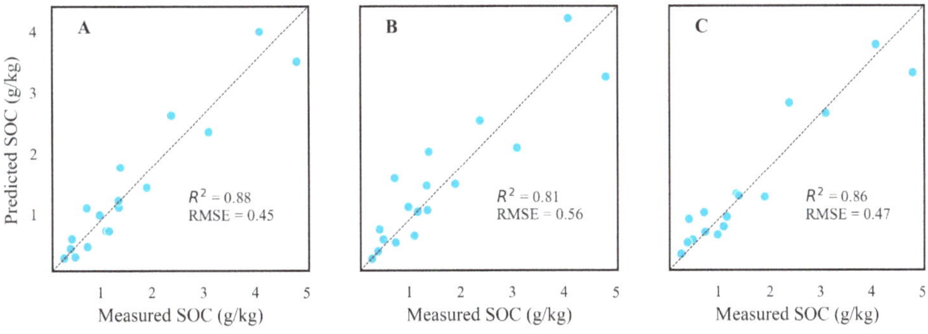

Figure 7. Scatter plots of the measured vs. predicted SOC values for different RF models based on combined datasets (spectral data and SIC). Black dotted lines represent the 1:1 lines.

Recognizing the precision of both the SOC prediction model and the instruments can inform targeted improvement strategies. For instance, if the instrument's precision surpasses that of the model, efforts should focus on enhancing the model's precision. This can be achieved by incorporating additional relevant features, refining existing algorithms, or exploring more sophisticated modeling techniques tailored to SOC data.

In summary, the precision of the prediction model, as depicted in Figures 6 and 7, reveals certain limitations when compared with the higher precision typically associated with instrumental measurements.

In summary, while prediction precision and instrumental measurement precision pertain to different domains, they share common principles, such as dependence on data quality, the necessity of validation and calibration, and the use of statistical methods to evaluate and enhance precision. Understanding their interconnection can lead to the better design and implementation of both predictive models and measurement systems, ensuring higher reliability and accuracy in various applications.

In light of the literature data, determining the precise wavelength at which a substance-specific signal is recorded can be debatable. For calcium carbonate, specific wavelengths have been determined: 1800 nm, 2350 nm, 2360 nm [45], 2325 nm [46], 2338 nm [47], 2340 nm [36], and 2341 nm (Gomez et al. 2008). However, in light of our analyses, the authors would suggest indicating wider ranges, related, on the one hand, to the diversity in soil samples and, on the other hand, the measurement method. The determination of a single length of the order of 1 nm, taking into account the spectral resolution of the apparatus at the level of 6 nm, seems methodologically unjustified. The reflectance/absorbance values at the suggested wavelengths may also be affected by the presence of certain clay minerals in the samples, such as chlorite or illite, which increase the absorbance of a given material in a similar range of the electromagnetic spectrum [47,48].

Organic carbon is one of the most commonly analyzed soil parameters using VIS-NIR spectrometry [7], considering the high importance of analyzing samples in the laboratory, with samples prepared in a specific way, which allows one to achieve more reliable results in contrast to analysis in the field. This is because in situ analysis must take into account local soil conditions, such as moisture content, structure, and, above all, the heterogeneity of the material [44]; the influence of the aforementioned factors is offset by the preparation of the material and its homogenization. The results of organic carbon prediction presented by many authors in available publications are highly promising and indicate the feasibility of using spectrometric techniques to analyze soil organic matter. R^2 values, indicating the accuracy of prediction, at levels exceeding 0.8–0.9, should be considered satisfactory.

Nevertheless, the selection of analytical material seems to be crucial in terms of prediction. Interfering factors are important, affecting the direct measurement of the spectral response of the soil in certain ranges or the specific "offset" of the entire spectral curve, due to high reflectance [36]. This is of great importance in the case of soils rich in calcium carbonate, such as those analyzed in this publication. The results obtained clearly indicate that it is necessary to use input fusion techniques, allowing researchers to take into account analytical laboratory results of calcium carbonate content. An increase in the quality of prediction using auxiliary data is also indicated in studies by other authors, who have taken into account, for example, soil moisture or temperature parameters [3].

One of the most important advantages of the VIS-NIR spectrometry method is that it facilitates and speeds up the analysis of soil materials. However, in the authors' opinion, it is necessary, at least at this stage, to take into account traditional techniques (laboratory analytics) to improve the quality of prediction. In any case, adopting a compromise—analyzing calcium carbonate in the laboratory and incorporating the results into a combined prediction model—represents a cost-effective solution in terms of labor input and analytical costs.

A methodological problem may be the relatively small database used for calibration, especially with a small total number of samples [49]. In the case of the present work, the number of samples used for calibration appeared to be sufficient [50]. On the other hand, increasing the measurement base would be difficult to achieve due to the uniqueness of the study material.

The comparison of the results obtained with those of other authors, in the case of VIS-NIR spectroscopy, often poses methodological problems. On the one hand, the analyzed material is highly diverse, including soils from different regions, with different basic properties (mineralogy, grain size, etc.). Nevertheless, it can be pointed out that most researchers obtain prediction results at an R^2 level in the range 0.6–0.9 [7]. However, it seems that the published values cannot be generalized due to the high variability in the research material. The soils used in the present study, with carbonate contents exceeding 40%, have not yet been analyzed in detail in terms of organic carbon prediction by VIS-NIR spectroscopy.

On the other hand, the processing of input data is performed in a differentiated way, which is due to the fact that authors are looking for the optimal solution (in the sense of the one producing the most reliable results). The use of diverse modeling techniques (e.g., Partial Least Squares Regression, cubist, Random Forest, Support Vector Machine, convolution neural network, XGBoost, memory-based learning, etc.) can lead to strongly divergent results in terms of prediction quality [4,51–53]. Consequently, there is no model solution that can be universally applied, but only a collection of individual case studies. Nevertheless, they provide an indispensable foundation for the creation of a library that takes into account different types and species of soils. The contribution of this publication in this regard is the inclusion of a particularly high carbonate content as an interfering factor in the organic carbon measurement range.

4. Conclusions

This study demonstrates the potential of VIS-NIR spectroscopy for SOC analysis in carbonate-rich soils. By integrating spectral data with SIC information and employing advanced modeling techniques, accurate predictions of SOC levels can be achieved, offering valuable insights for soil management and environmental monitoring.

In the case of some soils, it should be taken into account that, in certain ranges of the spectrum, reflectance may exceed the values for the reference materials.

Author Contributions: Conceptualization, P.B. and M.S.; methodology, A.S., P.B. and M.S.; software, A.S; formal analysis, A.S.; data curation and calculations, A.S. and P.B.; writing—original draft preparation, P.B. and A.S.; editing, P.B. and M.S.; supervision, P.B.; project administration, P.B.; visualization, A.S. All authors have read and agreed to the published version of the manuscript.

Funding: This research received no external funding.

Institutional Review Board Statement: Not applicable.

Informed Consent Statement: Not applicable.

Data Availability Statement: The data presented in this study are available upon request from the corresponding author.

Conflicts of Interest: The authors declare no conflicts of interest.

References

1. Ben-Dor, E. Quantitative Remote Sensing of Soil Properties. In *Advances in Agronomy*; Academic Press: Cambridge, MA, USA, 2002; Volume 75, pp. 173–243.
2. Ben-Dor, E.; Heller, D.; Chudnovsky, A. A Novel Method of Classifying Soil Profiles in the Field Using Optical Means. *Soil Sci. Soc. Am. J.* **2008**, *72*, 1113–1123. [CrossRef]
3. Debaene, G.; Bartmiński, P.; Siłuch, M. In Situ VIS-NIR Spectroscopy for a Basic and Rapid Soil Investigation. *Sensors* **2023**, *23*, 5495. [CrossRef]
4. Vestergaard, R.-J.; Vasava, H.B.; Aspinall, D.; Chen, S.; Gillespie, A.; Adamchuk, V.; Biswas, A. Evaluation of Optimized Preprocessing and Modeling Algorithms for Prediction of Soil Properties Using Vis-nir Spectroscopy. *Sensors* **2021**, *21*, 6745. [CrossRef]
5. Santana, F.B.; Otani, S.; De Souza, A.; Poppi, R. Comparison of PLS and SVM Models for Soil Organic Matter and Particle Size Using Vis-NIR Spectral Libraries. *Geoderma Reg.* **2021**, *27*, e00436. [CrossRef]
6. García-Sánchez, F.; Galvez-Sola, L.; Martínez-Nicolás, J.J.; Muelas-Domingo, R.; Nieves, M.; García-Sánchez, F.; Galvez-Sola, L.; Martínez-Nicolás, J.J.; Muelas-Domingo, R.; Nieves, M. Using Near-Infrared Spectroscopy in Agricultural Systems. In *Developments in Near-Infrared Spectroscopy*; IntechOpen: London, UK, 2017; ISBN 978-953-51-3018-5.
7. Ahmadi, A.; Emami, M.; Daccache, A.; He, L. Soil Properties Prediction for Precision Agriculture Using Visible and Near-Infrared Spectroscopy: A Systematic Review and Meta-Analysis. *Agronomy* **2021**, *11*, 433. [CrossRef]

8. Rosin, N.A.; Demattê, J.A.M.; Poppiel, R.R.; Silvero, N.E.Q.; Rodriguez-Albarracin, H.S.; Rosas, J.T.F.; Greschuk, L.T.; Bellinaso, H.; Minasny, B.; Gomez, C.; et al. Mapping Brazilian Soil Mineralogy Using Proximal and Remote Sensing Data. *Geoderma* **2023**, *432*, 116413. [CrossRef]
9. Ostovari, Y.; Ghorbani-Dashtaki, S.; Bahrami, H.-A.; Abbasi, M.; Dematte, J.A.M.; Arthur, E.; Panagos, P. Towards Prediction of Soil Erodibility, SOM and $CaCO_3$ Using Laboratory Vis-NIR Spectra: A Case Study in a Semi-Arid Region of Iran. *Geoderma* **2018**, *314*, 102–112. [CrossRef]
10. Reyes, J.; Ließ, M. On-the-Go Vis-NIR Spectroscopy for Field-Scale Spatial-Temporal Monitoring of Soil Organic Carbon. *Agriculture* **2023**, *13*, 1611. [CrossRef]
11. Rodionov, A.; Welp, G.; Damerow, L.; Berg, T.; Amelung, W.; Pätzold, S. Towards On-the-Go Field Assessment of Soil Organic Carbon Using Vis-NIR Diffuse Reflectance Spectroscopy: Developing and Testing a Novel Tractor-Driven Measuring Chamber. *Soil Tillage Res.* **2015**, *145*, 93–102. [CrossRef]
12. Nocita, M.; Kooistra, L.; Bachmann, M.; Müller, A.; Powell, M.; Weel, S. Predictions of Soil Surface and Topsoil Organic Carbon Content through the Use of Laboratory and Field Spectroscopy in the Albany Thicket Biome of Eastern Cape Province of South Africa. *Geoderma* **2011**, *167–168*, 295–302. [CrossRef]
13. Liu, Y.; Liu, Y.; Chen, Y.; Zhang, Y.; Shi, T.; Wang, J.; Hong, Y.; Fei, T.; Zhang, Y. The Influence of Spectral Pretreatment on the Selection of Representative Calibration Samples for Soil Organic Matter Estimation Using Vis-NIR Reflectance Spectroscopy. *Remote Sens.* **2019**, *11*, 450. [CrossRef]
14. Liu, J.; Zhang, D.; Yang, L.; Ma, Y.; Cui, T.; He, X.; Du, Z. Developing a Generalized Vis-NIR Prediction Model of Soil Moisture Content Using External Parameter Orthogonalization to Reduce the Effect of Soil Type. *Geoderma* **2022**, *419*, 115877. [CrossRef]
15. Cozzolino, D.; Cynkar, W.U.; Dambergs, R.G.; Shah, N.; Smith, P. In Situ Measurement of Soil Chemical Composition by Near-Infrared Spectroscopy: A Tool Toward Sustainable Vineyard Management. *Commun. Soil Sci. Plant Anal.* **2013**, *44*, 1610–1619. [CrossRef]
16. Shi, T.; Wang, J.; Chen, Y.; Wu, G. Improving the Prediction of Arsenic Contents in Agricultural Soils by Combining the Reflectance Spectroscopy of Soils and Rice Plants. *Int. J. Appl. Earth Obs. Geoinf.* **2016**, *52*, 95–103. [CrossRef]
17. Chatrenour, M.; Landi, A.; Bahrami, H.; Mirzaei, S. Dust Source Clay Content and Salinity Estimation Using VNIR Spectrometry. *Arid Land Res. Manag.* **2023**, *37*, 369–388. [CrossRef]
18. Gholizadeh, A.; Žižala, D.; Saberioon, M.; Borůvka, L. Soil Organic Carbon and Texture Retrieving and Mapping Using Proximal, Airborne and Sentinel-2 Spectral Imaging. *Remote Sens. Environ.* **2018**, *218*, 89–103. [CrossRef]
19. Viscarra Rossel, R.A.; Walvoort, D.J.J.; McBratney, A.B.; Janik, L.J.; Skjemstad, J.O. Visible, near Infrared, Mid Infrared or Combined Diffuse Reflectance Spectroscopy for Simultaneous Assessment of Various Soil Properties. *Geoderma* **2006**, *131*, 59–75. [CrossRef]
20. Debaene, G.; Bartmiński, P.; Niedźwiecki, J.; Miturski, T. Visible and Near-Infrared Spectroscopy as a Tool for Soil Classification and Soil Profile Description. *Pol. J. Soil Sci.* **2017**, *50*, 1–10. [CrossRef]
21. Biney, J.K.M.; Borůvka, L.; Chapman Agyeman, P.; Němeček, K.; Klement, A. Comparison of Field and Laboratory Wet Soil Spectra in the Vis-NIR Range for Soil Organic Carbon Prediction in the Absence of Laboratory Dry Measurements. *Remote Sens.* **2020**, *12*, 3082. [CrossRef]
22. Ji, W.; Li, S.; Chen, S.; Shi, Z.; Viscarra Rossel, R.A.; Mouazen, A.M. Prediction of Soil Attributes Using the Chinese Soil Spectral Library and Standardized Spectra Recorded at Field Conditions. *Soil Tillage Res.* **2016**, *155*, 492–500. [CrossRef]
23. Wetterlind, J.; Stenberg, B.; Söderström, M. Increased Sample Point Density in Farm Soil Mapping by Local Calibration of Visible and near Infrared Prediction Models. *Geoderma* **2010**, *156*, 152–160. [CrossRef]
24. Alomar, S.; Mireei, S.A.; Hemmat, A.; Masoumi, A.A.; Khademi, H. Prediction and Variability Mapping of Some Physicochemical Characteristics of Calcareous Topsoil in an Arid Region Using Vis–SWNIR and NIR Spectroscopy. *Sci. Rep.* **2022**, *12*, 8435. [CrossRef] [PubMed]
25. Feyziyev, F.; Babayev, M.; Priori, S.; L'Abate, G. Using Visible-Near Infrared Spectroscopy to Predict Soil Properties of Mugan Plain, Azerbaijan. *Open J. Soil Sci.* **2016**, *6*, 52–58. [CrossRef]
26. Tümsavaş, Z.; Tekin, Y.; Ulusoy, Y.; Mouazen, A. Prediction and Mapping of Soil Clay and Sand Contents Using Visible and Near-Infrared Spectroscopy. *Biosyst. Eng.* **2019**, *177*, 90–100. [CrossRef]
27. Hong, Y. Application of Fractional-Order Derivative in the Quantitative Estimation of Soil Organic Matter Content through Visible and near-Infrared Spectroscopy. *Geoderma* **2018**, *337*, 758–769. [CrossRef]
28. Zeng, R.; Rossiter, D.G.; Zhao, Y.G.; Li, D.C.; Zhang, G.L. Forensic Soil Source Identification: Comparing Matching by Color, Vis-NIR Spectroscopy and Easily-Measured Physio-Chemical Properties. *Forensic Sci. Int.* **2020**, *317*, 110544. [CrossRef]
29. Zhou, M.; Chen, G.; Dong, Z.; Xie, B.; Gu, S.; Shi, P. Estimation of Surface Albedo from Meteorological Observations across China. *Agric. For. Meteorol.* **2020**, *281*, 107848. [CrossRef]
30. Lagacherie, P.; Baret, F.; Feret, J.-B.; Madeira Netto, J.; Robbez-Masson, J.M. Estimation of Soil Clay and Calcium Carbonate Using Laboratory, Field and Airborne Hyperspectral Measurements. *Remote Sens. Environ.* **2008**, *112*, 825–835. [CrossRef]
31. Melendez-Pastor, I.; Navarro-Pedreño, J.; Gómez, I.; Koch, M. Identifying Optimal Spectral Bands to Assess Soil Properties with VNIR Radiometry in Semi-Arid Soils. *Geoderma* **2008**, *147*, 126–132. [CrossRef]
32. Ladoni, M.; Bahrami, H.A.; Alavipanah, S.K.; Norouzi, A.A. Estimating Soil Organic Carbon from Soil Reflectance: A Review. *Precis. Agric.* **2010**, *11*, 82–99. [CrossRef]

33. Angelopoulou, T.; Tziolas, N.; Balafoutis, A.; Zalidis, G.; Bochtis, D. Remote Sensing Techniques for Soil Organic Carbon Estimation: A Review. *Remote Sens.* **2019**, *11*, 676. [CrossRef]
34. Wu, C.-Y.; Jacobson, A.R.; Laba, M.; Baveye, P.C. Accounting for Surface Roughness Effects in the Near-Infrared Reflectance Sensing of Soils. *Geoderma* **2009**, *152*, 171–180. [CrossRef]
35. Bellon-Maurel, V.; McBratney, A. Near-Infrared (NIR) and Mid-Infrared (MIR) Spectroscopic Techniques for Assessing the Amount of Carbon Stock in Soils—Critical Review and Research Perspectives. *Soil Biol. Biochem.* **2011**, *43*, 1398–1410. [CrossRef]
36. Rasooli, N.; Farpoor, M.H.; Mahmoodabadi, M.; Esfandiarpour-Boroujeni, I. Vis-NIR Spectroscopy as an Eco-Friendly Method for Monitoring Pedoenvironmental Variations and Pedological Assessments in Lut Watershed, Central Iran. *Soil Tillage Res.* **2023**, *233*, 105808. [CrossRef]
37. *ISO 10694:1995*; Soil Quality—Determination of Organic and Total Carbon after Dry Combustion (Elementary Analysis). International Organization for Standardization: Geneva, Switzerland, 1995.
38. *ISO 10693:2014*; Soil Quality—Determination of Carbonate Content—Volumetric Method. International Organization for Standardization: Geneva, Switzerland, 2014.
39. Favilla, S.; Durante, C.; Vigni, M.L.; Cocchi, M. Assessing Feature Relevance in NPLS Models by VIP. *Chemom. Intell. Lab. Syst.* **2013**, *129*, 76–86. [CrossRef]
40. Li, H.-D.; Xu, Q.-S.; Liang, Y.-Z. Random Frog: An Efficient Reversible Jump Markov Chain Monte Carlo-like Approach for Variable Selection with Applications to Gene Selection and Disease Classification. *Anal. Chim. Acta* **2012**, *740*, 20–26. [CrossRef] [PubMed]
41. Breiman, L. Random Forests. *Mach. Learn.* **2001**, *45*, 5–32. [CrossRef]
42. Grimm, R.; Behrens, T.; Märker, M.; Elsenbeer, H. Soil Organic Carbon Concentrations and Stocks on Barro Colorado Island—Digital Soil Mapping Using Random Forests Analysis. *Geoderma* **2008**, *146*, 102–113. [CrossRef]
43. Shen, L.-T.; Zhang, F.-M.; Huang, J.; Li, Y.-P. Climate characteristics of day and night precipitation during the growing season in Inner Mongolia from 1961 to 2018. *Arid Zone Res.* **2020**, *37*, 1519–1527. [CrossRef]
44. Piccini, C.; Metzger, K.; Debaene, G.; Stenberg, B.; Götzinger, S.; Borůvka, L.; Sandén, T.; Bragazza, L.; Liebisch, F. In-Field Soil Spectroscopy in Vis–NIR Range for Fast and Reliable Soil Analysis: A Review. *Eur. J. Soil Sci.* **2024**, *75*, e13481. [CrossRef]
45. Ben-Dor, E.; Banin, A. Near-Infrared Reflectance Analysis of Carbonate Concentration in Soils. *Appl. Spectrosc.* **1990**, *44*, 1064–1069. [CrossRef]
46. Summers, D.; Lewis, M.; Ostendorf, B.; Chittleborough, D. Visible Near-Infrared Reflectance Spectroscopy as a Predictive Indicator of Soil Properties. *Ecol. Indic.* **2011**, *11*, 123–131. [CrossRef]
47. Khayamim, F.; Wetterlind, J.; Khademi, H.; Robertson, A.H.J.; Cano, A.F.; Stenberg, B. Using Visible and near Infrared Spectroscopy to Estimate Carbonates and Gypsum in Soils in Arid and Subhumid Regions of Isfahan, Iran. *J. Near Infrared Spectrosc.* **2015**, *23*, 155–165. [CrossRef]
48. Clark, R.N.; King, T.V.V.; Klejwa, M.; Swayze, G.A.; Vergo, N. High Spectral Resolution Reflectance Spectroscopy of Minerals. *J. Geophys. Res. Solid Earth* **1990**, *95*, 12653–12680. [CrossRef]
49. Chen, S.; Xu, H.; Xu, D.; Ji, W.; Li, S.; Yang, M.; Hu, B.; Zhou, Y.; Wang, N.; Arrouays, D.; et al. Evaluating Validation Strategies on the Performance of Soil Property Prediction from Regional to Continental Spectral Data. *Geoderma* **2021**, *400*, 115159. [CrossRef]
50. Debaene, G.; Niedźwiecki, J.; Pecio, A.; Żurek, A. Effect of the Number of Calibration Samples on the Prediction of Several Soil Properties at the Farm-Scale. *Geoderma* **2014**, *214–215*, 114–125. [CrossRef]
51. Wang, Z.; Chen, S.; Lu, R.; Zhang, X.; Ma, Y.; Shi, Z. Non-Linear Memory-Based Learning for Predicting Soil Properties Using a Regional Vis-NIR Spectral Library. *Geoderma* **2024**, *441*, 116752. [CrossRef]
52. Zhou, Y.; Chen, S.; Hu, B.; Ji, W.; Li, S.; Hong, Y.; Xu, H.; Wang, N.; Xue, J.; Zhang, X.; et al. Global Soil Salinity Prediction by Open Soil Vis-NIR Spectral Library. *Remote Sens.* **2022**, *14*, 5627. [CrossRef]
53. Guerrero, C.; Wetterlind, J.; Stenberg, B.; Mouazen, A.M.; Gabarrón-Galeote, M.A.; Ruiz-Sinoga, J.D.; Zornoza, R.; Viscarra Rossel, R.A. Do We Really Need Large Spectral Libraries for Local Scale SOC Assessment with NIR Spectroscopy? *Soil Tillage Res.* **2016**, *155*, 501–509. [CrossRef]

sensors

MDPI

Article

An Autocollimator Axial Measurement Method Based on the Strapdown Inertial Navigation System

Wenjia Ma [1,2], Jianrong Li [1], Shaojin Liu [1], Yan Han [1,2], Xu Liu [1,2], Zhiqian Wang [1,*] and Changhong Jiang [3,*]

1 Changchun Institute of Optics, Fine Mechanics and Physics, Chinese Academy of Sciences, Changchun 130033, China; mawenjia21@mails.ucas.ac.cn (W.M.); lijianrong@ciomp.ac.cn (J.L.); liusj@ciomp.ac.cn (S.L.); hanyan23@mails.ucas.ac.cn (Y.H.); liuxu223@mails.ucas.ac.cn (X.L.)
2 University of Chinese Academy of Sciences, Beijing 100049, China
3 School of Electrical and Electronic Engineering, Changchun University of Technology, Changchun 130012, China
* Correspondence: zhiqian_wang@sina.com (Z.W.); jch@ccut.edu.cn (C.J.)

Abstract: Autocollimators are widely used optical axis-measuring tools, but their measurement errors increase significantly when measuring under non-leveled conditions and they have a limited measurement range due to the limitations of the measurement principle. To realize axis measurement under non-leveled conditions, this paper proposes an autocollimator axis measurement method based on the strapdown inertial navigation system (SINS). First, the measurement model of the system was established. This model applies the SINS to measure the change in attitude of the autocollimator. The autocollimator was then applied to measure the angular relationship between the measured axis and its own axis, based on which the angular relationship of the axis was measured via computation through signal processing and data fusion in a multi-sensor system. After analyzing the measurement errors of the system model, the Monte Carlo method was applied to carry out a simulation analysis. This showed that the majority of the measurement errors were within ±0.002° and the overall measurement accuracy was within ±0.006°. Tests using equipment with the same parameters as those used in the simulation analysis showed that the majority of the measurement errors were within ±0.004° and the overall error was within ±0.006°, which is consistent with the simulation results. This analysis proves that this method solves the problem of the autocollimator being unable to measure the axis under non-leveled conditions and meets the needs of axis measurement with the application of autocollimators under a moving base.

Keywords: axis measurement; data fusion; autocollimation; SINS; computational measurement

1. Introduction

Axis measurement is an essential method for determining relative position and attitude [1,2], and it is widely used in industrial production, military operations, aerospace, and other fields. It is also employed in scientific research, including in straightness calibration [3], photon energy detection [4], and surface measurement [5]. In the aspect of axis angle measurement, it can be divided into mechanical methods, electromagnetic methods, optical methods, and inertial methods [6]. Among them, the mechanical and electromagnetic methods are more mature and less expensive to measure. However, the accuracy of electrical methods is easily affected by the environment, while mechanical methods are mostly contact measurements, which are limited in many fields. Optical measurement is a non-contact measurement method with high accuracy and sensitivity, which is widely used.

As a type of optical axis-measuring equipment, autocollimators benefit from advantages such as a high measuring accuracy, wide range, and non-contact measurement. The working principle of autocollimator is to use the orientation of the reflected beam from the target for pose calculation, which can realize precise single- and multi-axis angle measurements.

As technology has developed, so have autocollimators, from the traditional optical autocollimators to photoelectric autocollimators. The latter replaces the human eye with sensors such as charge-coupled devices (CCDs), quadrant photodiodes (QPDs), position sensitive detectors (PSDs), or complementary metal oxide semiconductors (CMOSs) for measurement, which improves the resolution and measurement accuracy [7–10]. The current research on measurement methods based on autocollimators mainly includes several aspects such as improving the measurement accuracy, increasing the measurement range, and increasing the number of measurement targets. For example, the use of new sensors improves the measurement accuracy [11–13], the design of the optical system improves the measurement accuracy and range [3,14], and the design of cooperative targeting achieves the measurement of the target's angle in three directions: yaw, pitch, and roll [15,16].

However, due to the limitation of the autocollimation measurement principle, its measurement range is mainly determined by the field of view of the optical system, the type of light source, and the image sensor, and usually the angular measurement range of the high-precision autocollimator is less than 1°, and the measurement distance is less than 50 m. In addition, in order to ensure measurement accuracy, the autocollimator needs to be roughly leveled with a geodetic coordinate system as a reference before use to avoid causing more significant measurement errors or making measurement impossible. However, in many measurement scenarios, it is impossible to level the autocollimator, which needs to remain stationary during the measurement process, making it impossible to carry out dynamic measurement. Therefore, autocollimators are usually used in the laboratory or after leveling on a stable platform. It is not possible to measure across long distances and on a large scale, such as, for example, in the case of ship installations, where the axes between the upper and lower layers of the hull are measured; in the case of large airplanes, where the axes are calibrated between each of the long-distance axes; or in the case of fast axes measurements of carrier vehicles in an off-site environment.

The principle of the strapdown inertial navigation system (SINS) is based on inertial characteristics; through the fusion data of the internal gyroscope and accelerometer sensors, it can realize accurate measurement of its angle and that of its strapdown equipment relative to the geodetic coordinate system. Accelerometers usually use micro-electromechanical system technology, and the displacement and angle can be inferred by integrating the acceleration in three directions [17–19]. Currently, the laser gyro and fiber-optic gyro measurement principles are based on the Sagnac effect, i.e., the beam propagation time slightly differs with rotation, and by measuring the time difference, the rotational speed and direction of an object can be obtained [20,21]. Compared with traditional gyroscopes, they have the advantages of high precision and shock resistance, so they are widely used in navigation systems [22,23].

At present, precision measurement is usually not limited to one kind of equipment, and multi-device cooperative work is one of the hot topics in current research [24,25]. With the improvement of the accuracy of laser gyro, fiber-optic gyro, and micro-electromechanical system (MEMS) inertial guidance technology, as well as the decrease in the cost and volume, inertial guidance has been employed in a large number of applications for the solution of position and navigation under multi-sensors. For example, inertial guidance is usually combined with a global navigation satellite system (GNSS) for integrated navigation [26–28], with a Doppler velocity log (DVL) for underwater navigation [29], and with a radar or camera for simultaneous localization and mapping (SLAM) algorithms [30].

Based on SINS characteristics and the defects that mean the autocollimator cannot measure under long-distance, wide-angle, or non-leveled conditions, it is of great practical significance and application value to proposes a non-leveled dynamic axis measurement method based on an SINS and autocollimator.

This paper is organized as follows: In Section 2, the system composition, measurement modeling, and experimental setups are described. Section 3 shows the simulation results and experimental results. Section 4 explains the experimental results, the potential limitations of this study, and how the system could be improved in future work. Section 5 presents the conclusions.

2. Methodologies

2.1. System Composition

Figure 1a shows the system composition, which includes a dual-axis photoelectric autocollimator and a strapdown inertial guide. The SINS consists of three fiber-optic gyros; it is a customized version purchased by the laboratory. The measurement errors of the inertial guide were within ±0.001° over a short time. The autocollimator consists of an optical system, a light source, and a CMOS sensor. The optical system was made by our lab and is designed for a focal length of 60 mm, an aperture of 25 mm, and a measuring range of 5 m. The model of the CMOS sensor is NOIP1SN5000A, made by ONSEMI, Scottsdale, AZ, USA; the sensor utilizes 4.8 μm × 4.8 μm pixels that support low-noise "pipelined" and "triggered" global shutter readout modes with 2592 × 2048 active pixels, with a plane mirror as the measurement target. The autocollimator has a measurement accuracy of ±0.001° in yaw and pitch under horizontal conditions, theoretically. The SINS has a measurement accuracy of ±0.01° theoretically, and had a measurement accuracy of ±0.001° in a short time proven by tests in the yaw, pitch, and roll directions. The measurement principle is shown in Figure 1b. In the measurement system, the inertial guide is used to measure the angular information between the system and O-XYZ relative to the geodetic coordinate system (i.e., northeast sky coordinate system), and the autocollimator is used to measure the angular relationship between the equipment and the Y_B axis of the target being measured (planar mirror). The fusion of the two data points is used to obtain the relationship between the axis of the target being measured relative to the geodetic coordinate system for the yaw and pitch angles.

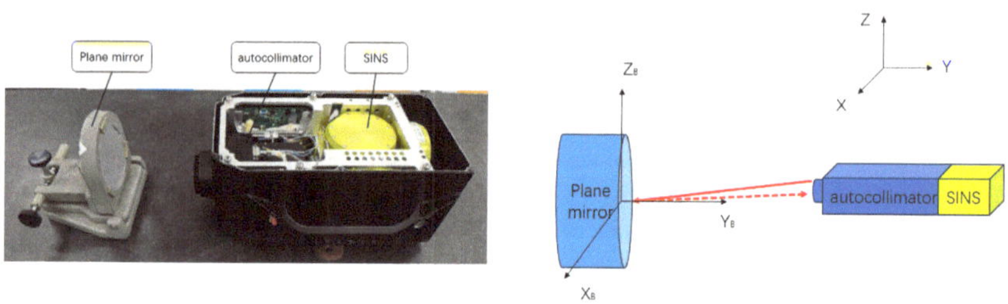

(**a**) (**b**)

Figure 1. System composition and principle. (**a**) The system includes an autocollimator and a strapdown inertial guide as the measurement equipment and a plane mirror as the target; (**b**) the autocollimator measures the plane mirror's axis and the SINS measures the geodetic coordinate system. The red arrows represent the return of beam from the collimator after reflection by the plane mirror.

2.2. Measurement Modeling

2.2.1. Coordinate System Establishment

The first step in establishing the system measurement model is to establish the measurement coordinate system, as shown in Figure 2, with the plane mirror as the measurement target. O-XYZ is the geodetic coordinate system (i.e., the northeast celestial coordinate system), O-$X_AY_AZ_A$ is the SINS coordinate system, O-$X_CY_CZ_C$ is the lens coordinate system, O-$X_PY_PZ_P$ is the camera coordinate system, and O-$X_BY_BZ_B$ is the plane mirror coordinate system. Among them, the inertial coordinate system, the lens coordinate system, and the camera coordinate system are all based on the Earth's level; the coordinate axes of these three coordinate systems are in the same direction, and the orientation is based on the Y_C direction along the optical axis.

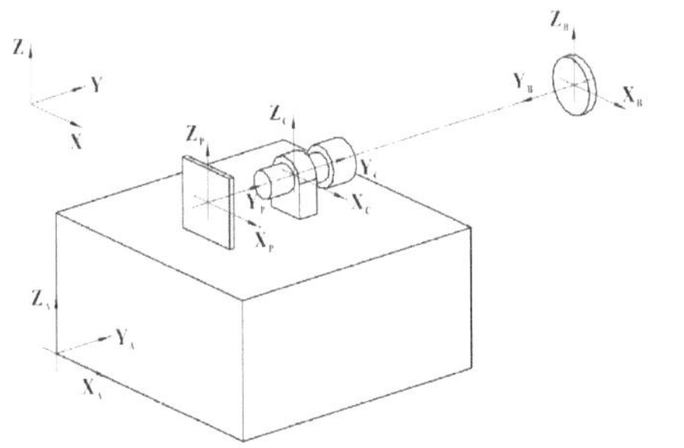

Figure 2. Measurement coordinate system. The measurement model is constructed using the geodetic coordinate system O-XYZ, the SINS coordinate system O-$X_AY_AZ_A$, the lens coordinate system O-$X_CY_CZ_C$, the camera coordinate system O-$X_PY_PZ_P$, and the plane mirror coordinate system O-$X_BY_BZ_B$.

2.2.2. Measurement Modeling of the Autocollimator

First, we consider the measurement model of the autocollimator in a leveled case. Projection along the opposite direction to the X-axis in Figure 2 yields the system shown in Figure 3. For this, we let the focal length of the autocollimator be *f*. The plane mirror reflects the emitted target, point P, to the image plane, point P (x_p,z_p). Then, the yaw angle α and pitch angle β of the plane mirror concerning the autocollimator can be computed using Equation (1).

$$\begin{aligned} \alpha &= \tfrac{1}{2} * \tan^{-1}\left(\tfrac{x_p}{f}\right) \\ \beta &= \tfrac{1}{2} * \tan^{-1}\left(\tfrac{z_p}{f}\right) \end{aligned} \quad (1)$$

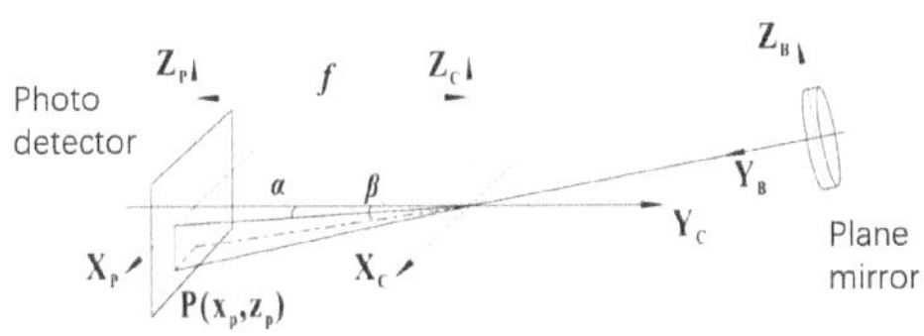

Figure 3. Autocollimator measurement coordinate system. By calculating the position of the reflected point P (x_p,z_p), the yaw angle α and pitch angle β of the plane mirror concerning the autocollimator can be obtained.

A further analysis of the measurement system under non-leveled dynamic conditions is shown in Figure 4 for the O-XZ planar projection obtained by projecting the measurement coordinate system in Figure 2 along the positive direction of the Y-axis. When the roll angle between the autocollimator and the horizontal plane is γ_A, the target is set to image point P (x_p,z_p) on the image plane of the camera and converted from the O-X_pZ_p coordinate system to the O-XZ coordinate system. The position coordinate conversion equations are as follows:

$$\begin{aligned} x &= x_p * \cos\gamma_A - z_p * \sin\gamma_A \\ z &= x_p * \sin\gamma_A + z_p * \cos\gamma_A \end{aligned} \quad (2)$$

Substituting Equation (2) into (1) yields the relative value of the measurement target to the axis of the measurement system in the non-leveled state:

$$\begin{aligned} \alpha &= \tfrac{1}{2} * \tan^{-1}((x_p * \cos\gamma_A - z_p * \sin\gamma_A)/f) \\ \beta &= \tfrac{1}{2} * \tan^{-1}((x_p * \sin\gamma_A + z_p * \cos\gamma_A)/f) \end{aligned} \quad (3)$$

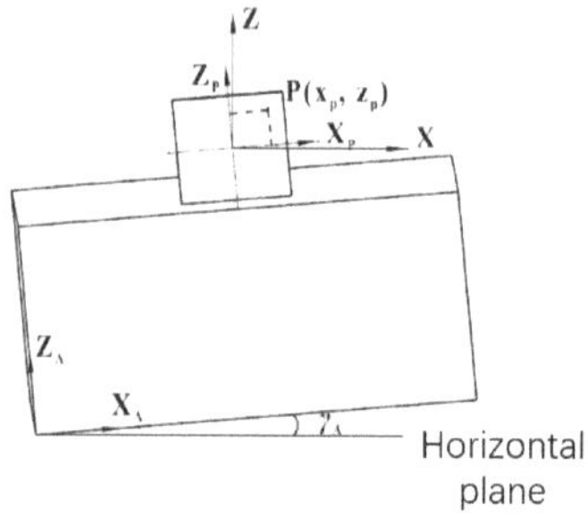

Figure 4. Autocollimator angle measurement under non-leveled conditions. γ_A is the roll angle between the autocollimator and the horizontal plane, which causes measurement errors.

2.2.3. Measurement Modeling of the SINS

As shown in Figure 5, the inertial guidance measurement angles are defined, respectively, as follows:

1. Yaw angle, α_A: the horizontal angle between the projection of the Y_A-axis onto the horizontal plane and the actual north direction;
2. Pitch angle, β_A: the angle between the Y_A-axis in the vertical projection plane and the horizontal plane;
3. Roll angle, γ_A: the angle between the X_A-axis in the vertical projection plane and the horizontal plane.

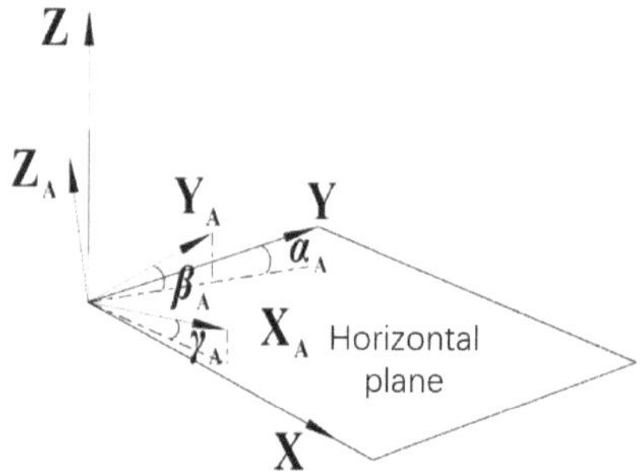

Figure 5. SINS measurement coordinate system. The yaw angle α_A, pitch angle β_A, and roll angle γ_A can be measured using the SINS.

Combined with Equation (3), the orientation and pitch angle of the measurement target relative to the geodetic coordinate system can be obtained as follows:

$$\begin{aligned} \alpha_0 &= \tfrac{1}{2} * \tan^{-1}\left(\left(x_p * \cos\gamma_A - z_p * \sin\gamma_A\right)/f\right) + \alpha_A \\ \beta_0 &= \tfrac{1}{2} * \tan^{-1}\left(\left(x_p * \sin\gamma_A + z_p * \cos\gamma_A\right)/f\right) + \beta_A \end{aligned} \tag{4}$$

2.3. Measurement Error Analysis

Based on the measurement model and the calculation formulae provided in Section 2, the sources of errors in the measurement system are further analyzed to prove the validity of the measurement system and its measurement accuracy.

The measurement errors of the system include systematic errors and random errors. Using Equation (4), the systematic errors of the system include the installation errors and focal length errors, and the random errors include the pixel errors and SINS measurement errors.

2.3.1. Installation Errors

Due to the virtual axis of the SINS, it is difficult to achieve parallelism between the coordinate system of the SINS and the autocollimator's coordinate system through installation and adjustment, which leads to a small deviation between the two coordinate systems. If we let the angle of the autocollimator rolling direction relative to the horizontal plane be γ_A, the angle between the autocollimator rolling axis and the inertial guide rolling axis be γ, and the angle between the inertial guide rolling axis and the horizontal plane be γ_0, then we have

$$\gamma_A = \gamma_0 + \gamma \tag{5}$$

Substituting Equation (5) into (4), we obtain

$$\begin{aligned} \alpha_0 &= \tfrac{1}{2} * \tan^{-1}\left(\left(x_p * \cos(\gamma_0 + \gamma) - z_p * \sin\,(\gamma_0 + \gamma)\right)/f\right) + \alpha A \\ \beta_0 &= \tfrac{1}{2} * \tan^{-1}\left(\left(x_p * \sin(\gamma_0 + \gamma) + z_p * \cos(\gamma_0 + \gamma)\right)/f\right) + \beta A \end{aligned} \tag{6}$$

From Equation (6), the roll angle γ between the SINS and the autocollimator affects the measurement results, and this error cannot be eliminated in the relative measurement.

This systematic error can usually be measured and corrected. The measurement of angle γ is usually realized through methods such as optical–mechanical calibration. However, the SINS cannot be measured accurately, because of its virtual axis. According to engineering experience, this measurement error can be controlled at about 10″, so the cross-roll error in the subsequent simulation is taken as $\gamma = \pm 0.003°$.

2.3.2. Focal Length Errors

The lens's focal length is usually determined by the optical design, but manufacturing and mounting errors will cause the focal length to be inconsistent with the design value. According to engineering experience, the actual focal length of the lens can usually be controlled within 0.1–0.5% of the theoretical value. After calibration, the actual focal length value can be controlled within 0.1% [31]. In this study, the theoretical focal length of the self-collimator is 60 mm and the actual value after calibration measurement is 59.64 mm. Accordingly, the following simulation takes the focal length error value as ±0.05 mm.

2.3.3. Pixel Errors

In Equation (5), x_p and z_p are the coordinates of the target's imaging position on the camera at the time of measurement. The errors in x_p and z_p are determined by the resolution of the camera image element. According to research on the pixel subdivision algorithm, the pixel errors can usually be reduced to within 0.1 pixels using differential calculation and other methods [32]. The actual camera pixel resolution used in this system is 4.8 μm. In order to ensure that the simulation was consistent with the actual test, the pixel error value was taken to be 0.5 pixels in the simulation process, which is 2.4 μm.

2.3.4. SINS Measurement Errors

From Section 2.2.2, the measurement error of the SINS is determined by its measurement accuracy. This system uses the fiber-optic inertial guide for measurement. For the measurement system in this study, the measurement is usually completed in a short period of time in a single power-up, so the effects of system errors can be ignored. The random errors of fiber-optic inertial guides include zero bias and noise. However, random errors usually have a negligible effect on the measurement results when measuring in a short period of time. After testing the inertial guide used in the system, it was proven that the measurement errors of this inertial guide were within ±0.001° over a short period of time, so this value was taken for the simulation.

2.4. Experiments

Experimental validation was carried out using laboratory equipment to verify the validity of the measurement model and simulation results. The instrumentation used in this testing included the SINS, an autocollimator, a roll adjustment stage, a plane mirror with a 2D adjustment stage, and a parallel light tube for measurements with the following accuracies:

- The installation error was taken as $\gamma = \pm 0.003°$;
- The focal length error was taken as ± 0.05 mm;
- The CMOS sensor measurement error was taken as ± 0.1 pixel;
- The SINS measurement error was taken as $\pm 0.001°$ in yaw, pitch, and roll;
- The parallel light pipes had a measurement accuracy of $\pm 0.2''$ in yaw and pitch;
- The roll adjustment table had a $\pm 15°$ adjustment range.

The experimental setup is shown in Figure 6. The experimental procedure was as follows: The measuring system (SINS with the autocollimator) was placed on the roll adjustment stage, and the plane mirror with the 2D adjustment stage was placed between the measuring system and the parallel light tube. Adjusting the 2D adjustment table changed the plane mirror's axis, and its axis changes were monitored with the parallel light tube. Measurement started from the position when the SINS was horizontal, and the range of roll adjustment was $\pm 10°$, with an interval of $2°$ between each adjustment. After each set of the system's roll values was adjusted, its yaw and pitch values were adjusted using the initial angle of the plane mirror as a reference. Each yaw and pitch adjustment interval was $0.2°$, with a range of $\pm 1°$. The yaw and pitch values of each set of the plane mirror adjustments were recorded, as well as the coordinate points of the image obtained from the autocollimator in the image plane (x_p,z_p). A total of 121 sets of corresponding data values were obtained.

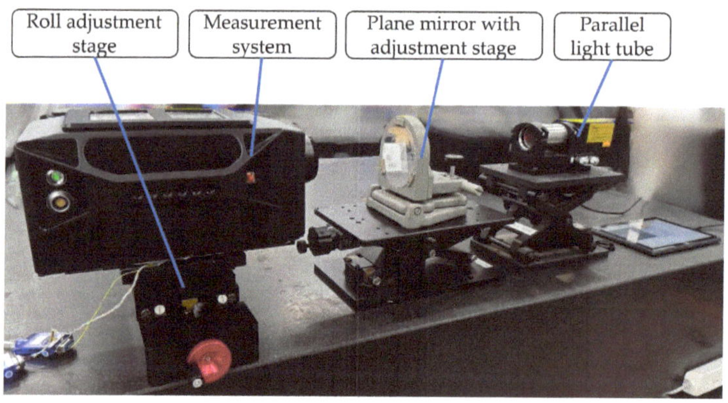

Figure 6. Experimental setup, including a measurement system, a plane mirror (measurement target), and a parallel optical tube to measure the true value of the amount of change in the target. The measurement accuracy of the system is verified by changing the angle of the plane mirror.

3. Results

3.1. Simulation Results

To ensure the accuracy and validity of the measurement model presented in Section 2.2, as well as the error analysis provided in Section 2.3, a Monte Carlo analysis based on Equation (6) was conducted. The Monte Carlo method, also known as the statistical test method, is a numerical simulation technique that focuses on probabilistic phenomena. Its fundamental principle is random sampling. By constructing a probabilistic model that closely represents the measurement system's performance and running random trials, the simulation can replicate the system's random measurement characteristics. The simulation results can be considered as actual measurement outcomes when the number of simulations is large enough. Substituting the four measurement error values from the error analysis in Section 2.3, all of the parameters used in the simulation were the same as those used in the experiments, specifically including the following:

- The installation error was taken as $\gamma = \pm 0.003°$;
- The focal length error was taken as ± 0.05 mm;
- The CMOS sensor measurement error was taken as ± 0.1 pixel;
- The SINS measurement error was taken as $\pm 0.001°$ in yaw, pitch, and roll.

The specific calculation steps were as follows:

1. Randomly generate an initial set of truth data, including the measurement of the system's roll angle (within $\pm 5°$) and the yaw and pitch angles of the plane mirror axis measured with the autocollimator (within $\pm 4.5°$);
2. Based on the generated truth data, back-project the theoretical truth data measured with the sensor, and randomly add the error data in Section 3.1 to them as the measurement data of the sensor;
3. Substituting the sensor measurement data into Equation (4), recalculate the yaw and pitch angles of the plane mirror as measurements;
4. Calculate the difference between the measured data and the true value data as the measurement error.

Scatter plots of the measurement errors of the system and the distribution of the errors when repeating the calculation 10,000 times are shown in Figure 7. The figures show that the system's measurement errors are overwhelmingly within the range of $\pm 0.002°$ and the overall measurement accuracy is within the range of $\pm 0.006°$. The mean square deviations of the yaw and pitch errors are $\sigma_{\_yaw} = 0.0020°$ (1σ) and $\sigma_{\_pitch} = 0.0019°$ (1σ).

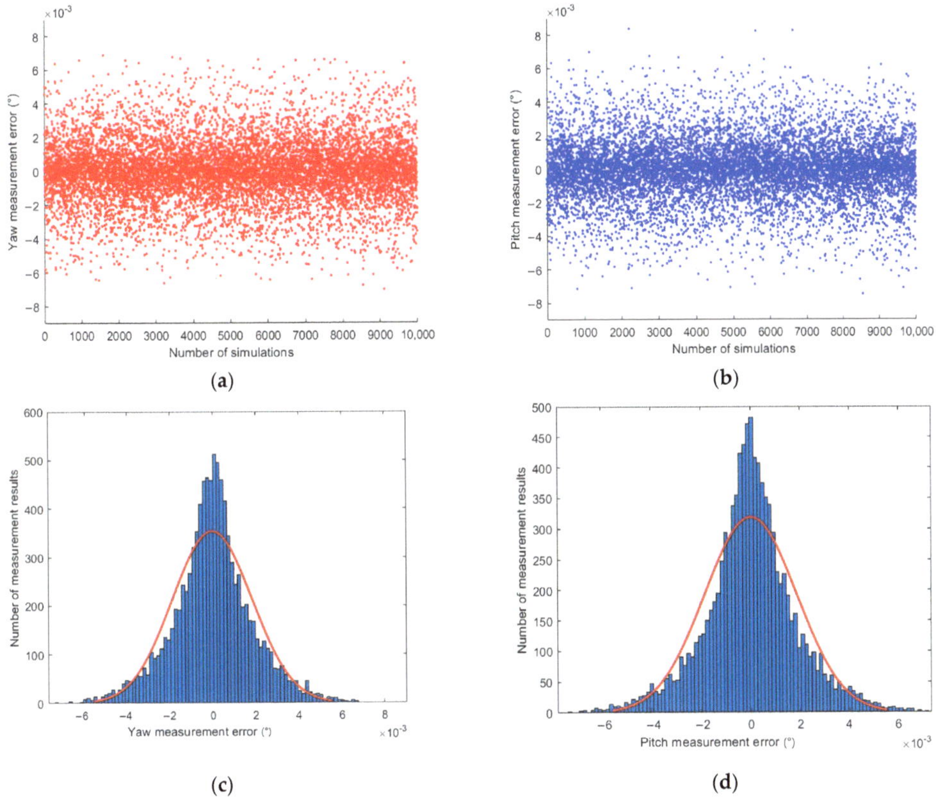

Figure 7. Simulation results. The red line refers to the probability distribution curve fitted to the simulated data. (**a**) Yaw error scatter plot; (**b**) pitch error scatter plot; (**c**) yaw error histogram; (**d**) pitch error histogram.

3.2. Experimental Results

The data measured in Section 2.4 are listed in Table 1. The measurement group 0 includes the initial yaw and pitch values of the mirror, as well as the position of the reflected target. For measurement groups 1 to 5, we sequentially increased the yaw and pitch of the plane mirror by about 0.2° in the same direction, using measurement group 0 as a reference. For measurement groups 6 to 10, we sequentially increased the yaw and pitch of the plane mirror by about 0.2° in the other direction, using measurement group 0 as a reference. The exact amount of change was measured by means of a parallel light tube, and the corresponding x_p and z_p values were also recorded.

Table 1. Experimental data.

Roll Angle (°)	Data Type	Measurement Groups										
		0	1	2	3	4	5	6	7	8	9	10
−0.001	Mirror yaw (°)	58.9891	59.1843	59.3862	59.5881	59.7882	59.9858	58.7882	58.5838	58.3834	58.1838	57.9819
	Mirror pitch (°)	90.2941	90.4921	90.6908	90.8952	91.0932	91.2998	90.0947	89.8982	89.6941	89.4959	89.2971
	x_p in CMOS	996.21	1082.25	1171.07	1259.81	1347.67	1434.39	908.08	818.45	730.23	642.92	554.45
	z_p in CMOS	985.64	1072.45	1159.24	1248.72	1334.92	1425.21	898.62	812.91	723.92	637.33	550.85
2.010	Mirror yaw (°)	58.9342	59.1346	59.3338	59.5343	59.7375	59.9371	58.7369	58.5367	58.3347	58.1356	57.9309
	Mirror pitch (°)	90.2945	90.4973	90.6951	90.8961	91.0947	91.2951	90.0933	89.8943	89.6949	89.4949	89.2946
	x_p in CMOS	996.78	1081.68	1166.11	1251.16	1337.37	1421.56	912.98	828.17	742.57	658.29	571.78
	z_p in CMOS	985.57	1077.33	1166.81	1257.67	1347.51	1437.87	894.77	805.01	715.01	624.82	534.54
4.002	Mirror yaw (°)	58.9664	59.1654	59.3642	59.5692	59.7686	59.9669	58.7627	58.5658	58.3646	58.1605	57.9634
	Mirror pitch (°)	90.2976	90.4936	90.6922	90.8951	91.0959	91.2957	90.0921	89.8914	89.6961	89.4982	89.2925
	x_p in CMOS	996.29	1077.64	1158.83	1242.53	1323.64	1404.31	913.21	832.94	750.86	667.69	587.58
	z_p in CMOS	985.66	1077.11	1170.01	1264.51	1358.31	1451.31	889.87	796.47	705.27	613.01	517.77
6.003	Mirror yaw (°)	58.9835	59.1883	59.3879	59.5843	59.7841	59.9892	58.7867	58.5819	58.3824	58.1865	57.9832
	Mirror pitch (°)	90.2931	90.4925	90.6961	90.8969	91.0935	91.2999	90.0967	89.8933	89.6947	89.4956	89.2954
	x_p in CMOS	996.43	1076.82	1154.83	1231.56	1309.67	1389.84	919.34	839.02	761.07	684.71	605.31
	z_p in CMOS	985.58	1081.81	1179.34	1275.66	1370.41	1469.31	891.17	793.71	698.31	603.02	507.11
8.007	Mirror yaw (°)	59.0043	59.2053	59.4065	59.6031	59.8038	60.0064	58.8056	58.6067	58.4061	58.2053	58.0048
	Mirror pitch (°)	90.2923	90.4947	90.6924	90.8946	91.0965	91.2957	90.0921	89.8955	89.6918	89.4951	89.2913
	x_p in CMOS	996.53	1071.88	1147.14	1220.53	1295.46	1371.41	922.08	847.62	772.76	697.43	622.66
	z_p in CMOS	985.35	1085.26	1182.96	1282.49	1382.28	1480.58	886.62	789.21	689.19	592.02	492.45
10.005	Mirror yaw (°)	59.0086	59.2059	59.4079	59.6074	59.8075	60.0061	58.8081	58.6015	58.4021	58.2022	58.0018
	Mirror pitch (°)	90.2878	90.4854	90.6857	90.8844	91.0879	91.2847	90.0821	89.8816	89.6849	89.4802	89.2815
	x_p in CMOS	996.58	1066.86	1139.08	1210.33	1281.35	1352.29	924.97	850.85	779.46	708.71	637.19
	z_p in CMOS	985.35	1085.76	1187.43	1288.21	1391.17	1490.91	882.17	780.19	680.59	577.41	477.05
−2.006	Mirror yaw (°)	58.9537	59.1531	59.3571	59.5554	59.7552	59.9555	58.7547	58.5509	58.3521	58.1568	57.9529
	Mirror pitch (°)	90.2958	90.4984	90.6992	90.8985	91.0972	91.2952	90.0911	89.8948	89.6965	89.4907	89.2906
	x_p in CMOS	996.41	1087.01	1179.91	1270.05	1360.52	1451.51	905.81	813.51	723.12	634.23	542.01
	z_p in CMOS	985.56	1071.19	1155.72	1239.63	1323.51	1406.63	899.65	817.05	733.63	647.11	563.26
−4.004	Mirror yaw (°)	59.0021	59.2047	59.4041	59.6004	59.8015	60.0088	58.8028	58.6031	58.4021	58.2037	58.0012
	Mirror pitch (°)	90.2947	90.4979	90.6974	90.8953	91.0954	91.2957	90.0922	89.8922	89.6928	89.4938	89.2909
	x_p in CMOS	996.21	1091.18	1184.89	1276.97	1371.23	1468.07	902.77	809.02	715.01	622.04	527.32
	z_p in CMOS	985.37	1067.94	1148.86	1229.01	1310.05	1390.91	903.54	822.61	742.01	661.53	579.94
−6.000	Mirror yaw (°)	58.9945	59.1977	59.3999	59.5983	59.7976	59.9975	58.7913	58.5925	58.3916	58.1939	57.9945
	Mirror pitch (°)	90.2922	90.4963	90.6957	90.8931	91.0932	91.2976	90.0955	89.8923	89.6914	89.4948	89.2926
	x_p in CMOS	996.27	1094.52	1191.97	1287.85	1383.99	1480.52	898.61	802.44	705.53	610.31	514.13
	z_p in CMOS	985.32	1064.99	1142.36	1218.89	1296.79	1376.34	909.37	830.68	752.81	676.65	598.34

Table 1. *Cont.*

Roll Angle (°)	Data Type	Measurement Groups										
		0	1	2	3	4	5	6	7	8	9	10
−8.001	Mirror yaw (°)	58.9714	59.1781	59.3785	59.5754	59.7751	59.9758	58.7727	58.5719	58.3701	58.1722	57.9702
	Mirror pitch (°)	90.2911	90.4971	90.6935	90.8979	91.0926	91.2933	90.0906	89.8917	89.6966	89.4977	89.2969
	x_p in CMOS	996.14	1098.83	1197.91	1296.01	1394.55	1494.01	897.39	798.02	698.53	600.47	500.78
	z_p in CMOS	985.68	1062.31	1135.11	1211.58	1283.52	1358.11	911.36	837.71	765.87	692.09	617.83
−10.006	Mirror yaw (°)	58.9665	59.1671	59.3617	59.5681	59.7611	59.9665	58.7616	58.5667	58.3666	58.1639	57.9635
	Mirror pitch (°)	90.2917	90.4933	90.6942	90.8944	91.0939	91.2957	90.0971	89.8946	89.6902	89.4913	89.2969
	x_p in CMOS	996.36	1098.48	1198.05	1302.56	1401.21	1505.27	893.01	793.38	691.28	588.71	487.71
	z_p in CMOS	985.79	1057.29	1129.02	1199.38	1270.61	1341.91	917.94	845.85	773.25	703.45	635.44

We substituted the x_p and z_p data in Table 1 into Equation (6) to calculate the measurement values of the mirror's change in yaw and pitch. The measurement group 0 was also used as the initial value in these calculations, and the measured values were compared with the true values to obtain the errors. From these calculations, the measurement errors are shown in Table 2, and each measurement error matches those of the measurement groups in Table 1.

Table 2. Measurement errors.

Roll Angle (°)	Data Type	Measurement Groups									
		1	2	3	4	5	6	7	8	9	10
−0.001	Yaw error (°)	0.0016	0.0029	0.0040	0.0048	0.0053	−0.0007	−0.0014	−0.0027	−0.0028	−0.0031
	Pitch error (°)	0.0013	0.0018	0.0028	0.0026	0.0031	−0.0004	−0.0007	−0.0008	−0.0013	−0.0008
2.010	Yaw error (°)	0.0011	0.0021	0.0033	0.0043	0.0042	−0.0016	−0.0025	−0.0034	−0.0041	−0.0042
	Pitch error (°)	0.0009	0.0016	0.0021	0.0026	0.0026	−0.0004	−0.0005	−0.0006	−0.0006	−0.0003
4.002	Yaw error (°)	0.0013	0.0026	0.0036	0.0042	0.0047	−0.0012	−0.0024	−0.0030	−0.0034	−0.0041
	Pitch error (°)	0.0004	0.0015	0.0016	0.0025	0.0026	−0.0006	−0.0009	−0.0013	−0.0012	−0.0006
6.003	Yaw error (°)	0.0011	0.0023	0.0035	0.0040	0.0047	−0.0012	−0.0024	−0.0031	−0.0036	−0.0037
	Pitch error (°)	0.0010	0.0013	0.0020	0.0028	0.0028	−0.0007	−0.0005	−0.0010	−0.0010	−0.0005
8.007	Yaw error (°)	0.0015	0.0020	0.0033	0.0040	0.0046	−0.0014	−0.0022	−0.0031	−0.0038	−0.0042
	Pitch error (°)	0.0006	0.0010	0.0015	0.0024	0.0022	−0.0004	−0.0015	−0.0012	−0.0012	0.0004
10.005	Yaw error (°)	0.0011	0.0022	0.0033	0.0041	0.0046	−0.0010	−0.0019	−0.0029	−0.0034	−0.0033
	Pitch error (°)	0.0008	0.0017	0.0023	0.0030	0.0037	0	−0.0002	−0.0006	−0.0004	0.0002
−2.006	Yaw error (°)	0.0009	0.0025	0.0035	0.0037	0.0046	−0.0013	−0.0019	−0.0030	−0.0039	−0.0039
	Pitch error (°)	0.0011	0.0017	0.0021	0.0031	0.0029	−0.0003	−0.0003	−0.0006	−0.0003	−0.0001
−4.004	Yaw error (°)	0.0010	0.0025	0.0035	0.0044	0.0050	−0.0009	−0.0022	−0.0028	−0.0036	−0.0040
	Pitch error (°)	0.0011	0.0020	0.0023	0.0028	0.0030	0.0001	−0.0002	−0.0004	−0.0005	−0.0010
−6.000	Yaw error (°)	0.0013	0.0023	0.0037	0.0044	0.0049	−0.0008	−0.0020	−0.0029	−0.0036	−0.0041
	Pitch error (°)	0.0014	0.0020	0.0023	0.0030	0.0032	−0.0002	0.0003	0.0002	0.0002	0.0007
−8.001	Yaw error (°)	0.0015	0.0024	0.0033	0.0039	0.0046	−0.0014	−0.0022	−0.0029	−0.0035	−0.0035
	Pitch error (°)	0.0010	0.0018	0.0025	0.0027	0.0032	0	−0.0003	−0.0003	−0.0003	−0.0005
−10.006	Yaw error (°)	0.0010	0.0021	0.0031	0.0038	0.0045	−0.0018	−0.0028	−0.0037	−0.0041	−0.0046
	Pitch error (°)	0.0014	0.0021	0.0028	0.0036	0.0038	0.0010	−0.0005	0.0005	0.0003	0.0008

Analyzing the measurement errors in Table 2 and Figure 8 reveals the following:

When the system measures within ±10° of the roll angle, most of the measurement errors are within ±0.004° and the overall error is within ±0.006°. The root mean square (RMS) values of the measurement errors are $\sigma_{\_yaw}$ = 0.003° (1σ) and $\sigma_{\_pitch}$ = 0.0018° (1σ). The measurement consistency is good, which is consistent with the simulation results.

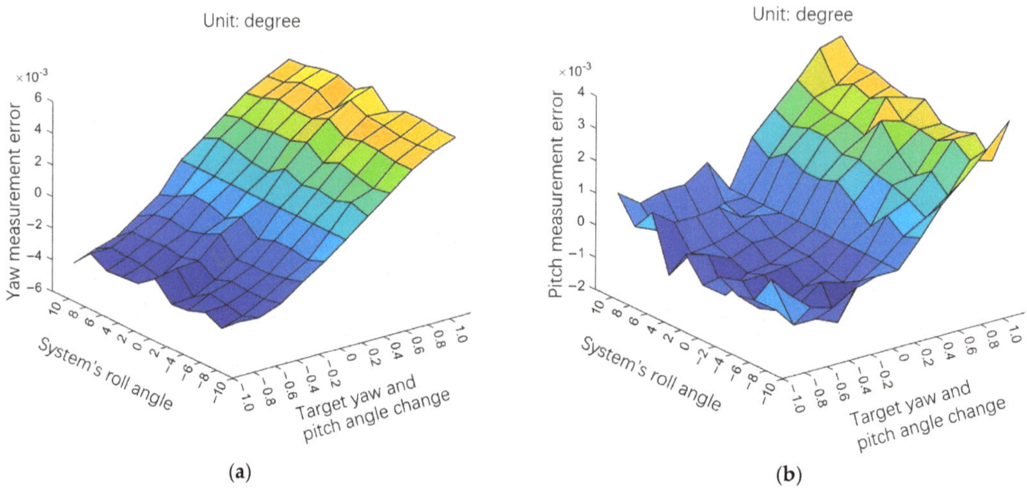

Figure 8. Experimental error distribution. Different color represents the amount of the surface in the z-axis direction. (**a**) Yaw error distribution; (**b**) pitch error distribution.

4. Discussion

Through Monte Carlo simulation of this measurement model, the variance of the yaw and pitch is obtained as σ_yaw = 0.0020° (1σ) and σ_pitch = 0.0019° (1σ) after running the simulation 10,000 times. It can be deduced from the simulation results that the measurement error of this system is stable and consistent for the target yaw and pitch angles.

As can be seen from the experimental results, when the system roll angle is unchanged and the angle of the measurement target changes, the measurement error increases with the change in angle of the measurement target, and the sign of the measurement error is related to the sign of the change in angle of the target. Also, within a specific range, the system's measurement error does not change due to the change in roll angle.

Comparing the experimental results with the simulation results, it can be seen that the yaw measurement error of the experimental results is larger than that of the simulation results. The reason for these different results may be the fact that the measurement group is too small to reflect the actual measurement capability of the system, or the fact that the CMOS sensor in the autocollimator does not coincide with the axis of the optical system when it is installed, or other reasons, which need to be further explored.

For axis measurements under non-leveled dynamic conditions, the accurate measurement of angles is usually achieved by optimizing the measurement environment, e.g., by designing vibration damping and leveling planes. This study started without optimizing the measurement equipment, which has an advantage in terms of preparation time, volume, and weight, although the accuracy is slightly lower in comparison. This measurement accuracy can still be suitable for vehicle-related axis measurement, aircraft axis measurement target calibration, and naval weapon axis measurement. Since the work presented in this paper involves only a measurement model and its validation under laboratory conditions, subsequent field tests need to be carried out to explore its engineering practicality further.

In this study, the data between the two sensors were only calculated and processed using the measurement model, but there may have been a time delay in the two sensors, which could have led to a decrease in the measurement accuracy due to the unsynchronized data when measuring on a dynamic platform; so, further exploration of data fusion and synchronization methods is required subsequently.

5. Conclusions

In order to realize axis measurement under non-leveled dynamic conditions using an autocollimator and extend the measuring range, an autocollimator axis measurement method based on the SINS is proposed. This article demonstrates the system model and measurement calculations, simulation analyses, and experimental verification of the model were carried out. The latter demonstrated that the majority of the method's measurement errors were within $\pm 0.002^\circ$ and the overall measurement error was within $\pm 0.006^\circ$. The measurement system was tested over a roll angle range of $\pm 10^\circ$, showing that most of the measurement errors were within $\pm 0.004^\circ$ and the overall measurement error was within $\pm 0.006^\circ$. This was consistent with the simulation results, showing a good measurement consistency.

Our system has certain advantages over other measurement methods, and the measurement accuracy can be further improved from the point of view of data fusion and synchronization of the two sensors in future work. According to the proposed study, a new measurement method can be provided for axis measurement in the case of a moving base.

Author Contributions: Conceptualization, W.M., C.J. and Z.W.; methodology, W.M., C.J. and Z.W.; software, S.L. and Z.W.; validation, W.M., Z.W. and J.L.; formal analysis, W.M. and Y.H.; investigation, Y.H.; resources, Z.W. and J.L.; data curation, W.M.; writing—original draft preparation, Z.W., X.L. and Y.H.; writing—review and editing, Z.W., Y.H., X.L., C.J. and J.L.; visualization, Z.W.; supervision, Z.W.; project administration, J.L. and Z.W.; funding acquisition, J.L. All authors have read and agreed to the published version of the manuscript.

Funding: This research was funded by the Scientific and Technological Development Program of Ji Lin Province, China (Grant No. 20210201048GX).

Institutional Review Board Statement: Not applicable.

Informed Consent Statement: Not applicable.

Data Availability Statement: The data presented in this study are available on request from the corresponding author.

Conflicts of Interest: The authors declare no conflicts of interest.

References

1. Kumar, A.S.A.; George, B.; Mukhopadhyay, S.C. Technologies and Applications of Angle Sensors: A Review. *IEEE Sens. J.* **2021**, *21*, 7195–7206. [CrossRef]
2. Gao, W.; Kim, S.W.; Bosse, H.; Haitjema, H.; Chena, Y.L.; Lu, X.D.; Knapp, W.; Weckenmann, A.; Estler, W.T.; Kunzmann, H. Measurement technologies for precision positioning. *CIRP Ann.-Manuf. Technol.* **2015**, *64*, 773–796. [CrossRef]
3. Zheng, F.; Long, F.; Zhao, Y.; Yu, C.; Jia, P.; Zhang, B.; Yuan, D.; Feng, Q. High-Precision Small-Angle Measurement of Laser-Fiber Autocollimation Using Common-Path Polarized Light Difference. *IEEE Sens. J.* **2023**, *23*, 9237–9245. [CrossRef]
4. Mandal, L.; Singh, J.; Ganesan, A.R. Michelson Interferometer with mirrored right-angled prism for measurement of tilt with double sensitivity. *J. Opt.* **2023**, *53*, 1545–1550. [CrossRef]
5. Meng, F.; Yang, G.; Yang, J.; Lu, H.; Dong, Z.; Kang, R.; Guo, D.; Qin, Y. Error Analysis of Normal Surface Measurements Based on Multiple Laser Displacement Sensors. *Sensors* **2024**, *24*, 2059. [CrossRef] [PubMed]
6. Wang, S.; Ma, R.; Cao, F.; Luo, L.; Li, X. A Review: High-Precision Angle Measurement Technologies. *Sensors* **2024**, *24*, 1755. [CrossRef] [PubMed]
7. Zou, Y.; Zhang, N.; Yuan, H.; Liu, C.; Liu, D. High precision deflection angle measurement of irregular light spot. *Chin. Space Sci. Technol.* **2022**, *42*, 107–113.
8. Guo, Y.; Cheng, H.; Liu, G. Three-degree-of-freedom autocollimation angle measurement method based on crosshair displacement and rotation. *Rev. Sci. Instrum.* **2023**, *94*, 0126806. [CrossRef] [PubMed]
9. Yan, L.; Yan, Y.; Chen, B.; Lou, Y. A Differential Phase-Modulated Interferometer with Rotational Error Compensation for Precision Displacement Measurement. *Appl. Sci.* **2022**, *12*, 5002. [CrossRef]
10. Kim, K.; Jang, K.W.; Ryu, J.K.; Jeong, K.H. Biologically inspired ultrathin arrayed camera for high-contrast and high-resolution imaging. *Light Sci. Appl.* **2020**, *9*, 28. [CrossRef]
11. Guo, J.; Lin, L.; Li, S.; Chen, J.; Wang, S.; Wu, W.; Cai, J.; Zhou, T.; Liu, Y.; Huang, W. Ferroelectric superdomain controlled graphene plasmon for tunable mid-infrared photodetector with dual-band spectral selectivity. *Carbon* **2022**, *189*, 596–603. [CrossRef]
12. Guo, J.; Lin, L.; Li, S.; Chen, J.; Wang, S.; Wu, W.; Cai, J.; Liu, Y.; Ye, J.; Huang, W. WSe_2/MoS_2 van der Waals Heterostructures Decorated with Au Nanoparticles for Broadband Plasmonic Photodetectors. *ACS Appl. Nano Mater.* **2022**, *5*, 587–596. [CrossRef]

13. Guo, J.; Liu, Y.; Lin, L.; Li, S.; Cai, J.; Chen, J.; Huang, W.; Lin, Y.; Xu, J. Chromatic Plasmonic Polarizer-Based Synapse for All-Optical Convolutional Neural Network. *Nano Lett.* **2023**, *23*, 9651–9656. [CrossRef]
14. Kong, Q.; Wen, L. Fast tilt measurement based on optical block. *Opt. Laser Technol.* **2022**, *148*, 107731. [CrossRef]
15. Du, M.-x.; Yan, Y.-f.; Zhang, R.; Cai, C.-l.; Yu, X.; Bai, S.-p.; Yu, Y. 3D position angle measurement based on a lens array. *Chin. Opt* **2022**, *15*, 45–55. [CrossRef]
16. Ren, W.; Cui, J.; Tan, J. Precision roll angle measurement system based on autocollimation. *Appl. Opt.* **2022**, *61*, 3811–3818 [CrossRef] [PubMed]
17. Dorofeev, N.V.; Kuzichkin, O.R.; Tsaplev, A.V. Accelerometric Method of Measuring the Angle of Rotation of the KinematicMechanisms of Nodes. *Appl. Mech. Mater.* **2015**, *770*, 592–597. [CrossRef]
18. Gao, Y.; Zhang, X.; Mai, G.; Hu, R. Measurement Method for Angle Acceleration of Turn-table Based on Accelerometer. *J. Astronaut. Metrol. Meas.* **2015**, *35*, 14–16.
19. Lu, Y.-l.; Pan, Y.-J.; Li, L.-l.; Liu, Y.; Peng, H. Measurement method of projectile's heading and pitching angle velocities based on biaxial accelerometer. *J. Chin. Inert. Technol.* **2015**, *23*, 160–164.
20. Ren, C.; Guo, D.; Zhang, L.; Wang, T. Research on Nonlinear Compensation of the MEMS Gyroscope under Tiny Angular Velocity. *Sensors* **2022**, *22*, 6577. [CrossRef]
21. Lijia, Y.; Qiang, L.; Guochen, W.; Hualiang, S.; Qingzhi, Z. An on-orbit high-precision angular vibration measuring device based on laser gyro. *Proc. SPIE* **2022**, *12169*, 1216943.
22. Liu, Y.; Luo, X.; Chen, X. Improvement of angle random walk of fiber-optic gyroscope using polarization-maintaining fiber ring resonator. *Opt. Express* **2022**, *30*, 29900–29906. [CrossRef] [PubMed]
23. Ma, R.; Yu, H.; Ni, K.; Zhou, Q.; Li, X.; Wu, G. Angular velocity measurement system based on optical frequency comb. In Proceedings of the Conference on Semiconductor Lasers and Applications IX, Hangzhou, China, 21–23 October 2019.
24. Ren, Y.; Zhang, F.; Liu, F.; Hu, X.; Fu, Y. Accuracy test method of multi-instrument cooperative measurement. *Proc. SPIE* **2022**, *12282*, 1228219.
25. Kim, J.-A.; Lee, J.Y.; Kang, C.-S.; Woo, J.H. Hybrid optical measurement system for evaluation of precision two-dimensional planar stages. *Measurement* **2022**, *194*, 111023. [CrossRef]
26. Suvorkin, V.; Garcia-Fernandez, M.; González-Casado, G.; Li, M.; Rovira-Garcia, A. Assessment of Noise of MEMS IMU Sensors of Different Grades for GNSS/IMU Navigation. *Sensors* **2024**, *24*, 1953. [CrossRef] [PubMed]
27. Chang, L.; Bian, Q.; Zuo, Y.; Zhou, Q. SINS/GNSS-Integrated Navigation Based on Group Affine SINS Mechanization in Local-Level Frame. *IEEE/ASME Trans. Mechatron.* **2023**, *28*, 2471–2482. [CrossRef]
28. Zhao, L.; Kang, Y.; Cheng, J.; Fan, R. A MFKF Based SINS/DVL/USBL Integrated Navigation Algorithm for Unmanned Underwater Vehicles in Polar Regions. In Proceedings of the 2019 Chinese Control Conference (CCC), Guangzhou, China, 27–30 July 2019; pp. 3875–3880. [CrossRef]
29. Liu, X.; Liu, X.; Wang, L.; Huang, Y.; Wang, Z. SINS/DVL Integrated System with Current and Misalignment Estimation for Midwater Navigation. *IEEE Access* **2021**, *9*, 51332–51342. [CrossRef]
30. Feng, L.; Qu, X.; Ye, X.; Wang, K.; Li, X. Fast Feature Matching in Visual-Inertial SLAM. In Proceedings of the 2022 17th International Conference on Control, Automation, Robotics and Vision (ICARCV), Singapore, 11–13 December 2022; pp. 500–504. [CrossRef]
31. Cheng, Q.; Hu, H.; Li, L.; Wang, X.; Luo, X.; Zhang, X. Distortion analysis and focal length testing of off-axis optical system. *Opt. Precis. Eng.* **2022**, *30*, 2839–2846. [CrossRef]
32. Wei, M.S.; Xing, F.; You, Z. A real-time detection and positioning method for small and weak targets using a 1D morphology-based approach in 2D images. *Light Sci. Appl.* **2018**, *3*, 18006. [CrossRef]

sensors

Review

Advances in Aeroengine Cooling Hole Measurement: A Comprehensive Review

Shuyan Yan [1,2,3], Junkai Shi [1,*], Guannan Li [1], Can Hao [1,2], Ying Wang [1], Hao Yu [3] and Weihu Zhou [1,2,*]

1 Institute of Microelectronics of the Chinese Academy of Sciences, Beijing 100029, China; yanshuyan22@mails.ucas.ac.cn (S.Y.)
2 University of Chinese Academy of Sciences, Beijing 100049, China
3 China Aviation Planning and Design Institute (Group) Co., Ltd., Beijing 100120, China
* Correspondence: shijunkai@ime.ac.cn (J.S.); zhouweihu@ime.ac.cn (W.Z.); Tel.: +86-010-82178675 (J.S. & W.Z.)

Abstract: Film cooling technology is of great significance to enhance the performance of aero-engines and extend service life. With the increasing requirements for film cooling efficiency, researchers and engineers have carried out a lot of work on the precision and digital measurement of cooling holes. Based on the above, this paper outlines the importance and principles of film cooling technology and reviews the evolution of cooling holes. Also, this paper details the traditional measurement methods of the cooling hole used in current engineering scenarios with their limitations and categorizes digital measurement methods into five main types, including probing measurement technology, optical measurement technology, infrared imaging technology, computer tomography (CT) scanning technology, and composite measurement technology. The five types of methods and integrated automated measurement platforms are also analyzed. Finally, through a generalize and analysis of cooling hole measurement methods, this paper points out technical challenges and future trends, providing a reference and guidance for forward researches.

Keywords: film cooling; cooling hole; optical measurement; online digital measurement; simultaneous measurement of multiple parameters

1. Introduction

With the continuous development of the aviation industry, the improvement of aero-engine performance has become a crucial issue with great attention in the modern aeronautical field. The turbine inlet temperature is one of the key technical indicators of aero-engines, and improve of turbine inlet temperature is an effective way to increase thrust and thrust-to-weight ratio. According to calculations, for each 55 °C increase in total turbine temperature, the engine thrust will increase by approximately 10% [1]. Nowadays, the turbine inlet temperature of the fourth-generation aero-engine, which has a thrust-to-weight ratio of around 10, has reached approximately 2000 K [2]. A proposed method from the Beihang University Comprehensive Thermal Management Team indicated that the turbine inlet temperature has even achieved 2400 k. This advancement can increase the theoretical speed range by 156%, reduce fuel consumption by 15%, and enable higher Mach numbers and longer flight range [3].

The extremely high temperature means hot section components face harsh operating conditions, especially for turbine blades. To address this issue, the advanced alloy material with thermal barrier coating has been developed to process better heat resistance capacity [4]. However, fourth-generation single-crystal alloy material has a maximum temperature tolerance of about 1450 K [5], which cannot meet the required operational condition of approximately 2000 K. To overcome the heat resistance limitations of blade materials and avoid blade failure due to excessive operating temperature, thermal protection technology should be integrated in the blade's design and manufacturing. This integration fills the temperature gap and ensures blades maintain excellent reliability and service life in harsh operating conditions. Figure 1 illustrates the development of blade materials in response to the increase turbine temperatures.

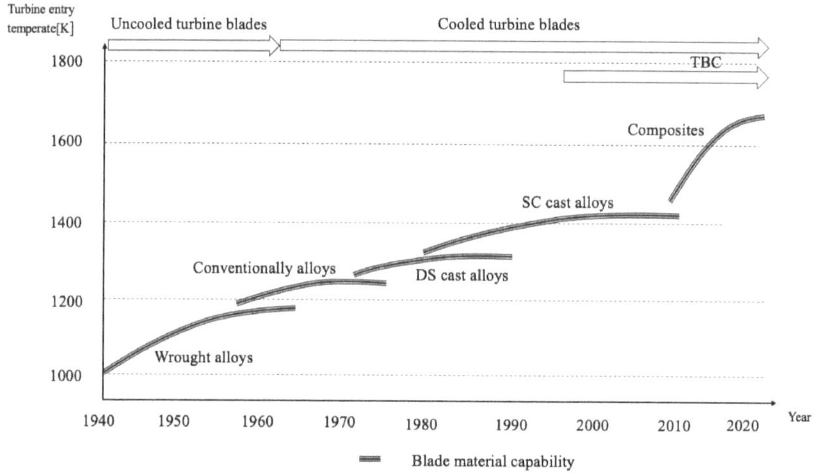

Figure 1. Development trend of aero-engine blade material.

In addition to high temperature-resistant materials and thermal barrier coating, thermal protection technologies mainly include air cooling technology which can be generally divided into two types: internal cooling and film cooling (external cooling). Figure 2 presents a schematic diagram of turbine blade film cooling technology [6], including external and internal cooling. In internal cooling, the cool airflow is guided through the internal passage to enhance the heat transfer. In film cooling, cool airflow expels from the internal passage and cooling holes and is applied to the blade surface as the protection layer, shielding the blade from the impact of high-temperature air. And Figure 3 presents a schematic diagram of principle of film cooling [7].

Moreover, the essential of the film cooling technology is to have numerous cooling holes (with diameters about 0.3 to 0.5 mm and depth-to-diameter ratios up to 10:1) distributed as rows along the leading edge, pressure and suction side, and blade tip region to generate air circulation [8,9]. According to the literature reports, thermal barrier coatings can reduce the temperature of turbine blades' surface by around 100 to 150 K. In comparison, film cooling technology can decrease the temperatures from around 400 to 500 K with more effective cooling performance [10]. Since 1970, film cooling technology has been designed into turbine blades and has become an effective method of thermal protection. Also, this method can be used in conjunction with other cooling methods. Figure 4 presents the development of blade cooling methods since 1960 [11].

The geometrical structures of cooling holes, including axial angle, diameter, orifice shape, spacing, and hole positional accuracy, are key indicators of film cooling technology which influence the effectiveness of cooling performance [12–17]. To meet measurement requirements for cooling hole geometrical parameters, it is vital to quantitatively assess all cooling hole characteristics by an efficient and precise method. Measurement results of critical indicators are helpful to understand the relationship between the geometrical structure of cooling holes and the effectiveness of film cooling technology. As well as this, researchers can further ensure safe, reliable operation and extend the service life of aero-engines according to the measurement results.

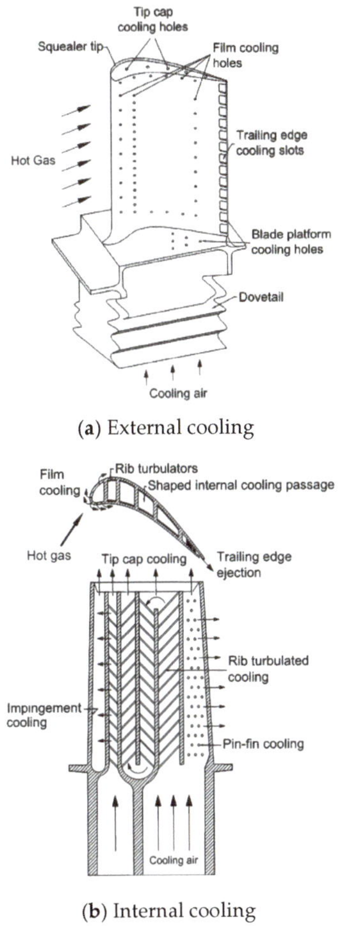

Figure 2. Schematic diagram of turbine blade cooling technologies [6].

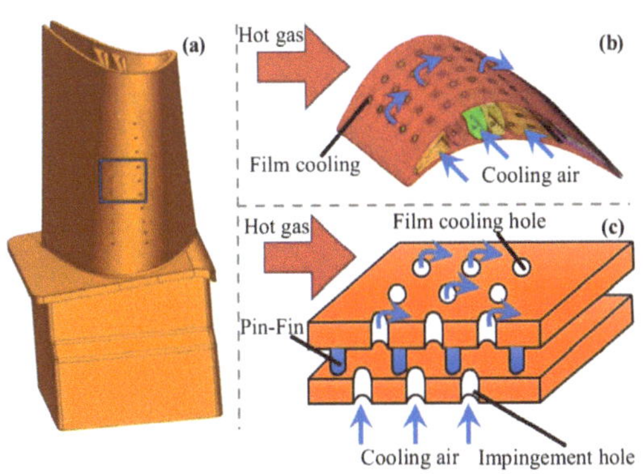

Figure 3. Schematic diagram of film cooling principle [7]. (**a**) Turbine blade model; (**b**) Schematic diagram of film cooling principle for blade; (**c**) Schematic diagram of film cooling principle for cooling hole.

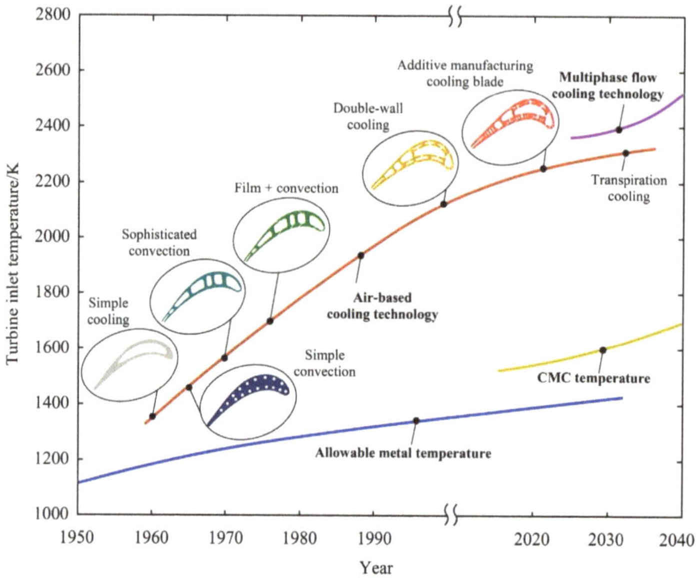

Figure 4. Development trend of cooling methods against the turbine inlet temperature [11].

In this paper, some of the literature related to the cooling hole measurement are reviewed. The overall structure of this paper is presented as follows. Section 2 briefly describes the development of shaped cooling holes with their geometrical features and corresponding enhanced cooling performance. Section 3 summaries and compares common manufacturing processes of the cooling hole. Section 4 focuses on the measurement technologies for the cooling holes. The key measurement indicators, traditional measurement methods, and principles of digital measurement technology are detailed, respectively. Further, the advantages, disadvantages and applied scenarios of each digital measurement technology are analyzed and compared. Finally, Section 5 gives the conclusion and outlook for future work.

2. Development of Shaped Cooling Hole Geometrical Design

The most common type of cooling hole is the cylindrical hole, which can be manufactured by mature and low-cost methods, such as laser processing and electrical discharge machining (EDM). With the development of film cooling technology, more and more novel shaped cooling holes have been utilized to enhance the efficiency of cooling performance when compared to the cylindrical hole [18–23].

In 2005, Bunker [19] categorized the geometrical features of early shaped cooling holes into four types, as shown in Figure 5. The projection of the hole axis on the blade surface is considered as the longitudinal direction, and then its orthogonal direction is considered as the lateral direction. The cooling hole of shape A is the classic fan-shaped form, with both lateral and longitudinal expansion on the outlet area (angle β and angle δ). The cooling hole of shape B contains only lateral expansion on the outlet area (angle β), while the cooling hole of Shape C contains only longitudinal expansion on the outlet area (angle δ). Shape D is the conical shape hole, expanding equally in all directions from the inlet to the outlet, centered around the hole axis.

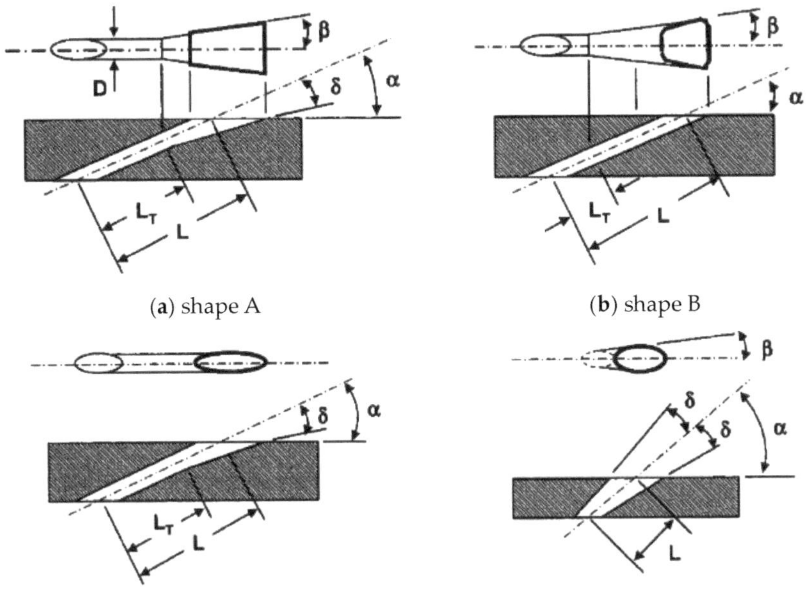

(a) shape A (b) shape B (c) shape C (d) shape D

Figure 5. Geometries for 4 types of shaped cooling hole [19].

Shaped cooling holes mainly adopt the design concept involving expanded outlets. The expansion of the outlet area facilitates the diffusion of the airflow to achieve a lower blowing ratio, reduce aerodynamic loss, and a larger cooling air coverage area, which is beneficial to enhancing cooling performance [20].

References [21–23] evaluate the cooling performances of various shaped cooling holes. Liu [23] concluded the improvements in the cooling performance of various shaped cooling holes against cylindrical cooling holes in Table 1. However, shaped cooling holes require more complex manufacturing processes, demanding higher precision and more advanced manufacturing technologies. Also, compared with the cylindrical hole, shaped hole involves more geometrical features can result in measurement difficulties. Finally, as the outlet area of cooling holes expands, the spacing between each hole gets larger. This means that there are fewer shaped holes for the definite blade surface area than normal cylindrical holes. The reformation of shape poses a challenge for the arrangement of cooling holes.

Table 1. Improvement in cooling performance against shaped cooling holes [23].

Hole Shape	% Improvement of Cooling Performance
Fan shape	10–40
Conical	15
Console	20
Sister hole	15–23
Compound angle	4–10
Trench shape	15–20

3. Development of Cooling Hole Manufacturing Processes

3.1. Cooling Hole Manufacturing Processes

The main manufacturing processes for cooling holes include electrical discharge machining (EDM), electrohydrodynamic (EHD) jet drilling, and laser drilling.

Currently, EDM is the most widely used process with mature technology. The principle of EDM is removing material through electro-erosion which is caused by spark discharge between the positive and negative electrodes of the tool and the workpiece, whereas due to the limitation of the material properties, the ceramic material has poor electrical conductivity. It is particularly challenging to process ceramic thermal barrier coatings. Also, EDM is characterized by low efficiency and low precision. The noise, smoke, and harmful gases generated in manufacturing are potentially hazardous to operators [24].

The principle of EHD jet drilling method is based on the micro holes manufacturing using a metal tube electrode. The workpiece is connected to the positive electrode, while a metal tube in a glass nozzle is connected to the negative electrode, and then electrolyte solution is ejected from the glass tube through a high-voltage electric field. Eventually, excess material from the positive electrode metal workpiece is removed under the influence of an electric field [25]. This method is not a thermal manufacturing method. Therefore, it can lead to better surface quality and lower roughness without microcracks or recast layers on the inner walls of the cooling holes. However, a drawback of this method is the difficulty in controlling the shape of cooling holes during the electrolytic etching process. Compared to the other two methods, its application is less widespread.

The principle of the laser drilling method is removing excess material by heating and melting with a high-energy light beam [26]. This method is known for its high precision, high efficiency, and wide applicability of materials, making it a key area of current research for the cooling hole manufacturing process.

Considering the trade-off against economic benefits and production quality, nanosecond laser drilling and electrical discharge machining (EDM) are common choices in engineering production. However, both methods are thermal processes, which create heat-affected zones on the inner walls of cooling holes, leading to microcracks and recast layers. These effects can significantly impair the cooling efficiency, overall reliability and service life of the turbine [27]. To mitigate thermal damage, important progress has been made in cooling technologies, specifically for cooling hole manufacturing. General Electric (GE) company in the USA, in collaboration with SYNOVA corporation in Switzerland, developed the water-assisted laser method. Water helps to lower the temperature in the manufacturing area, wash away debris, and guide the laser beam to the manufacturing target [28]. Similar research on water-assisted laser methods has been conducted in China, with Zhang Wenwu of the Chinese Academy of Sciences proposing a method [29]. These technologies effectively reduce thermal damage and enable high-quality manufacturing of cooling holes.

On the other hand, drilling processes continuously evolve, with researchers developing methods such as femtosecond laser drilling [28,30] and compound laser drilling with varying pulse width [31]. To some extent, these methods have improved issues related to microcracks and recast layers, reduced the harm caused by the heat-affected zone, and enhanced precision and efficiency in processing. However, these drilling processes cannot completely eliminate the effects of thermal manufacturing, indicating that manufacturing processes of cooling holes should be optimized further [32]. Figure 6 displays three types of micro holes drilled by nanosecond, picosecond, and femtosecond lasers [33].

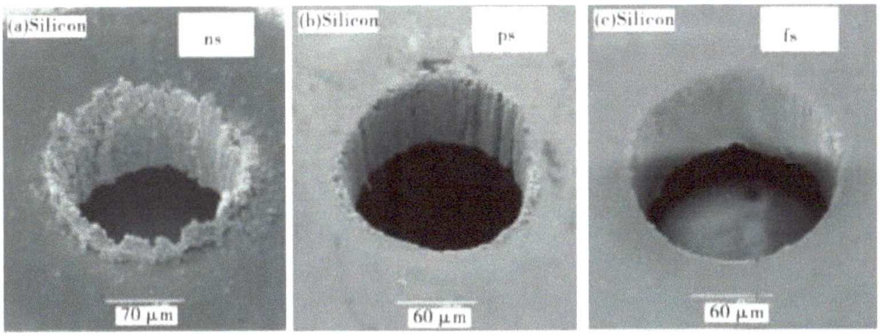

Figure 6. Manufacturing quality against three types laser drilling method [33]. (**a**) Micro hole drilled by nanosecond laser. (**b**) Micro hole drilled by picosecond laser. (**c**) Micro hole drilled by femtosecond laser.

3.2. Difficulties in Cooling Hole Manufacturing

Currently, there are several technical challenges that exist in the manufacturing of cooling holes.

Back strike is a common issue during the laser drilling process. Due to immature manufacturing parameters, hollow and complex structure of blades, and other factors, the control system is unable to accurately identify the location and timing of drilling. These factors may result in incomplete perforation, creating blind cooling holes, or over-penetration even after the hole is made, leading to melting or scorching on the inner wall of cooling hole and damage to the internal cooling paths of blade [34], as shown in Figure 7 [35]. Due to the complexity of internal cooling paths in the blade, conventional methods cannot observe the internal condition, making it difficult to detect back strike. Hence, it is necessary to implement quality control during the manufacturing of the cooling holes to improve manufacturing precision.

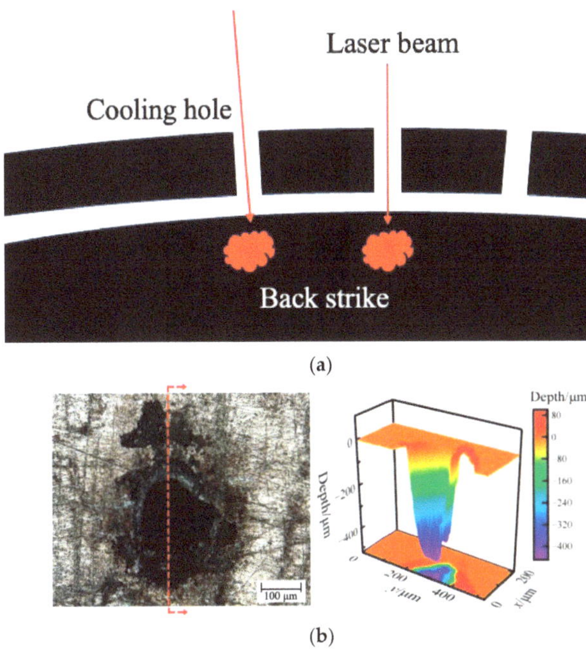

Figure 7. Schematic diagram of back strike. (**a**) Principle of back strike. (**b**) Example of back strike [35].

Drilling holes in blades covered with a thermal barrier coating presents another challenge in manufacturing cooling holes. For the blades with a thermal barrier coating, the conventional sequence of the hole manufacturing process is "drilling before coating". This method can result in cooling hole shrinkage, where the coating may obstruct or cover the orifices, causing a reduction in orifices area, diminishing the airflow capability, and even leading to the clogging of the cooling holes. According to the literature reports, the shrinkage rate of this method can be 15% to 20%, with the hole diameter reducing from 0.35–0.4 mm (before coating) to 0.25–0.3 mm (after coating). Additionally, the drilling process may damage the material structure around the edges of the orifice, resulting in fatigue cracks, which ultimately affect the bonding strength between the thermal barrier coating and the blade surface, as well as their impact resistance capabilities [36].

"Drilling after coating" is a new developing trend. The SYNOVA corporation [37] proposed a method that combines abrasive water jet machining and EDM, applying the "drilling after coating" technology to alloy blades. But differences in thermal expansion coefficients, toughness, stiffness, interface geometric structures and other material properties between the thermal barrier coatings and the alloy materials can result in potential issues such as interlayer tearing and cracking [38], impacting the safe service of the blades.

In response to these challenges in the cooling hole manufacturing processes, engineered solutions are unavailable nowadays. It is necessary to improve manufacturing precision to avoid or mitigate the risks associated with the above issues.

4. Development of Cooling Hole Measurement Technology

4.1. Key Quality Indicators for Cooling Hole

The key quality indicators for cooling holes primarily include:

1. Hole Diameter: The diameter of hole orifice, with a general tolerance of 0.10 mm.
2. Hole Positional Accuracy: the hole geometric position and hole spacing. The geometric position is the coordinate value of the point where the hole's axis intersects the blade profile in the blade coordinate system. The tolerance for position accuracy is generally between 0.10–0.15 mm.
3. Hole Axis Orientation: The angular tolerance of the hole axis is generally within $\pm 1^\circ$ [39]. Currently, there are no explicit design standards or technical requirements for the axis angle [40].
4. Orifice Shape: orifice shapes have two main types: circular orifices and shaped orifices.
5. Internal Surface Roughness: The surface quality of inner wall is influenced by the manufacturing process. Presently, there are no specific numerical requirements.
6. Maximum Thickness of recast layer: thermal manufacturing processes induce thermal effects, forming a recast layer on the hole's inner walls. The thickness of recast layer needs to be controlled. Presently, there are no specific numerical requirements.
7. Blind Hole Rate: Design specifications require that the blind hole (blocked hole) rate should be 0%, ensuring that each hole is fully open and functional for proper airflow distribution.

The schematic diagram of discrete cooling hole quality indicators is demonstrated in Figure 8.

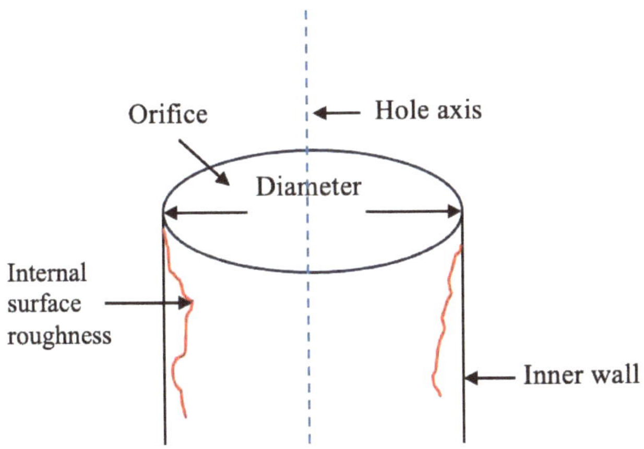

Figure 8. Schematic diagram of discrete cooling hole quality indicators.

4.2. Cooling Hole Measurement Technology

4.2.1. Traditional Measurement Method

Currently, the measurement of cooling holes often relies on manual inspection by quality checkers, including plug gauges, visual comparison, and water flow tests.

1. Plug gauge method

Quality checkers use plug gauges with different diameters, inserting them into a cooling hole to approximate the internal diameter and thus measure the diameter of the hole. This method is extremely slow and risks the plug gauge breaking inside the hole if mishandled, potentially resulting in scrap. Due to the manufacturing process of cooling holes, the actual internal surface has high roughness, poor roundness, and taper shape. The plug gauge can only measure the maximum ideal diameter in such cases, the schematic diagram is shown in Figure 9.

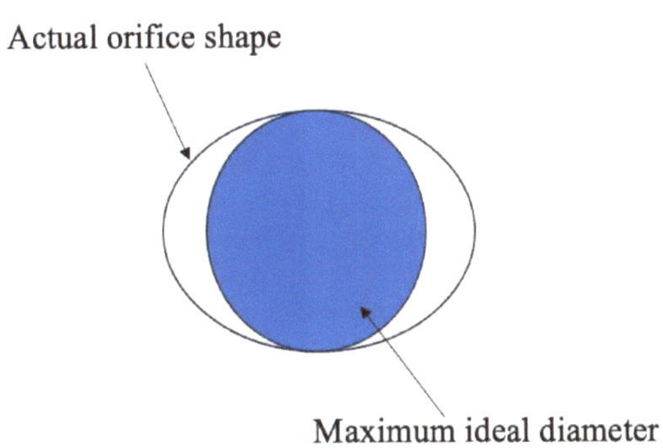

Figure 9. Schematic diagram of plug gauge measurement.

2. Sample Visual Comparison Method

This method relies on the visual judgment of quality checkers. They compare the blade with the standard sample to judge the positional accuracy, orifice shape, and diameter of the cooling hole. Presently, there is no unified national calibration standard for calibrating the standard sample blades used in visual comparisons, nor are there instruments to calibrate the positional accuracy of cooling holes in standard samples. It is challenging to use standard blade samples for quantitative traceability or data transfer [40].

3. Water Flow Method

This method involves injecting water inside the blade and visually observing whether the cooling holes allow water to pass through and whether the flow value is similar. This method depends on the blade's internal cooling path design and does not apply to all blades, making it less common than the other methods.

The measurement methods commonly used in engineering scenarios rely heavily on manual operation and invasive inspection techniques. Subjective manual judgments can only provide qualitative indicators. They cannot quantitatively evaluate critical quality indicators simultaneously, such as positional accuracy, hole axis orientation, hole diameter, and blind hole verification. As for the manufacturing quality of the internal surface of cooling holes, including issues like microcracks, recast layers, and thermal barrier coating defects, no corresponding measurement methods are currently available.

The automation and intelligence levels of traditional measurement methods urgently need improvement. By utilizing automated data collection and processing, it is possible to reduce manual intervention, thereby increasing efficiency and the accuracy and precision of measurement results.

4.2.2. Digital Measurement Method

1. Probing Measurement Technology

Researchers have taken advantage of miniature-sized probes for in-depth micro-hole measurements, including fiber probe technology [41–45] and capacitive probe technology [46–51]. The working mechanism of this method is presented in Figure 10. Salah crafted a rotational wire probe using stainless steel wire and a microtube, employing an acoustic emission device to perform contact detection by approaching and impacting the inner walls of cooling hole [41]. This method is utilized for measuring the diameter and roundness of the holes. Their experiment successfully measured the micro holes with diameters less than 1 mm and depth-to-diameter ratios of approximately 10:1, obtaining 3D profiles of the inner walls. Cui Jiwen [42] developed a twin Fiber Bragg Grating (FBG) probe for measuring large depth-to-diameter ratio micro holes. This design achieved multidimensional tactile perception along the X-axis and Z-axis while guiding the optical signal through the probe [43]. The design mitigates shadow effects to some extent, and the probe, with a diameter of less than 100 μm, is suitable for measuring micro holes in various industries. Building on this research, Feng Kunpeng [44] integrated the FBG probe with a measuring machine and introduced a data processing method with transformation of the signal domain and multiple fitting, enhancing the measurement accuracy of micro-hole diameters. Muralikrishnan [45] employed fiber deflection probing technology (FDP), integrating the fiber probe with a coordinate measuring machine to measure the diameter and shape of micro holes, achieving a measurement uncertainty of 0.07 μm. This technology can measure micro holes with depth-to-diameter ratios up to 20:1. However, the accuracy depends on the alignment of the probe with the hole axis and the machine axis, as there is no established reference standard for measurement. Figure 11 displays the measurement principle.

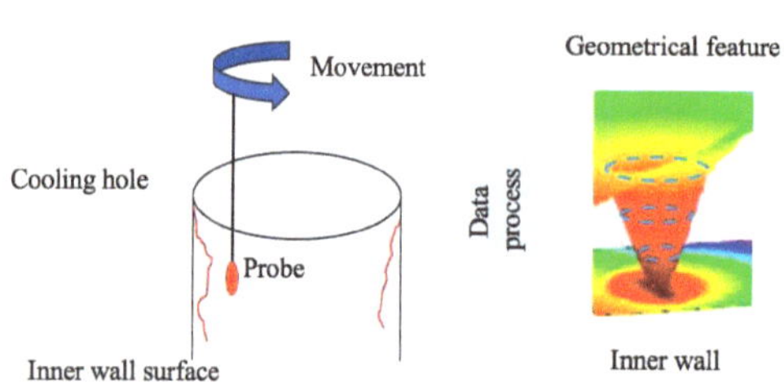

Figure 10. Working mechanism of probing measurement method.

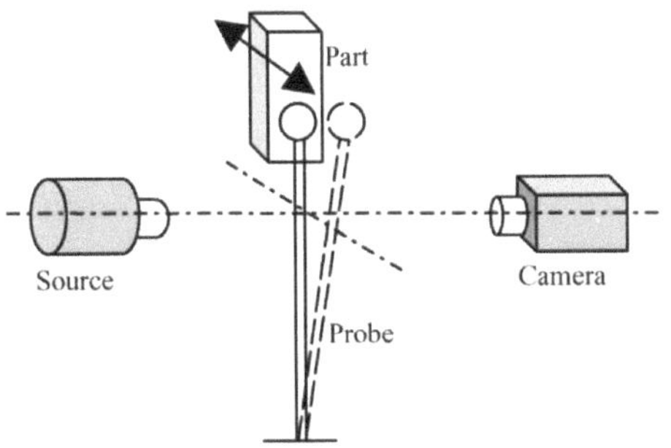

Figure 11. Measurement principle [45].

Ma Yuzhen [46,47] researched the capacitive probe measurement method for micro holes. To measure the diameter of deep and angled holes, they developed a non-contact capacitive probe that measures the gap between the probe electrode and the hole's inner wall. They proposed a hole axis fitting algorithm that combines the projection and least squares fitting. Experiment results confirm that the measurement data maintains consistent accuracy for holes with a depth-to-diameter ratio exceeding 10:1 and is not affected by where the probe enters the hole. Sun Xuan [48] established a micro hole diameter measurement system based on a coaxial cylindrical capacitive sensor. By identifying the central axis of the sensor and the micro hole using a charge-coupled device (CCD) camera and aligning the probe with the hole center using a movement system, then driving the capacitive probe into the hole and measuring micro holes with a depth-to-diameter ratio of 13:1. This system can measure the internal diameter at any depth within the hole, with a standard deviation 0.167 μm. Lee Neville [49] introduced a low-cost capacitive probe hole measuring system, determining the center position of the signal by locating the position of the minimum capacitive signal. Bian [50] developed a specialized hole diameter measurement system based on spherical scattering electric field technology. This system converts the tiny gap between a detection sphere on the probe and the test piece into an electrical signal, enabling non-contact, nanoscale resolution measurements of hole diameters.

Li Qi [8] utilized the principle of laser interferometry to design a cooling hole measurement system based on a laser rangefinder sensor. Guided by a digital blade model, the coordinate measuring machine (CMM) drives the probe into the cooling hole at various depths. The sensor acquires measurement data, which is processed by specialized software to determine the diameter of cooling holes and any position deviations.

In addressing the micro-probe measurement methods, researchers predominantly integrate CMMs with probes, controlling the probe to follow a planned path and scan the inner wall of cooling holes. These approaches offer high reliability and are not influenced by the hole depth or the inner wall's characteristics, allowing precise measurements for critical areas. However, these methods cannot provide information on the positional accuracy or minor damages. These notable limitations include low efficiency, not being applicable for measuring curved holes, the potential for causing damage to the sample surface, and the inability to meet the demands of large-scale industrial measurements.

2. Optical Measurement Technology

Optical measurement technology represents a principal approach for non-contact measurement of cooling holes, encompassing machine vision, 3D reconstruction, stripe pattern projection, and luminous flux methods. The non-contact character of optical methods enables in situ measurement capabilities. Additionally, by not physically interacting with the object, these technologies offer significant advantages in preventing any potential damage to cooling holes during the measurement process. The working mechanism of this method is presented in Figure 12.

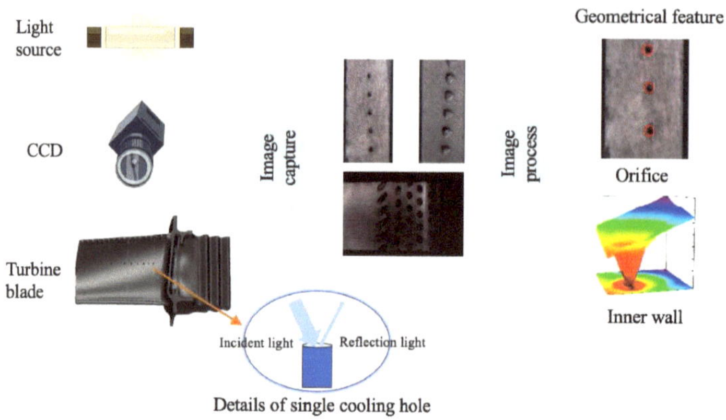

Figure 12. Working mechanism of optical measurement method.

Cheng Yuqi from Huazhong University of Science and Technology utilized a stereo vision 3D reconstruction technology to measure the diameter of cooling holes. The experiment results indicate that the evaluation error in measuring diameters is within 0.05 mm. This method allows simultaneous measurement of multiple cooling holes, significantly enhancing measurement efficiency [51]. Li Lei from Xi'an Jiaotong University [52] developed a cooling hole measurement method based on microscopic image sequence topographical reconstruction. A new measuring operator was established to measure the focus of the cooling hole image sequence. The reconstructed models obtained by this method show a standard deviation ranging from 0.007 mm to 0.018 mm. For cooling holes with a depth-to-diameter ratio close to 5:1, the absolute error in diameter is less than 0.01 mm. Figure 13 demonstrated measurement system. Zhao Yuanyuan from Shanghai Jiaotong University [53] employed a light field camera to capture sub-aperture images of cooling holes. Epipolar plane images (EPI) was generated from sub-aperture images and convert depth information into 3D point cloud data through EPIs. This method captures the 3D point cloud of cooling holes in a single exposure, greatly enhancing the measurement efficiency of cooling holes and demonstrating the potential of light field cameras in the micro hole measurement area. Figure 14 presents the original light field image and 3D cloud point of cooling hole.

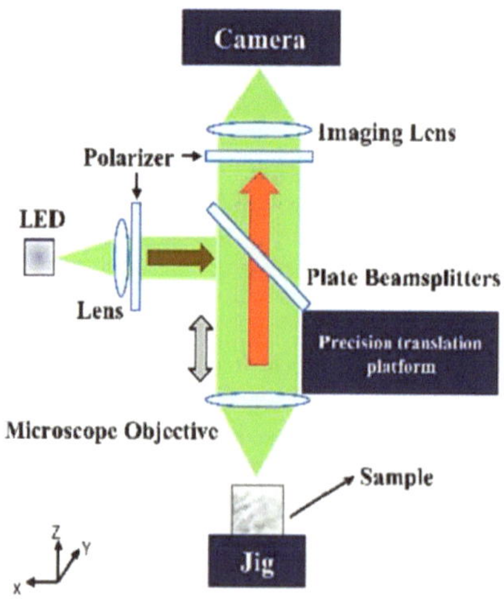

Figure 13. Principle of optical measuring system [52].

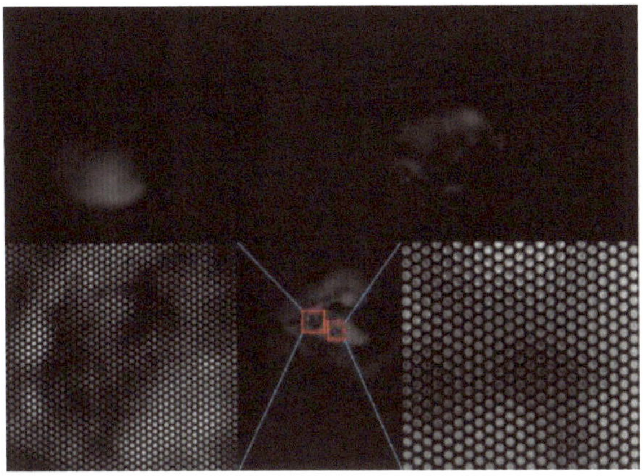

(**a**) Original light field of cooling hole

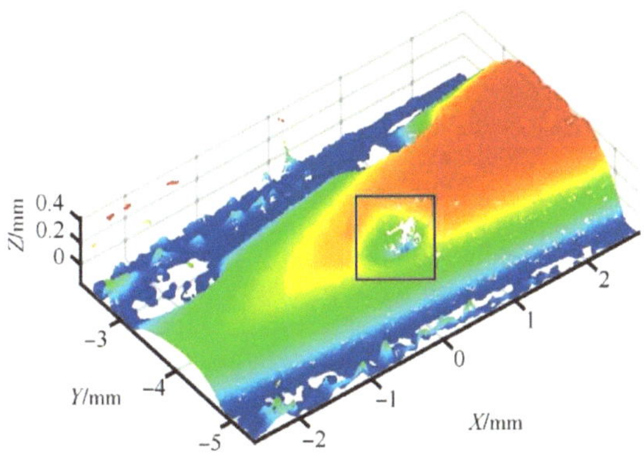

(**b**) 3D cloud point of cooling hole

Figure 14. Experimental results of cooling hole [53].

Munkelt [54] addressed the issue of thermal barrier coating potentially covering or filling cooling holes. An optical 3D scanning method based on the fringe projection principle was utilized to scan the blades, enabling the automatic detection of covered cooling holes and precise guidance for the laser drilling process. Xu Dongjing at Nanjing University of Aeronautics and Astronautics [55,56] proposed a method for measuring the geometric parameters of micro holes based on luminous flux. They established a mathematical model correlating the area of an orifice with the emitted luminous flux. A measured hole is non-compliant if the measured luminous flux does not align with the luminous flux values for standard holes. This method can meet the industrial measurement requirements, which is a variation rate of 5%. Still, it only provides qualitative assessments, lacking the capability for quantitative measurement of hole geometry parameters. Jin, from South Korea [57], studied through silicon via (TSV) used in semiconductor device packaging. TSVs have an aspect ratio of 20:1 and diameters ranging from 50 to 200 μm [58]. Jin used an optical comb of femtosecond pulse laser in the infrared range as a light source and, based on spectral resolved interferometry, achieved measurement for micro holes with a depth-to-diameter ratio of 7:1. However, this method cannot provide information on the 3D profile of inner walls [57]. Wu Chunxia [59] developed a near-infrared microscopic interferometry technology with aberration compensation for TSV inspection. This method enabled micro hole measurement with a depth-to-diameter ratio up to 6:1, including the depth and bottom surface morphology.

The above measurement methods primarily focus on finished cooling holes. However, in situ measurements during the manufacturing process of cooling holes can timely detect errors, allowing for adjustment and optimization of manufacturing parameters, enhancing manufacturing quality to its maximum and avoiding defective workpieces. Weifang Sun [7] implemented an in-situ measurement method during the drilling process by integrating an image-capturing device into a laser drilling machine with an image edge feature extraction algorithm, enabling the measurement of cooling holes' diameter and roundness. The experimental results indicate that the absolute errors of the diameter and roundness are 0.05 μm and 11.13 μm, respectively. This method demonstrates the potential and feasibility of in situ measurements in the domain of cooling holes. Shetty [60] utilized a vision system to acquire the diameter and orifice shape of cooling holes, coupled with a collimating tube to illuminate the inner wall to determine the presence of hole bottom. This method enables real-time in situ measurement of the drilling depth and drilling speed.

3. Infrared Imaging Technology

Besides optical measurement technologies, researchers have exploited the principles of infrared imaging, generating heat within the blade and identifying the geometric characteristics of cooling holes through infrared images. The working mechanism of this method involves detecting and measuring the infrared radiation emitted by object. By using a detector to measure the difference in infrared radiation between the object and background, infrared images can be obtained [61]. The detailed working mechanism of this method is presented in Figure 15. Rosemau [62] developed a measurement system for cooling holes based on infrared imaging. This system cyclically heats and cools the blade, capturing infrared images of the thermal airflow effusing from the cooling holes. The temperature intensity change rate during the heating and cooling processes is utilized to assess hole quality. Experiment results indicate that the system achieved a 98.3% recognition rate for defective holes and a 99.7% recognition rate for qualified holes. This method allows for preliminary filtering of cooling hole quality through qualitative assessment, although it is limited in measuring precise geometric parameters due to its measuring principle. He Qing [63] proposed a high-pressure turbine cooling hole testing method based on infrared imaging principles, capable of identifying whether the holes are clear or blocked. Xia Kailong [36] optimized He Qing's work by constructing an infrared measurement system, including a thermal imager, heat excitation source, turntable, and movement system. The system applies thermal excitation to cooling holes, captures infrared image sequences, and uses the canny algorithm and Hough circle transform function for image processing to calculate the diameters of cooling holes. Experiment results demonstrate that average deviations between the horizontal and vertical row of cooling holes and plug gauge are 4.40% and 2.32%, respectively. The infrared map of cooling hole is shown in Figure 16.

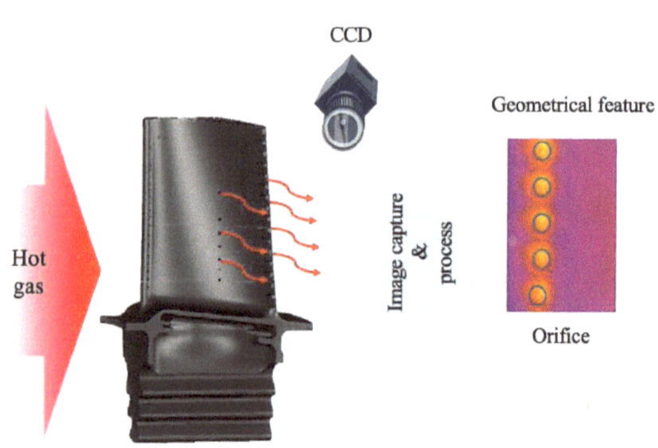

Figure 15. Working mechanism of infrared imaging measurement method.

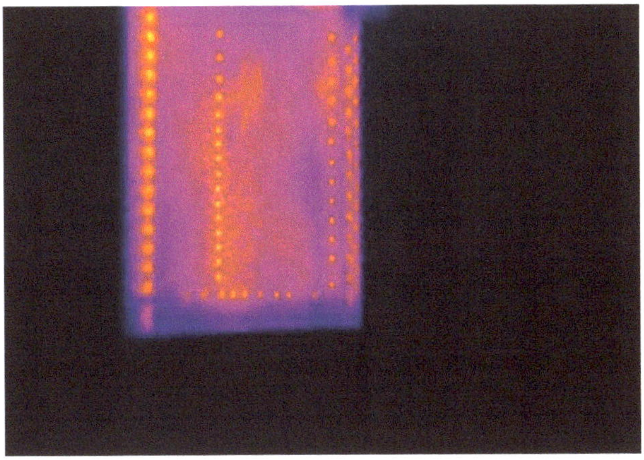

Figure 16. Infrared image of cooling hole [36].

4. CT Scanning Technology

Industrial CT technology is a widely used non-destructive testing method that involves a series of X-ray measurements taken from different angles to generate cross-sectional images and 3D profiles of the scanned object, allowing users to see inside of the object without cutting. This technology applies to various aerospace components [64]. The working mechanism of this method is presented in Figure 17. Wang Wenhu [65] conducted measurements on blades using industrial CT scanner. They extracted a 3D point cloud of shaped cooling holes as key features. Through computational processing, they obtained critical parameters such as the size, shape, and contour of cooling holes.

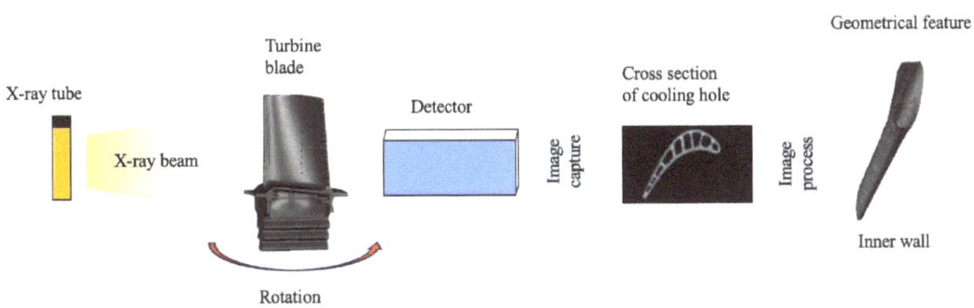

Figure 17. Working mechanism of computer tomography measurement method.

Jiang Qilin [66] conducted quality inspections on blades, using industrial CT to measure the cooling hole profiles. By examining cross-sectional scans, any internal wall intersections can be observed clearly. Figure 18 demonstrates a blade cross-section by industrial CT scan. Yang Zenan [67] utilized cone-beam CT technology to compare cooling holes manufactured with different manufacturing parameters and analyze their geometrical features and manufacturing quality. This work demonstrates the feasibility of using cone-beam CT technology to evaluate the quality of cooling holes.

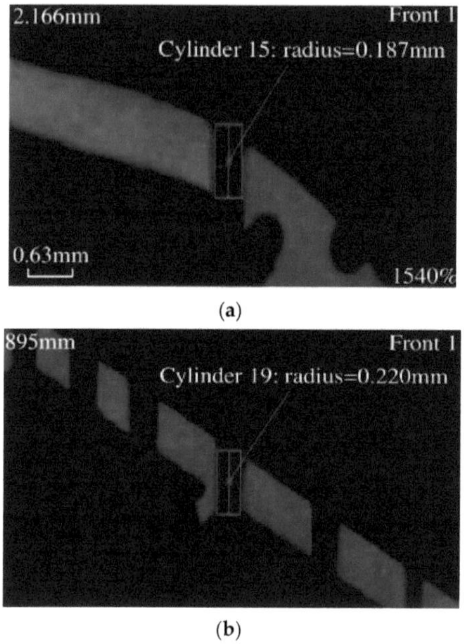

Figure 18. Blade profile image from industrial CT [66]. (**a**) Measurement results of squealer tip from industrial computer tomography (CT). (**b**) Measurement results of 2nd row cooling hole from industrial computer tomography (CT).

5. Composite Measurement Technology

Due to the limitations of single-method measurements in obtaining comprehensive quality indicators of cooling holes, multi-sensor composite measurement methods have garnered attention.

Chen Xiaomei [68] researched measuring micro hole positions on complex curved surfaces. Traditional methods typically employ a single optical vision sensor, but the optical focusing function faces limitations due to the complexity of surfaces. Consequently, they proposed a dual-sensor autofocus method combining vision with tactile sensing, and experiment results indicate that for measuring micro holes with a diameter of 0.5 mm distributed on an elliptical cylinder, the focusing deviation ranges from −23 μm to +95 μm. This dual-sensor autofocus method proves to be a more accurate and reliable method for measuring micro holes on complex surfaces, detailed diagram is shown in Figure 19. Sui Xin from Changchun University of Science and Technology [69,70] developed a multi-sensor technology combining contact and non-contact measurements. They utilized a fiber probe to touch the inner wall along the hole, with a CCD recording the relative position of the inner wall and the probe. This method achieves the measurement of five parameters for micro holes, including cylindricity, diameter, roundness, taper, and straightness. Additionally, measurement results for cylindricity and taper demonstrate repeatability of 1.36 μm and 1.51 μm, respectively.

Stimpson [71] conducted research on the cooling performance of cooling holes produced via additive manufacturing. To assess these cooling holes, they employed a combination of Industrial CT and Scanning Electron Microscope (SEM), which allowed them to measure the geometric features of cooling holes and the roughness of inner walls. The integration of these two technologies provided a comprehensive analysis of both the macro-structural and micro-structural characteristics of cooling holes, offering insights into how additive manufacturing technology affects these critical components.

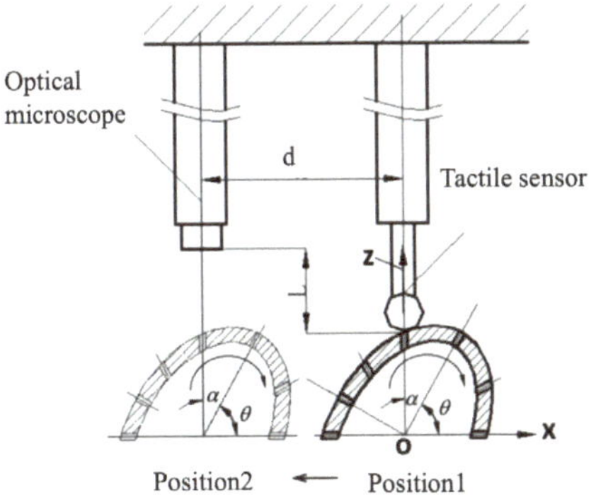

Figure 19. Dual-sensor autofocusing configurations [69].

NOVACAM company from Canada [72] developed the cooling hole inspection system, EDGEINSPECTTM, based on low coherence interferometry measurement technology. This non-contact inspection system is capable of acquiring high-precision 3D point cloud at speeds from 2100 to 100,000 points per second. The measurement results provided by this system include parameters such as the orifice shape, inner diameter, and the axial orientation of cooling holes. SURVICE Metrology, a company based in the United States, have integrated blade surface images from optical scanners with internal 3D profile from industrial CT, and analyzed the manufacturing accuracy of cooling holes [73]. The technology roadmap is illustrated in Figure 20. Additionally, General Electric (GE)company proposed a composite method combining CMMs with optical scanners to inspect the distribution of cooling holes in blades [74].

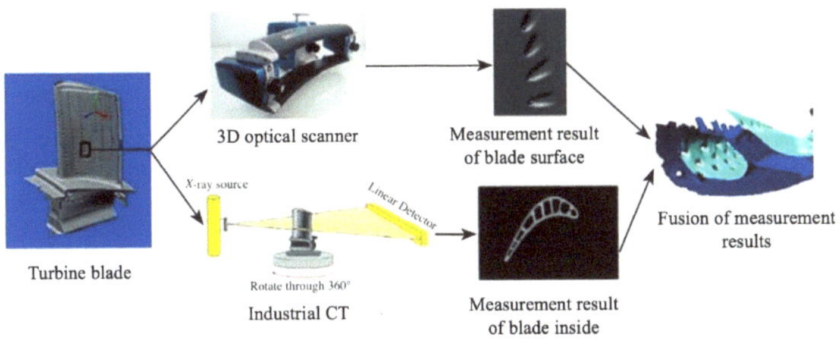

Figure 20. Technology roadmap of composite measurement system from SURVICE Metrology corporation [32].

6. Automated Measurement Platform

To meet the measurement requirements of cooling holes, researchers have constructed an automated measurement platform that integrates machine vision and image processing methods to address practical inspection challenges and to study problems encountered in engineering applications. Bao Chenxing [75] designed and developed a 4-axis cooling hole measurement system based on CCD. This system uses a turntable to rotate the blade along a specific axis, and the CCD captures images of cooling holes during rotation without aligning a CCD with a hole. Further, the standard for cooling hole alignment is where the image of the cooling hole is a perfect circle. The hole axis and diameter are identified and calculated with Halcon17.12, an open-source image processing software. This system demonstrates the repeatability error of 0.2° for the hole axis and 0.1 mm for the hole diameter. Figure 21 illustrates the detection device and imaging system used in the experiment.

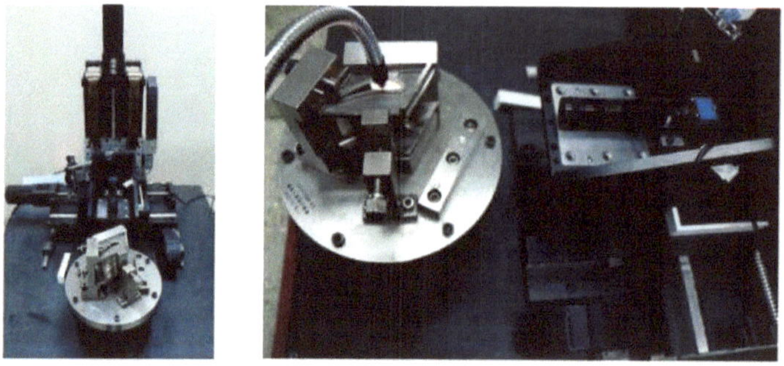

Figure 21. Diagram of detection device and imaging system [75].

Bi Chao and team from the Precision engineering Institute for aircraft industry conducted a series of studies [76–82] focused on cooling hole measurement and construction and design of the measurement system. These studies include establishing coordinate systems within the measurement systems and transforming measurement data from 2D coordinates in the image coordinate systems to 3D coordinates in global coordinate systems by mathematical approach [76]. Additionally, Bi constructed 4-axis and 5-axis visual measurement platforms using CMM, high-precision turntables, and CCD [39,77–80]. The 4-axis visual measurement platform captures sequential images of the inner wall of cooling holes with different depths by axially moving the CCD, using depth from focus methods to get the 3D profile of the inner wall of cooling holes, the detailed measurement system is illustrated in Figure 22, whereas it cannot provide specific dimensions of the inner wall. The 5-axis visual measurement platform organizes the measurement trajectory based on the 3D digital model of blades, achieving a repeatability accuracy for hole diameter within -10 μm to $+10$ μm and for hole center coordinates within -12 μm to $+12$ μm. Building on this research, Bi [81] proposed a method for synthesizing axis of cooling hole using a 3D point cloud. By fitting the annular point cloud of inner wall to obtain the center coordinates and then fitting the center coordinates at different depths into a straight line, the axis of cooling hole is finally established. To simplify, convert the calculated direction vector of axis into the angle between the vector and the coordinate axis, as well as the angle between the vector projection and the coordinate axis. The angles show a repeatability error within 0.3°.

Wang Cheng [40] utilized a 5-axis optical CMM to measure cooling holes. They employed a CCD to capture images of cooling holes and calculate the diameter and center coordinates of the holes. To validate the measurement accuracy, the measuring platform was used to measure a simulation specimen (standard disc), and limit error is 0.024 mm for diameter and 0.042 mm for positional accuracy. Nevertheless, since the measurement process requires continuous adjustments relative position of the CCD and cooling hole to align the hole axis, it relied on manual experience and subjective visual judgment, which can potentially affect the precision of the measurement results.

Liao Tao [82] modelled a digital virtual specimen of blades by extracting blade profile parameters, including the number and position information of cooling holes. They developed specialized software based on the digital virtual specimen, enabling real-time prediction and compensation of drilling positions during the manufacturing, according to the extent of blade deformation. The simulation results and experiment results demonstrate that the positioning errors of cooling holes are 1.34 μm and 4.25 μm, respectively, proving this prediction method can satisfy the cooling efficiency requirements. Zhang Min [83] researched the issue of axial inspection of cooling holes. They proposed an inspection scheme for the axial direction based on an improved Gaussian mapping algorithm. Using a line laser scanner, the high-precision point cloud is obtained to extract the axial feature parameters of cooling holes. The experiment results confirm that this method achieved an extraction precision of 0.6290°.

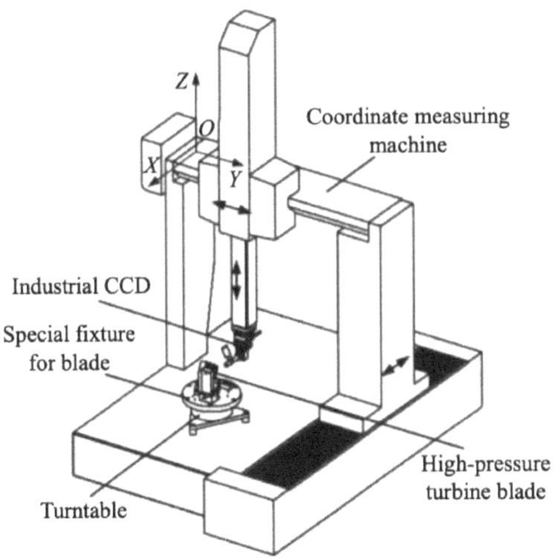

Figure 22. Schematic diagram of measurement a 4-axis vision coordinate measuring machine (CMM) [80].

Based on the above studies, Table 2 organizes and summarizes the main measurement technologies for cooling holes, outlining the measurement parameters of each technology along with their advantages and disadvantages.

Table 2. Comparison of main measurement technology for cooling hole.

Main Measurement Technology		Measured Features	Advantage	Disadvantage
Optical measurement	Light field	Geometrical feature of hole inner wall (partial) Orifice shape and diameter	High efficiency	Limited data
	Image recognition	Orifice shape and diameter	High efficiency	Limited data
	luminous flux	Orifice diameter	High efficiency	Limited data
	3D reconstruction	Geometrical feature of hole inner wall (partial) Orifice shape and diameter	High efficiency	Limited data
Industrial CT		Geometrical feature of hole inner wall	Generalized measurement result	High cost, low efficiency
Infrared imaging		Orifice diameter	High efficiency convenient construction of measurement system	Limited data
Probing measurement	Capacitive probe	Geometrical feature of hole inner wall	Suitable for high depth to diameter ratio hole	Low efficiency, low resolution, limited data
	Fiber probe	Geometrical feature of hole inner wall	Suitable for high depth to diameter ratio hole	Low efficiency, low resolution, limited data
	Laser interferometry	Geometrical feature of hole inner wall	Suitable for high depth to diameter ratio hole	Low efficiency, low resolution, limited data

4.3. Difficulties in Application of Digital Measurement Technology for Cooling Hole

Based on the background mentioned above, this section summarizes the technological difficulties in applying digital measurement technologies to cooling holes:

- Microscale and Complex Geometries

Cooling holes typically have small sizes, large depth-to-diameter ratios, and complex axial angles. It is challenging for light to illuminate the cooling holes and reflect to the camera, which restricts ordinary visual observation and visual inspection of orifice areas. These geometric characteristics result in significant inconvenience to the measurement.

- Requirements for High Precision and High Resolution

In addition to vital quality indicators such as the diameter and axis orientation of cooling holes, which evidently impact cooling efficiency, some minute damages like recast layers and microcracks are hard to measure accurately. Traditional measurement equipment and technologies often struggle to achieve the aim of precise measurement, failing to meet the requirements for high precision and resolution.

- Standardization and Uniformity

The absence of uniform quality assessment standards and standardized measurement methods can influence the accuracy and reliability of cooling hole measurement results. Different measurement devices and technologies may generate varying data types, posing an adverse to ensuring data consistency and comparability across different measurement platforms and methodologies.

- Time Efficiency

While ensuring measurement accuracy, the time cost of measurement must also be considered. Given that a single blade has hundreds of cooling holes and an aero-engine requires several hundred blades, manual inspection methods are impractical for large-scale production. Therefore, efficient measurement methods are crucial to meet large-scale production requirements without sacrificing high precision.

4.4. Shortcoming of Current Measurement Technology for Cooling Hole

The shortcomings of current measurement technologies are analyzed in this section.

- Single Measurement Quality Indicators

Research on non-contact measurement primarily focused on the geometric shape of the orifice, with less emphasis on positional accuracy and inner wall manufacturing quality. Contact measurement can measure the geometry of the inner walls, but fails to reflect manufacturing quality and micro defects accurately. So far, single-measurement methods can only assess the geometric shape, positional accuracy, or inner wall quality of cooling holes and cannot measure all quality indicators in a single operation.

- Absence of Digital Measurement Strategy

Most current research primarily focuses on individual cooling hole measurement methods and enhancing measurement precision. The absence of comprehensive measurement planning, position and orientation modelling of blades and measuring equipment, digital modelling of cooling hole geometric features, and the construction of cooling hole measurement platforms are notable. It is urgent to fulfil the aim of efficient and precise automated measurement.

- Absence of Error Analysis

In the actual measurement process of cooling holes, due to the wide distribution of holes and significant variation in axial orientation, measuring each cooling hole individually requires the measurement devices to vary the position and orientation continuously. This process may induce cumulative errors, affecting the final measurement results. Currently, there is no specialized research addressing this issue.

- Absence of Comprehensive Multi-Hole Measurement

Discrete hole characteristics cannot estimate the influence of positional accuracy on cooling performance from the perspective of cooling hole distribution. At the moment, there is limited research on combining multiple discrete hole measurement data to obtain a comprehensive cooling hole distribution pattern on the blade.

5. Summary and Outlook

As the requirements of aero-engine performance continuously increase, the development of cooling technologies, including the optimization and improvement of shaped cooling holes and manufacturing processes, presents problems for precise measurement. This paper focuses on analyzing various cooling hole measurement methods with different principles, comparing their advantages and appropriate applied scenarios, and identifying the difficulties and current technological drawbacks. This paper is of significant value for future research.

Future work in this field should be developed in the following directions:

- Establishing a Comprehensive Technical Framework for Cooling Hole Measurement

This framework involves generating measurement planning strategies based on digital blade models, implementing automated digital measurement based on intelligent equipment, and conducting data analysis and quality assessment. The ultimate goal is to enhance the level of automation and efficiency in cooling hole measurement.

- Optimizing Vision Measurement Methods

For vision-based measurement technologies, the design of the optical path and the selection of measurement equipment and construction of the measurement system should be refined. These optimizations point to improving the precision of measurement results and operation efficiency. Enhancements should include advanced image processing algorithms, calibration technologies, and the utilization of higher resolution CCD.

- Focusing on Comprehensive Cooling Hole Quality Indicators

Future research should not only concentrate on measuring discrete hole geometrical features, but also pay attention to a comprehensive study of the overall distribution of cooling holes and the manufacturing quality of inner walls. By employing composite measurement methods and integrating results, a 3D digital model of the blade can be established. This model will facilitate analysis of the relationship between cooling performance and cooling hole distribution, as well as their geometrical features and manufacturing quality. Understanding these relationships is beneficial for optimizing blades' design and manufacturing processes to enhance cooling efficiency.

- Analyzing and Assessing Errors in Cooling Hole Measurement Systems

The sources and quantity of error in cooling hole measurement systems should be analyzed and evaluated in detail. Implementing software algorithms helps mitigate the impact of the errors and enhances measurement precision consequently. This aspect should be harmonized with the efficiency of automated measurement equipment to balance accuracy against efficiency. Application strategies should include developing advanced calibration technologies, improving data processing algorithms, and integrating real-time feedback mechanisms into the measurement process.

Author Contributions: Conceptualization, S.Y.; methodology, S.Y., J.S. and W.Z.; resources, S.Y., J.S., G.L., Y.W., H.Y. and C.H.; investigation, S.Y., J.S., G.L., Y.W. and C.H.; writing—original draft preparation, S.Y.; writing—review and editing, S.Y., J.S. and W.Z.; supervision, J.S., Y.W. and W.Z. All authors have read and agreed to the published version of the manuscript.

Funding: This research was funded by the National Key Research and Development Program of China, grant number 2023YFF0722803. This research was sponsored by Institute of Microelectronics of the Chinese Academy of Sciences, and all research results are owned by Institute of Microelectronics of the Chinese Academy of Sciences.

Conflicts of Interest: Authors Shuyan Yan and Hao Yu are employed by the China Aviation Planning and Design Institute (Group) Co., Ltd.

References

1. Han, J.C.; Dutta, S.; Ekkad, S. *Gas Turbine Heat Transfer and Cooling Technology*, 2nd ed.; CRC Press: Boca Raton, FL, USA, 201[illegible] pp. 1–27.
2. Liu, J.Y.; Quan, Y.K.; Xu, G.Q.; Chai, J.M.; Yin, Q.Y.; Liu, J.S. Research Progress on Phosphor Thermometry Technology for Aer[illegible] Engine Hot Section Components. *J. Aerosp. Power* **2023**, *38*, 2861–2871.
3. Ding, S.T.; Liu, C.K.; Qiu, T.; Li, G. A Combined Cycle Method of Primary and Secondary Flows in Turbine Engines. Paten[illegible] CN111237085B, 14 May 2021.
4. Chen, R.Z.; Wang, L.B.; Li, J.H. Review and Prospect of the Development of Casting Superalloys. *J. Aeronaut. Mater.* **2000**, *1*, 55–6[illegible]
5. Cumpsty, N.A. *Jet Propulsion: A Simple Guide to the Aerodynamic and Thermodynamic Design and Performance of Jet Engines*; Cambridg[illegible] University Press: Cambridge, UK, 1997.
6. Han, J.C. Recent Studies in Turbine Blade Cooling. *Int. J. Rotating Mach.* **2004**, *10*, 443–457.
7. Sun, W.; Yi, J.; Ma, G.; Li, F.; Liu, X.; Gao, F.; Lu, C. A Vision-Based Method for Dimensional in Situ Measurement of Coolin[illegible] Holes in Aero-Engines during Laser Beam Drilling Process. *Int. J. Adv. Manuf. Technol.* **2022**, *119*, 3265–3277. [CrossRef]
8. Unnikrishnan, U.; Yang, V. A Review of Cooling Technologies for High Temperature Rotating Components in Gas Turbin[illegible] *Propuls. Power Res.* **2022**, *11*, 293–310. [CrossRef]
9. Zhang, J.; Zhang, S.; Wang, C.; Tan, X. Recent Advances in Film Cooling Enhancement: A Review. *Chin. J. Aeronaut.* **2020**, *3*[illegible] 1119–1136. [CrossRef]
10. Li, Q.; Ma, Y.L.; Zhang, L. Discussion on Blade Film Hole Detection Technology Based on Laser Ranging Principle. *Sci. Informati*[illegible] **2021**, *1*, 94–96.
11. Clifford, R.J. Rotating heat transfer investigations on a multi-pass cooling geometry. *AGARD Heat Transf. Cool. Gas Turbines* **1985** *N86-29823*, 7–21.
12. Ni, M.; Zhu, H.R.; Qiu, Y. Overview of Cooling Technology for Aero Engine Turbine Blades. *Gas Turbine Technol.* **2005**, *4*, 25–33+38
13. Li, S.F.; Huang, K.; Ma, H.S.; Chen, D.Y. Research Progress on Design and Manufacture Technology of Film Cooling Holes for Aeroengine Turbine Blades. *J. Therm. Eng. Power* **2022**, *37*, 1–11.
14. Huang, K.N.; Zhang, J.Z.; Guo, W. Effect of Partial Blockage inside Film Hole on Film Cooling Characteristics. *J. Aerosp. Power* **2014**, *29*, 1330–1338.
15. Li, G.C.; Zhu, H.R.; Bai, J.T.; Xu, D.C. Experimental Investigation of Film Cooling Effectiveness on Leading Edge with Various Geometries. *J. Propuls. Technol.* **2008**, *2*, 153–157.
16. Liu, Z.; Ding, Y.Q.; Xie, Y.H.; Zhang, W.X.; Wei, Z.P. Study on Effects of Film Hole Arrangement on Turbine Endwall Film Cooling Characteristics and Aerodynamic Performance. *J. Xi'an Jiaotong Univ.* **2023**, *57*, 22–33.
17. Tang, X.Z.; Li, L.P.; Huang, Z.J.; Liang, Y.C.; Zhong, W.L. Influence of Hole Spacing on the Film Cooling Effectiveness of a Gas Turbine Moving Blade. *J. Power Eng.* **2018**, *38*, 105–113+131.
18. Goldstein, R.J.; Eckert, E.R.G.; Burggraf, F. Effects of Hole Geometry and Density on Three-Dimensional Film Cooling. *Int. J. Heat. Mass. Transf.* **1974**, *17*, 595–607. [CrossRef]
19. Bunker, R.S. A Review of Shaped Hole Turbine Film-Cooling Technology. *J. Heat Trans.* **2005**, *127*, 441–453. [CrossRef]
20. Wilson, D.J.; Goldstein, R.J. Effect of Film Cooling Injection on Downstream Heat Transfer Coefficients in High-Speed Flow. *J. Heat Trans.* **1973**, *95*, 505–509. [CrossRef]
21. Kim, S.; Lee, K.; Kim, K.-Y. A Comparative Analysis of Various Shaped Film-Cooling Holes. *Heat. Mass. Trans.* **2012**, *48*, 1929–1939. [CrossRef]
22. Zhai, Y.N.; Liu, C.L. Experimental Study on the Film Cooling Performance of Odd-Shaped Film Holes with Large Inclination Angle under High Turbulence Intensity. *J. Xi'an Jiaotong Univ.* **2017**, *51*, 16–23.
23. Liu, J.S.; Malak, M.F.; Tapia, L.A.; Crites, D.C.; Ramachandran, D.; Srinivasan, B.; Muthiah, G.; Venkataramanan, J. Enhanced Film Cooling Effectiveness with New Shaped Holes. In Proceedings of the ASME Turbo Expo, Glasgow, UK, 14–18 June 2010.
24. Kang, X.L. Principles and Applications of EDM Composite Machining. *Mech. Electr. Prod. Dev. Innov.* **2007**, *4*, 175–176.
25. Shi, W.X.; Zhang, M.Q.; Yin, M. Research and Development of Electro-Stream Machining Technology. *Aeronaut. Manuf. Technol.* **2001**, *1*, 25–27.
26. Shirk, M.D.; Molian, P. A Review of Ultrashort Pulsed Laser Ablation of Materials. *J. Laser Appl.* **1998**, *10*, 18–28. [CrossRef]
27. Xia, K.L.; Ge, C.; Wang, Q.T.; He, Q. Research Progress on Detection Technology for Film Cooling Holes and Coating Defects of Turbine Blades. *Aeronaut. Manuf. Technol.* **2022**, *65*, 13.
28. Song, J.B.; Yi, H.Y. Technology for Gas Film Hole of Aircraft Engine Turbine Blade. *Tool Technol.* **2020**, *54*, 82–86.
29. Zhang, W.W.; Wang, Y.F.; Wang, B.; Zhang, G.Y.; Zhang, T.R. A Composite Water-Assisted Laser Processing System and Its Processing Method. Patent CN112824004B, 22 November 2022.
30. Li, X.J.; Dong, Y.W.; Yin, C.P.; Zhao, Q.; You, Y.C. Geometric Parameters Evolution experiment of Hole during Femtosecond Laser helical drilling. *Chin. J. Lasers* **2018**, *45*, 102–111.
31. Zhang, X.B.; Sun, R.F. Sequential Laser Drilling Technology. *Acta Aeronaut.* **2014**, *35*, 894–901.

32. Dong, Y.W.; Wu, Z.P.; Li, X.J.; Yin, C.P.; You, Y.C. Current Situation and Development Trend of Processing and Measurement Technology for Turbine Blade Film Cooling Hole. *Aeronaut. Manuf. Technol.* **2018**, *61*, 16–25.
33. Liu, X.L.; Tao, C.H.; Liu, C.J.; Hu, C.Y.; Chen, X. Investigation of Processing Methods and Development of Gas Holes of Engine Blade. *Mater. Rep.* **2013**, *27*, 117–120.
34. Wang, J.Q.; Zhao, W.S. Breakthrough Detection Technology of High-Speed EDM Drilling Based on Support Vector Machine. *Electr. Mach. Moulds* **2017**, *4*, 56–59+67.
35. Zeng, Q.Y.; Wang, B.; Zhang, W.W. Monte Carlo simulation and experimental investigation on back strike protection in laser drilling. *Chin. J. Lasers* **2023**, *51*.
36. Xia, K.L.; He, Q.; Zhang, Y.S. Measurement Method of Turbine Blade Film Aperture Based on Infrared Thermal Imaging and Shrinkage Law. *Acta Aeronaut.* **2022**, *43*, 594–606.
37. Gao, C.; Zhuang, L.; Zhao, K.; Guo, C. Abrasive Water Jet Drilling of Ceramic Thermal Barrier Coatings. *Procedia CIRP* **2018**, *68*, 517–522. [CrossRef]
38. Gao, C.; Zhuang, L.; Qiu, Y.; Zhao, K. Feasibility of Drilling Holes on Thermal Barrier Coated Superalloy Using Electrical-Discharge Machining. *Procedia CIRP* **2020**, *95*, 522–526. [CrossRef]
39. Bi, C.; Hao, X.; Liu, M.C.; Fang, J.G. Design and Establishment of the Machine Vision Measuring System for Film cooling Holes. *Acta Metrol. Sin.* **2020**, *41*, 775–780.
40. Wang, C.; Liu, T.; Mu, X.; Liu, P.; Zhu, L.Z. Research on Aero Engine Blade Film Hole Measuring Technology. *Meas. Technol* **2012**, *32*, 27–30.
41. Elfurjani, S.; Ko, J.; Jun, M.B.G. Micro-Scale Hole Profile Measurement Using Rotating Wire Probe and Acoustic Emission Contact Detection. *Measurement* **2016**, *89*, 215–222. [CrossRef]
42. Cui, J.; Feng, K.; Hu, Y.; Li, J.; Tan, J. A Twin Fiber Bragg Grating Probe for the Dimensional Measurement of Microholes. *IEEE Photonics Technol. Lett.* **2014**, *26*, 1778–1781. [CrossRef]
43. Chen, C.; Li, D.G. Optical Properties and Applications of Fiber Bragg Gratings. *J. Naval Univ. Eng.* **2000**, *4*, 11–15+22.
44. Feng, K.; Cui, J.; Zhao, S.; Li, J.; Tan, J. A Twin FBG Probe and Integration with a Microhole-Measuring Machine for the Measurement of Microholes of High Aspect Ratios. *IEEE-ASME Trans. Mechatron.* **2016**, *21*, 1242–1251. [CrossRef]
45. Muralikrishnan, B.; Stone, J.A.; Stoup, J.R. Fiber Deflection Probe for Small Hole Metrology. *Precis. Eng.* **2006**, *30*, 154–164. [CrossRef]
46. Ma, Y.Z.; Yu, Y.X.; Wang, X.H. Diameter Measuring Technique Based on Capacitive Probe for Deep Hole or Oblique Hole Monitoring. *Measurement* **2014**, *47*, 42–44. [CrossRef]
47. Ma, Y.Z.; Ma, L.; Zheng, Y.Z. The Measurement Techniques for Angular 3-D Pinholes Based on Capacitive Probe. *Measurement* **2017**, *97*, 145–148. [CrossRef]
48. Sun, X.; Ma, Y.Z.; Yu, Y.X.; Zheng, Y.Z. Inner Hole Diameter Measuring Based on Capacitive Sensor. *Adv. Mater. Res.* **2013**, *739*, 596–601. [CrossRef]
49. Lee, N.K.-S.; Chow, J.; Chan, A.C.K. Design of Precision Measurement System for Metallic Hole. *Int. J. Adv. Manuf. Technol.* **2008**, *44*, 539–547. [CrossRef]
50. Bian, X.; Cui, J.; Lu, Y.; Tan, J. Ultraprecision Diameter Measurement of Small Holes with Large Depth-To-Diameter Ratios Based on Spherical Scattering Electrical-Field Probing. *Appl. Sci.* **2019**, *9*, 242. [CrossRef]
51. Cheng, Y.Q.; Li, W.L.; Jiang, C.; Wang, G.; Xu, W.; Peng, Q.Y. A Novel Cooling Hole Inspection Method for Turbine Blade Using 3D Reconstruction of Stereo Vision. *Meas. Sci. Technol.* **2021**, *33*, 015018. [CrossRef]
52. Li, L.; Li, B.; Zhang, R.; Xue, Z.; Wei, X.; Chen, L. Geometric Parameters Measurement for the Cooling Holes of Turbine Blade Based on Microscopic Image Sequence Topographical Reconstruction. *Measurement* **2023**, *210*, 112562. [CrossRef]
53. Zhao, Y.Y.; Zeng, F.; Li, Y.; Gan, M.Y.; Shi, S.X. 3D Measurement Technique for Film Cooling Holes Based on Light Field Imaging. *Acta Aeronaut.* **2021**, *42*, 416–426.
54. Munkelt, C.; Kühmstedt, P.; Aschermann, L.; Seidel, F. Automatic Complete High-Precision Optical 3D Measurement of Air Cooling-Holes of Gas Turbine Vanes for Repair. In Proceedings of the Optical Measurement Systems for Industrial Inspection IX, Munich, Germany, 22–25 June 2015.
55. Xu, D.J. Research on Micro-Aperture Rapid Measuring System Based on Luminous Flux. Master's Thesis, Nanjing University of Aeronautics and Astronautics, Nanjing, China, 2013.
56. Xu, D.J.; Ye, M.; Ni, Z.Q. The Study of the Micro-Aperture Rapid Measuring Theory Based on Luminous Flux. *J. Mech. Eng.* **2013**, *2*, 7–9.
57. Jin, J.; Kim, J.W.; Kang, C.-S.; Kim, J.-A.; Lee, S. Precision Depth Measurement of through Silicon Vias (TSVs) on 3D Semiconductor Packaging Process. *Opt. Exp.* **2012**, *20*, 5011. [CrossRef]
58. Wang, M.Y.; Zhang, H.B.; Hu, W.B.; Mei, Y.H. Review on the through Silicon Via Technology in the 3D System in Package (3D-SiP). *J. Mech. Eng.* **2023**, *59*, 1–16.
59. Wu, C.X.; Ma, J.Q.; Gao, Z.S.; Guo, Z.Y.; Yuan, Q. Measurement of through Silicon Via by Near-Infrared Micro Interferometry Based on Aberration Compensation. *Opt. Precis. Eng.* **2023**, *31*, 12. [CrossRef]
60. Shetty, D.; Eppes, T.; Campana, C.; Filburn, T.; Nazaryan, N. New Approach to the Inspection of Cooling Holes in Aero-Engines. *Opt. Laser. Eng.* **2009**, *47*, 686–694. [CrossRef]

61. Ciampa, F.; Mahmoodi, P.; Pinto, F.; Meo, M. Recent Advances in Active Infrared Thermography for Non-Destructive Testing of Aerospace Components. *Sensors* **2018**, *18*, 609. [CrossRef]
62. Rosemau, R.D.; Nawaz, S.; Niu, A.; Wee, W.G. Aircraft Engine Blade Cooling Holes Detection and Classification from Infrared Images. In Proceedings of the Nondestructive Evaluation of Aging Aircraft, Airports, and Aerospace Hardware III, Newport Beach, CA, USA, 3–5 March 1999.
63. He, Q.; Ge, C.; Wang, Q.T. Method and Platform for Detecting Cooling Film Holes on Working Blades of High-Pressure Turbines. Patent CN109751972B, 26 February 2021.
64. Li, X.Y.; Yang, L.; Sun, C.G.; Li, X.J.; Liu, Y. High-Pressure Turbine Blade Defect Technology Based on Micro-Focus Industrial CT. *Foundry* **2023**, *72*, 904–908.
65. Wang, W.H.; Zhang, Z.F.; Jiang, R.S.; Qin, C.C.; Zhu, X.X.; Huang, B. Method for Detecting and Evaluating the Geometric Structure of Shaped Film Holes Based on Industrial CT Scanning. Patent CN201910970716.7, 14 October 2019.
66. Jiang, Q.L.; Cao, K.Q.; Chen, L.; Feng, Z.P.; Jia, T.Q.; Sun, Z.R. Process of Turbine Blade Film Cooling Holes by Nanosecond and Femtosecond Laser Pulses. *Aeronaut. Manuf. Technol.* **2021**, *64*, 53–61.
67. Yang, Z.N.; Zha, H.Y.; Huang, Z.Y.; Xu, J.J.; Liang, W.; Zhang, Z.; Wang, Z.D.; Huang, H.D. Development and Prospect of the Drilling and Detection of Film Cooling Hole on Turbine Blade and Vanes. *Fail. Anal. Prev.* **2023**, *18*, 14–20.
68. Chen, X.; Longstaff, A.P.; Fletcher, S.; Myers, A. Analysing and Evaluating a Dual-Sensor Autofocusing Method for Measuring the Position of Patterns of Small Holes on Complex Curved Surfaces. *Sens. Actuators* **2014**, *210*, 86–94. [CrossRef]
69. Sui, X.; Xu, X.P.; Sun, J.; Zhang, L. Microhole Geometry Measurement Applying Multi-Sensor Technology. *Opt. Optoelectron. Technol.* **2009**, *7*, 79–82.
70. Sun, Q.S.; Sui, X.; Xu, H.F. Research on an Optoelectronic Detection System for Microhole Shape Measurement. *Sci. Technol. Inf.* **2008**, *14*, 14.
71. Stimpson, C.K.; Snyder, J.C.; Thole, K.A.; Mongillo, D. Effectiveness Measurements of Additively Manufactured Film Cooling Holes. *J. Turbomach.* **2017**, *140*, 011009. [CrossRef]
72. Cooling Holes Inspection & 3D Metrology—Novacam. Available online: https://www.novacam.com/applications/cooling-holes-inspection (accessed on 27 January 2024).
73. Survice_Measurement of Cooling Hole CMSC 2011 Sub. Available online: https://www.cmsc.org/stuff/contentmgr/files/0/2bdcf766d9d5daf6e892c46153c591d3/misc/cmsc2011_wed_gh_0800_survice.pdf (accessed on 27 January 2024).
74. Nigmatulin, T. Turbine Shroud Cooling Hole Diffusers and Related Method. Patent US20030082046A1, 26 October 2001.
75. Bao, C.X.; Wang, L.; Li, K.; Min, L.X.; Geng, C.K. Research on Rapid Detection Technology of Gas Film Hole Based on CCD. *Aviat. Precis. Manuf. Technol.* **2017**, *53*, 52–55.
76. Bi, C.; Hao, X.; Liu, M.C.; Liu, Y. Establishment and Conversion of Coordinate System in Visual Measurement of Film Cooling Holes. *J. Sens. Technol.* **2019**, *32*, 1515–1521.
77. Bi, C.; Hao, X.; Liu, M.; Ji, G.B. Design and Establishment of Five-Axis Visual Measuring System for Film cooling Holes. *Mach. Tool Hydraul.* **2021**, *49*, 123–127.
78. Bi, C.; Pi, K.S.; Sheng, B.; Long, K.; Hao, X. Measurement Technology of blade-Shaped Holes Based on Machine Vision. *Tool Technol.* **2022**, *56*, 147–151.
79. Bi, C.; Zhang, Y.; Zhang, C.; Zhou, P. Study on Automatic Focusing Strategy of Visual Coordinate Measuring System. *Tool Technol.* **2022**, *56*, 136–141.
80. Bi, C.; Hao, X.; Liu, M.C.; Fang, J.G. Study on Calibration Method of Rotary Axis Based on Visual Measurement. *Infrared Laser Eng.* **2020**, *49*, 167–174.
81. Bi, C.; Zhang, C.; Fan, C.Y.; Fang, J.G. Study on Visual Measuring Technology of Axis Direction of Film Cooling Holes. *J. Aerosp. Metro Meas.* **2022**, *42*, 37–42.
82. Liao, T.; Dong, Y.W.; Zhang, S.T.; Bi, C.; Fang, J.G. Error Analysis Method of Turbine Blade Film Hole Based on Virtual Measurement. *Aeronaut. Sci. Technol.* **2021**, *32*, 50–59.
83. Zhang, M.; Yan, X.S.; Xi, X.C.; Zhao, W.S. Measurement of Axial Orientation of Film Cooling Holes of Turbine Blade Based on Laser Sensor. *Procedia CIRP* **2022**, *113*, 172–177.

sensors

Article

Measurement Method for Contact Wire Wear Based on Stereovision

Wei Zhou *, Zhe Qin, Xinyu Du, Xiantang Xue, Haiying Wang and Hailang Li

Infrastructure Inspection Research Institute, China Academy of Railway Sciences Corporation Limited, Beijing 100081, China
* Correspondence: wzhou@rails.cn

Abstract: The contact wire wear is an important parameter to ensure the safety operation of electric railways. The contact wire may break if the wear is serious, which leads to transportation interruptions. This study proposes an optical measurement method of contact wire wear, using stereovision technology. The matching method of stereovision based on line-scan cameras is proposed. A lookup-table method is developed to exactly determine the image resolution caused by the contact wire being in different spatial positions. The wear width of the contact wire is extracted from catenaries' images, and the residual thickness of the contact wire is calculated. The method was verified by field tests. The round-robin tests of the residual thickness at the same location present excellent measurement repetitiveness. The maximum difference value between dynamic test results and ground measurement results is 0.13 mm. This research represents a potential way to implement condition-based maintenance for contact wire wear in the future in order to improve the maintenance efficiency and ensure the safety of catenary infrastructure.

Keywords: contact wire wear; residual thickness; optical measurement; stereovision

1. Introduction

The contact wire is one of the most important components in an electrified railway system. It makes direct contact with the pantograph of the electric locomotive and supplies current to the locomotive. Because of the continuous friction between the pantograph and the contact wire, the wire surface wears out, thus reducing the effective cross-section of the wire [1–3]. The reduction of the cross-section causes different problems. Firstly, the mechanical strength of the contact wire reduces. If the cross-section is too small, the contact wire may break, which might cause hazardous accidents. Secondly, the resistance increases as the cross-section becomes smaller, and the voltage reduces which may not be able to meet the operational needs of electric locomotives. As a result, the contact wire wear should be monitored regularly to ensure the safety of the catenary infrastructure. If a section of contact wire with serious wear is detected at an early stage, this section of contact wire could be strengthened or replaced by a new section. Therefore, the lifetime of contact wire could be extended. In the past few decades, the measurement of contact wire wear has mainly relied on the method of manual measurement. For the manual measurement method, the safety risk for the maintenance workers is high; the power on the catenary has to be cut off, which leads to the traffic interruption; and the inspection workload is heavy because of the point-by-point measurement way. With the development of the high-speed electrified railway, the requirements were put forward to develop the dynamic inspection method of contact wire wear. The automatic measurement method of contact wire wear on rolling stock improves the maintenance efficiency and reduces the maintenance cost, because of the timely and fast defect detection. Furthermore, the continuous wear measurement along the railway line helps the railway operators to diagnose the abnormal function state, evaluate the quality of catenary, and implement the condition-based maintenance of catenary.

The contact wire height, stagger, and wear are the primary geometrical properties of the contact wire. The height and stagger describe the spatial position of the contact wire relative to the rail surface. The residual thickness indicates the degree of contact wire wear. With the rapid development of sensor technology in the past few decades, various measurement methods for the geometry parameters of contact wire have been developed [4–7], such as the microwave reflection analysis and electrical and optical methods. Among them, the optical method has been the most successful one to perform the measurement of contact wire wear, which includes the residual thickness measurement method, the laser scanning method, and vision measurement method [6]. The residual thickness measurement method presents a few limitations, such as the necessity of modifying the pantograph, a limited measurement range, and the fact that the maximum number of detectable wires is only one [6,7]. The obstacles of the laser scanning method are the limitation of the total measurement points per second, and it is sensitive to the reflectivity of contact wires. For the vision measurement method, the wear surface of contact wire is illuminated, and the wear width is extracted by analyzing the image taken by the cameras. Several vision measurement methods were reported, such as structured light vision measurement [8–10] and stereovision measurement [11–13]. For example, researchers from RTRI developed a contact wire wear measurement system based on structured light vision measurement [8]. The RTRI's system was installed on a Shinkansen vehicle and was capable of an accuracy of within ±0.1 mm in residual thickness. Chugui et al. [9] developed a structured light vision measurement system with an RMS error of the residual thickness of 0.15 mm and a measurement rate of ~150 Hz. Nie [10] adopted a 2D laser sensor (ELAG Elektronik AG, Winterthur, Switzerland) to measure the wire wear of metro rigid catenary.

The limitations of structured light vision measurement are the low measurement frequency and the influence of sunlight in daytime inspection.

Stereovision measurement using two or more line-scan cameras is a better solution for the measurement of contact wire wear on rolling stock [7,14]. The advantages of stereovision measurement are the high scanning frequency and the high resolution because of the one-dimensional imaging chip of the line-scan camera. For this application, the scanning plane of the line-scan camera is perpendicular to the direction of the train. However, it is still a challenging work since it has to overcome several obstacles, such as the image quality during day and night, the measurement resolution, and the variations in both contact wire height and stagger [7,14]. Great efforts have been made to overcome the above challenges. Laser diodes combined with interference filters are used as the active light source, which improves the image SNR (signal-noise ratio) in daytime inspection [7,12]. Due to considerations of operational safety and laser temperature control, the power of the laser needs to be carefully designed. In order to cover the range of stagger with high resolution, multiple line-scan cameras were adopted by ATON and MEDES [7]. To minimize the influence of the variation in contact wire height on measurement, Van Gigch et al. [15] and Borromeo et al. [7] proposed a method of adjusting the focal lengths of cameras online separately, while Kusumi et al. [16] developed a method for keeping a constant distance from cameras to contact wire by controlling the mirrors. You et al. [12] derived the equation of calculating the wear width using the pinhole camera model, which involved the impact of height variation on resolution. On the other hand, stereo matching is another challenge in stereovision measurement. For a binocular vision system, stereo matching can be defined as finding the corresponding points between the stereo image pairs. Interest point detectors and feature descriptors are commonly used methods for solving the stereo-matching problem [14]. Zhou et al. [17] proposed a trinocular stereovision system that includes a projector and two area-scan cameras and utilized the phase information of pixels to complete the matching of homonymous points. Shu et al. [18] designed a trinocular stereovision system by using three area-scan cameras arranged in equilateral triangles and solved the matching problem by using the method of epipolar constraint with three intersecting cameras.

The author previously studied the method based on multi-view stereovision to measure the height and stagger of contact wire [19,20]. Four CCD line-scan cameras with a resolution of 4096 pixels and different angles and four high power spotlights (575 W for each spotlight) were adopted to form a robust system under most of the environmental conditions. If one of the four cameras was influenced by the direct sunlight, the other two cameras were used to perform the measurement. However, since the spectrum of the spotlight is wide and continuous, the strong sunlight is hard to weaken, and the grayscale feature of the wear surface could be hardly highlighted because of the poor image SNR. The contact wire in the image is usually a black object against a bright background in daytime inspection, thus making it difficult to measure the contact wire wear. Based on previous research, this paper proposes a contactless optical measurement method of contact wire wear, using the multi-view stereovision technology. A matching method of stereovision based on line-scan cameras is proposed. The third line-scan camera is adopted to overcome the correspondence problem. Compared with the matching method of binocular stereovision in the literature, this matching method is fast and robust, which satisfies the requirement of real-time measurement in day and night inspection. The wear width of the contact wire is extracted from catenaries' images. A lookup-table method is developed to exactly determine the image resolution in various sections of measurement plane. Compared with the methods reported in the literature [7,15,16], no additional adjustment device is required to adjust the focal lengths of cameras or maintain a constant working distance in this study. Currently, little attention is being paid to the impact of stagger variation on resolution. In most existing studies, the focus is on minimizing the influence of the variation in contact wire height on measurement [7,15,16] or on evaluating the impact of height variation on resolution [12]. In this study, the impact of both the height variation and the stagger variation on the image resolution has been taken into account by using the lookup-table method, which ensures the accuracy of the wear measurement of contact wires. Then, the residual thickness is calculated according to the rated cross-sectional parameters of contact wire. In order to highlight the grayscale feature of the wear surface and improve the image SNR in daytime inspection, a new kind of high-speed synchronized stroboscopic lighting technology using monochromatic LED lamps combined with narrow band-pass optical filters is developed as an alternative active lighting method. Compared with the laser diode lighting method reported in the literature [7,12], the stroboscopic LED lighting method avoids the risk of laser operation and is harmless to nearby people. In addition, there is no need to design a precise mechanical adjustment device to make the scanning plane of the line-scan camera coincide with the laser irradiation plane.

The remaining sections of this paper are organized as follows: Section 2 introduces the basic principle of contact wire wear measurement using the vision measurement method. The measurement model and implementation method are detailed in Section 3. The test results and a discussion are given in Section 4. Finally, the work of the full paper is summarized, and conclusions and suggestions are given.

2. Basic Principle

Figure 1 shows the cross-section of a new contact wire (left side) and the cross-section of a contact wire with wear (right side). The key feature to be measured may be classified as the residual thickness (h) or the width of the wear surface (w), as can be seen in Figure 1b. For most railway companies, the residual thickness (h) or the worn section area (A) of the contact wire is used as the indicator of the degree of wire wear.

As shown in Figure 1, the lower half part of the cross-section of the wire is arc-shaped. In the case of a circular contact wire with radius, r, the residual thickness (h) can be calculated from the width of the wear surface, w:

$$h = r + \sqrt{r^2 - \left(\frac{w}{2}\right)^2} \tag{1}$$

The relationship between the worn section area, A, and the width of the wear surface, w, can be described by [7,21]:

$$A = r^2 \sin^{-1}\left(\frac{w}{2r}\right) - \frac{w}{2}\sqrt{r^2 - \left(\frac{w}{2}\right)^2} \quad (2)$$

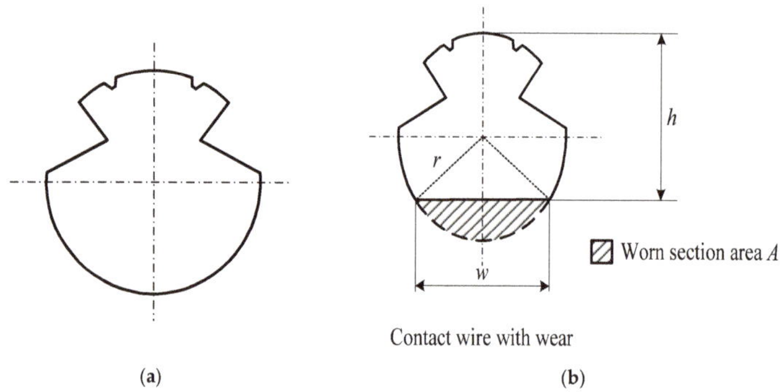

Figure 1. Cross-section of a contact wire: (**a**) a new contact wire and (**b**) a contact wire with wear.

The contact wire wear measurement using the vision measurement method is based on the fact that the wearing surface of the contact wire presents good light-reflecting characteristics. The wearing surface is illuminated by the active lighting source, i.e., spotlights, laser diodes, and LED. Because of the continuous friction between the pantograph and the contact wire, the wearing surface is flat and has much greater light-reflecting characteristics than the lateral surface of the contact wire. In common, the grayscale of the wear surface region is much higher than the rest of the region of the contact wire. The red lines in Figure 2 show the grayscale curve of a new contact wire (Figure 2a) and the grayscale curve of a contact wire with wear (Figure 2b). Appropriate image-processing techniques are adopted to extract the key points, including the edge points and the central point of the wear surface, such as the Sobel operator and cross-correlation template matching. The width of the wear surface is the distance between the two edge points (points A and B in Figure 2b), indicating the residual geometric thickness of the contact wire. The central point of the wear surface (point C in Figure 2b) is usually chosen to calculate the values of height and stagger.

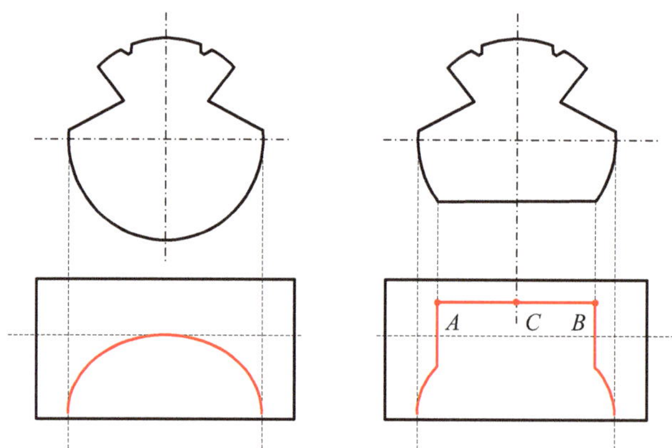

Figure 2. Grayscale curves (red lines) of cross-section of contact wires: (**a**) a new contact wire and (**b**) a contact wire with wear.

3. Materials and Methods

3.1. Stereovision Measurement Model

Since the image resolution varies with both the height and stagger values of the contact wire, the measurement of height and stagger is the basis of the measurement of contact wire wear. The measurement principle of the contact wire geometry parameters using the stereovision method is shown in Figure 3. In the figure, $O_W X_W Z_W$ is the world coordinate system. The origin of this coordinate system, O_W, is set at the midpoint of the train roof, and Y_W represents the direction of the train. O_1X_1 and O_2X_2 are the imaging coordinate systems of the left and right cameras, respectively. $O_{c1}X_{c1}Z_{c1}$ and $O_{c2}X_{c2}Z_{c2}$ are the camera coordinate systems of the left and right cameras, respectively. For these two camera coordinate systems, each origin is located at the principal point of the camera lens. The point P is projected onto the imaging chip of the line-scan camera through the optical lens. The imaging locations of the point P on the two cameras are P_1 and P_2, respectively. U_{01} and u_{02} are the image coordinates of the principal points of the two camera lenses.

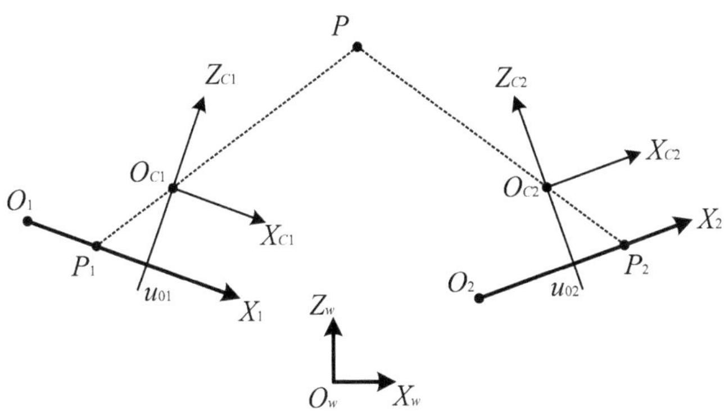

Figure 3. Measurement principle of the contact wire geometry parameters using stereovision method with line-scan cameras.

By using perspective projection transformation, the transformation relation between the camera imaging coordinate system and the camera coordinate system is shown in Equation (3):

$$s\begin{bmatrix} u \\ 1 \end{bmatrix} = \boldsymbol{M}_1 \begin{bmatrix} X_c \\ Z_c \end{bmatrix} \tag{3}$$

where s is the non-zero scale factor; u is the coordinate of image point of point P in any camera imaging coordinate system; $[X_c, Z_c]$ is the coordinate of point P in the corresponding camera coordinate system; and $\boldsymbol{M}_1$ is the internal parameter matrix of the camera and is given by the following:

$$\boldsymbol{M}_1 = \begin{bmatrix} f_e & u_0 & 0 \\ 0 & 1 & 0 \end{bmatrix} \tag{4}$$

In Equation (4), f_e is the normalized focal length of the camera lens, and u_0 is the image coordinate of the principal point of the camera lens.

By using Euclid-space transformation, the transformation relation between the camera coordinate system and the world coordinate system is as shown below [22]:

$$\begin{bmatrix} X_c \\ Z_c \end{bmatrix} = \boldsymbol{M}_2 \begin{bmatrix} X_W \\ Z_W \end{bmatrix} = \begin{bmatrix} \boldsymbol{R} & \boldsymbol{T} \\ 0 & 1 \end{bmatrix} \begin{bmatrix} X_W \\ Z_W \end{bmatrix} \tag{5}$$

where $[X_w, Z_w]$ is the coordinate of point P in the world coordinate system; $\boldsymbol{M}_2$ is the external parameter matrix of the camera; and $\boldsymbol{R}$ and $\boldsymbol{T}$ are the rotation matrix and translation vector, respectively.

Combining Equation (3) with (5), the line-scan camera model is given by the following:

$$s\begin{bmatrix} u \\ 1 \end{bmatrix} = \boldsymbol{M}_1 \begin{bmatrix} X_c \\ Z_c \end{bmatrix} = \boldsymbol{M}_1\boldsymbol{M}_2 \begin{bmatrix} X_w \\ Z_w \end{bmatrix} = \boldsymbol{M} \begin{bmatrix} X_w \\ Z_w \end{bmatrix} \quad (6)$$

where $\boldsymbol{M}$ is the intrinsic parameter matrix. $\boldsymbol{M}$ is described by the following:

$$\boldsymbol{M} = \boldsymbol{M}_1\boldsymbol{M}_2 = \begin{bmatrix} m_{11} & m_{12} & m_{13} \\ m_{21} & m_{22} & m_{23} \end{bmatrix} \quad (7)$$

By eliminating the variables in Equation (6), we can obtain the equation as follows:

$$u = \frac{m_{11}X_w + m_{12}Z_w + m_{13}}{m_{21}X_w + m_{22}Z_w + m_{23}} \quad (8)$$

Equation (8) is the perspective projection equation of the line-scan camera, which describes the relationship between the coordinate of point P in the world coordinate system and the image coordinate of point P in the camera imaging coordinate system.

Multiple sets of calibration point data are obtained within the measurement range, including the world coordinate data, $[X_w, Z_w]$, and the corresponding imaging coordinate data, u. Then, the parameters of the matrix, $\boldsymbol{M}$, in Equation (8) are calculated using mathematical procedures.

In principle, the coordinate of point P in the world coordinate system, $[X_w, Z_w]$, could be calculated by using the imaging coordinate data, u, of any two line-scan cameras. The subscripts a and b are used to represent the left and right cameras in Figure 3, respectively. By combining the perspective projection equations of the two cameras, X_w and Z_w are given by the following:

$$\begin{aligned} X_w &= \tfrac{q_2k_1 - q_1k_2}{p_1q_2 - p_2q_1} \\ Z_w &= \tfrac{p_1k_2 - p_2k_1}{p_1q_2 - p_2q_1} \end{aligned} \quad (9)$$

The parameters p_1, p_2, q_1, q_2, k_1, and k_2 in Equation (9) are given by the following:

$$\begin{aligned} p_1 &= m_{11a} - u_a m_{21a} \\ p_2 &= m_{11b} - u_b m_{21b} \\ q_1 &= m_{12a} - u_a m_{22a} \\ q_2 &= m_{12b} - u_b m_{22b} \\ k_1 &= u_a m_{23a} - m_{13a} \\ k_2 &= u_b m_{23b} - m_{13b} \end{aligned} \quad (10)$$

Equation (9) is the measurement model of the contact wire geometry parameters based on stereovision via triangulation. To perform the dynamic measurement, the pixel coordinates of the contact wire in the two line-scan cameras, u_a and u_b, are extracted through image processing. Then, the positions of the contact wire in the world coordinate system, $[X_w, Z_w]$, are calculated using Equation (9) and the calibrated parameters of the intrinsic parameter matrix, $\boldsymbol{M}_a$ and $\boldsymbol{M}_b$, of the two line-scan cameras.

In the overlapping section [1], not only the two contact wires, but also the messenger wires located above the contact wires, would be captured. As mentioned above, when there is one target in the FOV (field of view) of the line-scan camera, the world coordinate of the target could be calculated by using the imaging coordinate of the target in the images of two cameras. However, when the number of the targets is more than one, it is necessary to find the corresponding points of the same target between the stereo image pairs, that is, to solve the stereo-matching problem. Various research studies have been conducted to determine the accurate corresponding points, such as feature descriptors, interest point detectors, and epipolar constraint method [14,22].

This article proposes a matching method based on the position feature of the target. The third line-scan camera is adopted to overcome the correspondence problem. The target imaging coordinate information of the third line-scan camera is used to check the matching of corresponding points in the first two line-scan camera images, which could eliminate the uncertainty caused by the matching of binocular images.

The proposed matching method is illustrated in Figure 4. A, B, and C are three targets in the FOV of the stereovision system; O_1, O_2, and O_3 are the principal points of the three camera lenses; a_1, b_1, and c_1 are the imaging points of the three targets in the left camera; a_2, b_2, and c_2 are the imaging points of the three targets in the right camera; and a_3, b_3, and c_3 are the imaging points of the three targets in the middle camera. The matching procedure operates in three steps, which are described as follows.

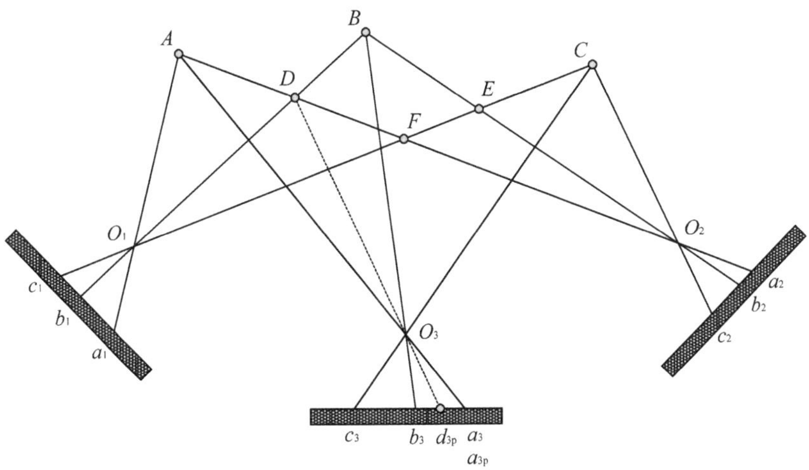

Figure 4. Stereovision matching method using the third line-scan camera.

Step 1: All possible spatial points are reconstructed based on the extracted image coordinate data of the left and the right cameras, using the enumeration method. Among the possible spatial points shown in Figure 4, A, B, and C are the real targets, while D, E, and F are the false targets. Then, the key to the matching procedure is to verify all of the reconstructed spatial points.

The verification of points A and D is illustrated bellow. Point A is reconstructed using the imaging point a_1 of the left camera and a_2 of the right camera. Point D is reconstructed using b_1 of the left camera and a_2 of the right camera.

Step 2: The reconstructed points A and D are re-projected onto the imaging chip of the middle line-scan camera and form the re-projected imaging points a_{3p} and d_{3p}. The re-projection imaging coordinates of points a_{3p} and d_{3p} are denoted as u_{a3p} and u_{d3p}, respectively.

Step 3: The imaging coordinates of a_3, b_3, and c_3 are denoted by u_{a3}, u_{b3}, and u_{c3}. The validity of the re-projected imaging points a_{3p} and d_{3p} is verified by calculating the distance between the extracted image coordinate data and the re-projected imaging coordinate data. An appropriate threshold, T, is adopted by taking into account the average deviation distance of re-projection of the line-scan camera. With regard to the re-projected imaging point a_{3p}, the distance between u_{a3} and u_{a3p} is less than the threshold (T), verifying the effectiveness of reconstructed point A. However, reconstructed point D is judged as a false target since there is no candidate imaging point near the re-projected imaging point d_{3p}.

The verification-procedure speed in this study is fast and satisfies the requirement of real-time measurement. In addition, the matching method is robust for both day and night inspection since the matching precision is not easily affected by the changes in ambient lighting.

3.2. Determining the Wear Width of the Contact Wire

Determining the wear width of the contact wire includes two steps, the extraction of wear width in pixels from the image and the calculation of the physical wear width in mm. The specific steps of extracting the wear width in pixels from the image are as shown below:

Step 1: The candidate wires are extracted by using the edge detection operator. The first-order difference of grayscale curve is designed to detect the candidate objects in different weather conditions. Figure 5 shows the original grayscale curve of catenary on a cloudy day in an overlapping section. Therefore, there are two peak-shaped regions and two valley-shaped regions in the background grayscale curve, where the peak-shaped regions are the two contact wires in the overlapping section and the valley-shaped regions are the two messenger wires. Figure 6 shows the processed images by using the first-order difference operator. The grayscale gradient of the contact wire is obviously higher than that of the messenger wire, which helps to minimize the interference from the messenger wire in matching.

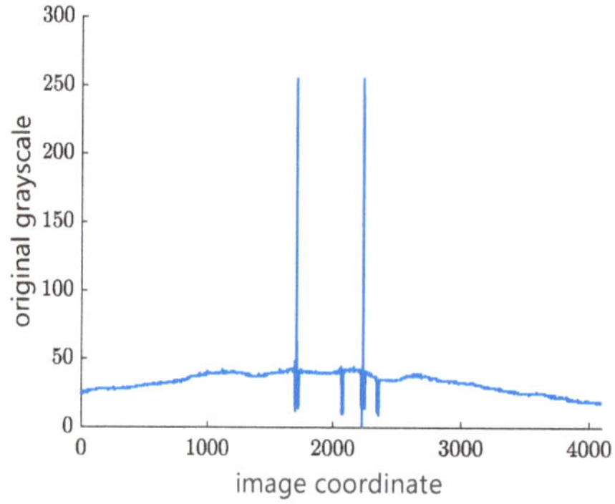

Figure 5. Original grayscale curve of catenary in an overlapping section.

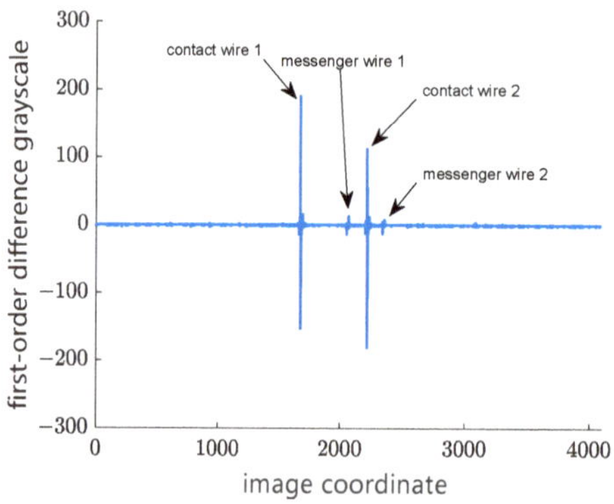

Figure 6. First-order difference of grayscale curve of catenary in an overlapping section.

Step 2: The peak-shaped regions and the valley-shaped regions are extracted from the fitting background grayscale curve. Mean filter operator is adopted to acquire the fitting background grayscale curve and minimize the influence of noise. Then, the boundary between the contact wire and the background is carefully determined. An empirical threshold of the pixel width of a contact wire is used to assist in determining boundaries. The threshold is determined by experimental statistics, and the value of the threshold is affected by the resolution of the line-scan camera, the FOV of the camera and the height variation in the contact wire.

Step 3: The left and right edge points of the wear surface are located within the peak-shaped region of a contact wire in the image. As mentioned above, the wearing surface has much higher light-reflecting characteristics than the lateral surface of the contact wire. As a result, the grayscale of the wear surface region is several times higher than that of the rest of regions of the contact wire. As shown in Figure 7, the grayscale of the lateral surface regions of the contact wire decreases rapidly from the wear edge point to the background. In this case, the second-order difference of the grayscale curve is adopted to locate the left and right wear edge points within the peak-shaped region of the local grayscale curve of the contact wire. The wear width in pixels is extracted by calculating the difference of the coordinates between the left and the right wear edge points. Meanwhile, the central point of the wear surface is determined.

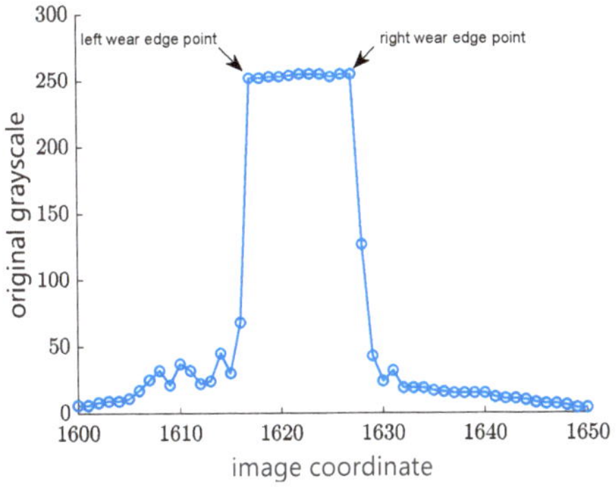

Figure 7. Original grayscale curve of a contact wire with a wear surface.

Step 4: The real targets in the FOV are determined by using the stereovision matching method proposed in this study. Several targets may be extracted in an overlapping section in daytime inspection, i.e., the operating contact wire, the non-operating contact wire, and the two messenger wires which support these two contact wires. The position of each target in the world coordinate system is calculated by triangulation, using Equation (9).

Step 5: The messenger wires are excluded based on the fact that the height of the contact wire is lower than that of the messenger wire. A target-tracking operator is used to exclude the other interference targets, such as the droppers and the cantilevers of the support system. This is due to the fact that the contact wire is continuous along the railway line, while the projections of both the droppers and the cantilevers on the image are discontinuous in the direction of the railway line.

The calculation of the physical wear width in mm relies on the determination of the image resolution in the measurement range. For different types of electrified railways in China, the height from the car roof to the contact wire would be in the range between 1300 mm and 2500 mm. The stagger would vary from −400 mm to 400 mm in an electrified railway to avoid the continuous friction at the same point of the pantograph. If the effect of the stagger variation is not taken into account, a significant error would be introduced when considering the image resolution. In this study, the image resolution in various sections of measurement range is carefully calibrated using a calibration tool, which takes into account the impact of both the height variation and the stagger variation. As shown in Figure 8a, a set of horizontally arranged targets is fixed on top of a slide bar, with a spacing of 100 mm between each target. A flat surface with a width of 6 mm is fabricated on the bottom of each target, which simulates a contact wire with a wear surface, as shown in Figure 8b. Each target in the horizontal array denotes a contact wire with a different stagger value. The height of the target array could be set in the measurement range by moving the slider bar up and down.

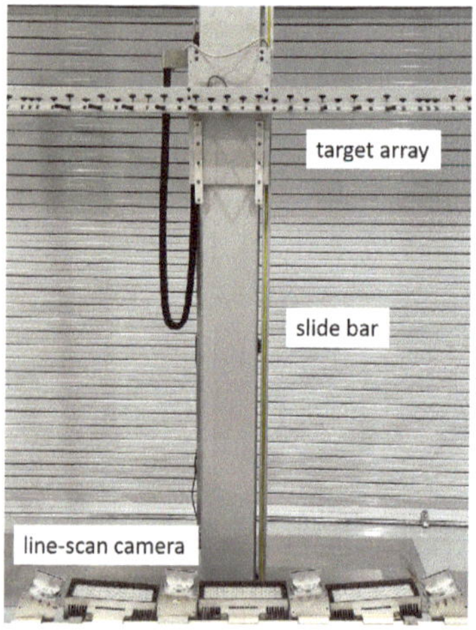

(a)

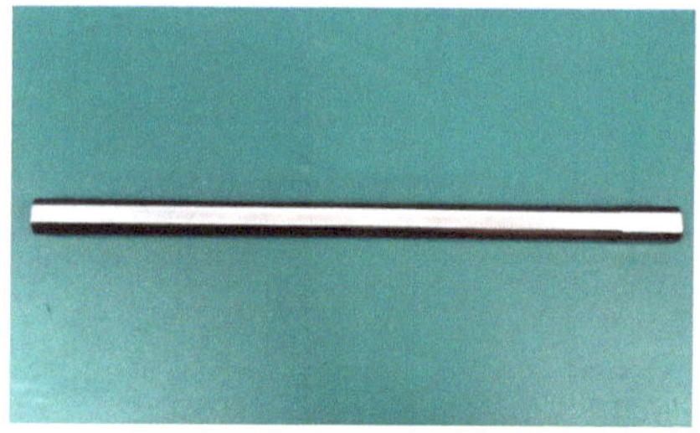

(b)

Figure 8. Calibration tool of image resolution: (**a**) calibration tool setup and (**b**) a target with a wear surface.

First, the parameters of the matrix (***M***) in Equation (8) of each line-scan camera are calculated using multiple sets of calibration data, including the world coordinate data, $[X_W, Z_W]$, and the corresponding imaging coordinate data, u. After that, the wear width of each target is extracted in pixels from the captured grayscale curve, using the above image-processing procedure, as shown in Figure 9.

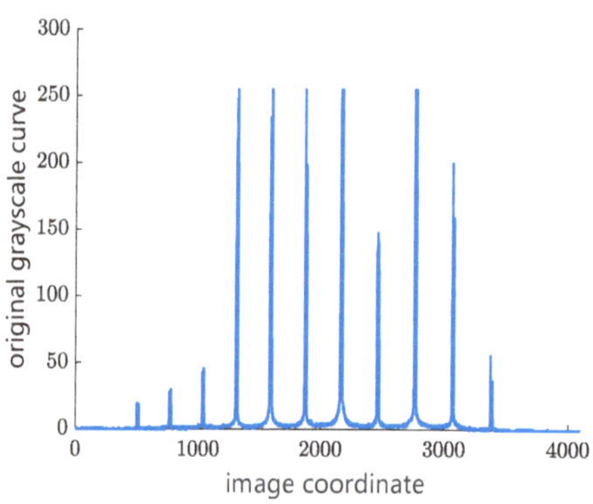

Figure 9. Grayscale curve of the arranged targets of the calibration tool.

The image resolution with a fixed position is calculated by dividing the preset physical width of 6 mm by the extracted image width of the wear surface, with the unit of mm/pixel. Table 1 illustrates the calibrated image resolutions of the second line-scan camera at different horizontal positions with a height of 1300 mm from the car roof. The image resolutions of each line-scan camera at different horizontal or vertical positions are determined by repeating the above calibration. Thus, the measurement range is divided into an array of small square areas. The image resolution within the same small square area is considered to be uniform. As a result, the distribution of the image resolution in the measurement range is determined by using this lookup-table method. Then, the physical wear width is calculated accordingly. Finally, the residual thickness, h, of the contact wire is calculated according to the wire type and the rated cross-section diameter of the contact wire by using Equation (1).

Table 1. Calibrated image resolution of the second line-scan camera at different horizontal positions with a height of 1300 mm from the car roof.

Horizontal Position (mm)	Image Resolution (mm/pixel)
−400	0.33
−300	0.32
−200	0.32
−100	0.31
0	0.30
100	0.30
200	0.29
300	0.28
400	0.27

3.3. Measurement Apparatus Design

The architecture diagram of the measurement apparatus based on the measurement method for contact wire wear in this study is shown in Figure 10. The measurement apparatus is composed of the stereovision measurement module on the train roof, the processing module inside the train, and the vehicle compensation measurement module under the train. The stereovision measurement module consists of four line-scan cameras with different angles and three LED lamps. The vehicle compensation measurement module includes three displacement sensors. When the train sways, two sensors are used to measure both the left and the right vertical displacement of the train relative to the rail surface, and then the roll angle of the train is calculated by using the displacement data of these two sensors and the width of the train; the third sensor is applied to measure the horizontal displacement of the train relative to the center line of the track. The processing module processes the image data of catenary captured by the line-scan cameras and the vehicle compensation data and calculates the geometry parameters (stagger and contact wire height) and the wear parameters (residual thickness, h; or worn area, A). The processing module receives the distance pulses produced by the photoelectric encoder, which is installed on the train wheel and triggers all the line-scan cameras to exposure simultaneously.

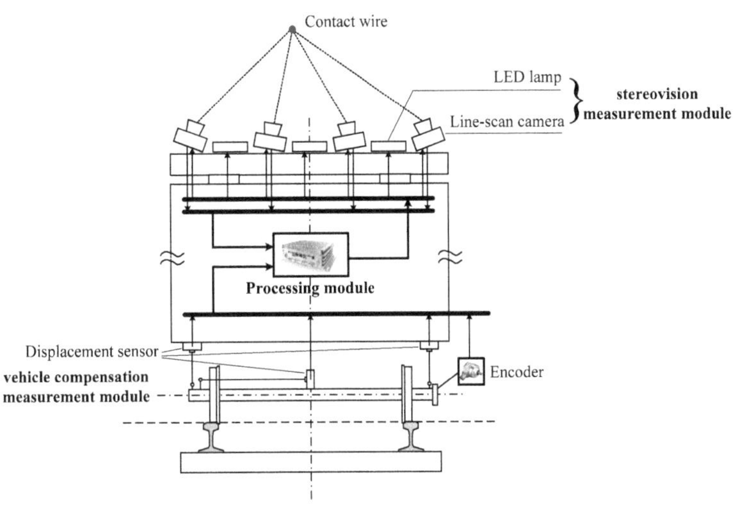

Figure 10. Architecture diagram of the measurement apparatus.

The flowchart of the processing module to perform the measurement of the contact wire wear is shown in Figure 11, which includes four steps:

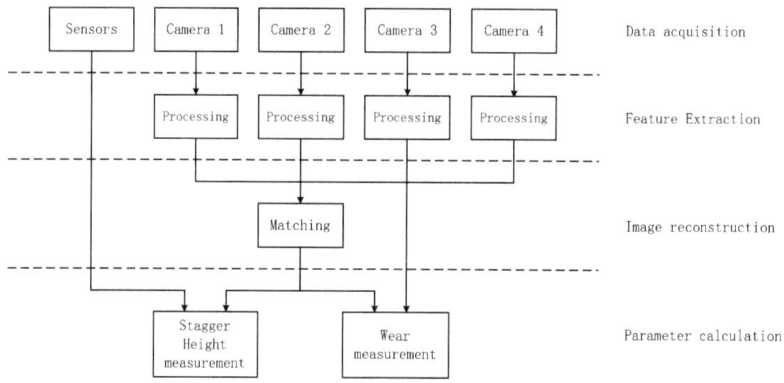

Figure 11. Architecture diagram of the processing module.

Step 1: Acquire the image data of the four line-scan cameras and the voltage data of the three displacement sensors.

Step 2: Extract the features from each line-scan camera's image, including the central point coordinate and the wear width of the candidate targets.

Step 3: Execute the stereovision matching algorithm to determine the corresponding points of the contact wire among the stereo images, and then calculate the position of the contact wire in the train roof coordinate system.

Step 4: Calculate the position of the contact wire in the rail surface coordinate system by using the calculated position data from Step 3 and the measurement data of the vehicle compensation measurement module based on the Euclid-space transformation. Then, the physical wear width of the contact wire is determined by using both the data of the wear width in pixels from Step 2 and the image resolution data corresponding to the current position.

The catenary image quality is essential to the wear measurement of the contact wire. In this study, a new kind of high-speed synchronized stroboscopic lighting technology is developed as an alternative active light source. The high-speed synchronization of lighting and camera exposure is realized by a same-trigger pulse. The imaging and lighting devices are able to work stably at 1000 Hz in the pulsed mode, which ensures the dynamic measurement of contact wire wear on the inspection train. Since the duty cycle of the pulsed lighting is low (less than 0.1), the power consumption of the light source is greatly reduced, which is no more than 5% of the spotlights' total power in our previous research [19,20]. Compared with the continuous LED lighting, the thermal performance of the LED chip in stroboscopic mode is significantly improved, which greatly extends the lifetime of the light source. Moreover, compared with the continuous intensity lighting with rated current, an over-driven pulsed current is used to obtain a light beam with higher luminous flux, which highlights the image features of the wear surface of the contact wire.

Blue LED lamps combined with band-pass optical filters are adopted to minimize the influence of sunlight. A comparison test in daytime was performed on two light sources, that is, the white LED lamp and the blue stroboscopic LED lamp with the optical filter, as shown in Figure 12. In Figure 12a, the light zigzag lines are the contact wires due to the good light-reflecting characteristics, while the dark zigzag lines are the messenger wires due to the poor reflectivity. In Figure 12b, the grayscale of the sky is greatly reduced, and the image SNR in the daytime environment is significantly improved by using the proposed lighting technology. The messenger wires disappear in Figure 12b because of the little difference in grayscale between the messenger wires and the sky. In this case, the number of the extracted candidate targets is reduced, which accelerates the stereo matching.

(a) (b)

Figure 12. Comparison of catenary images illuminated by two types of light sources in daytime inspection: (**a**) white LED lamps and (**b**) blue stroboscopic LED lamps with band-pass optical filters.

4. Results and Discussion

4.1. Experimental Result

The measurement apparatus based on the proposed measurement method was fixed on an inspection train. The four line-scan cameras are the Eliixa+ multi-line high-speed CMOS cameras (Teledyne E2V company, Thousand Oaks, CA, USA), with a resolution of 4096 pixels. Compared with the CCD camera used in the previous research [19,20], the CMOS camera used in this study eliminates the blooming effect and smear effect under strong light illumination. The lens is an Interlock C 35 mm lens (Carl Zeiss AG, Oberkochen, Germany), with robust full-metal construction and low distortion, and then the angle of FOV for each camera is about 60°. The parameters of the intrinsic parameter matrix, $\boldsymbol{M}$, are calculated using the proposed calibration tool. Figure 13 shows the re-projection error diagram of the four cameras. The average re-projection error for all calibration points of the four line-scan cameras is 0.42, 0.55, 0.55, and 0.52 pixels, respectively. The measurement accuracy of both stagger and height is within ±3 mm in the train roof coordinate system.

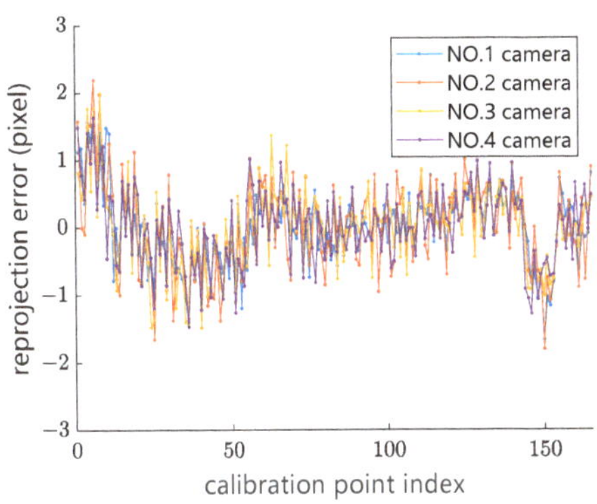

Figure 13. Re-projection error distribution of four line-scan cameras.

Field tests were performed in National Railway Track Test Center of China Academy of Railway Sciences. The following three groups of experiments were carried out to verify the availability of the proposed measurement method for contact wire wear.

The first group of experiments is functional testing. The type of contact wire in this test is CTA110, and the rated diameter of the contact wire is 12.34 mm. Figure 14 shows the measurement results of residual thickness, stagger, and contact wire height in a straight-line section, while Figure 15 shows the results of these parameters in a curve-line section. The blue and orange lines in Figure 15 represent two contact lines, which belong to two adjacent anchor segments. As shown in Figure 14, the variation in the residual thickness in the straight-line section is relatively small, probably due to the well-distributed elasticity within a span of the straight-line section. In contrast, the minimum residual thickness of each span (except the overlapping section) locates around the catenary mast (refer to the three blue squares in Figure 15) in the curve-line section, which indicates that the distribution of catenary elasticity within a span in the curve-line section is not uniform. This is probably due to the large pull-off force on the contact wire at the mast in the curve-line section with a small curve radius, resulting in better elasticity mid-span than around the mast. The local minimum residual thickness in Figure 15 is located in the overlapping section (refer to the yellow ellipse), probably because of the poor catenaries' elasticity around the conversion point of the two operating contact wires.

The second group of experiments is the repeatability verification experiment. The round-robin tests of the residual thickness in the same line section present good measurement repetitiveness, which is shown in Figure 16. According to the statistical results of the difference between the two tests shown in Figure 16b, for more than 90% of the measuring points, the absolute difference between the two tests is less than 0.1 mm, indicating that the method presents high repeatability.

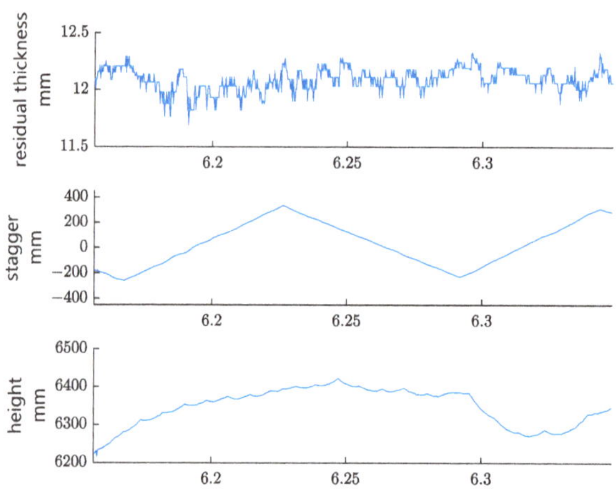

Figure 14. Test results in a straight-line section.

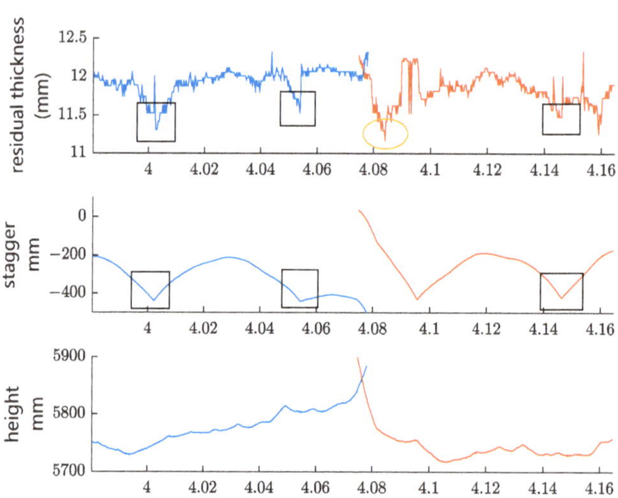

Figure 15. Test results in a curve-line section.

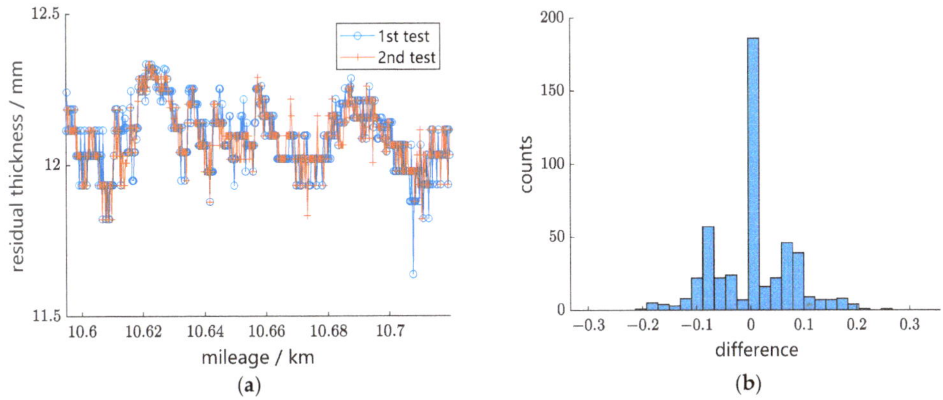

Figure 16. Repeatability verification experiment: (**a**) measurement curves of residual thickness of two tests and (**b**) statistical results of difference between the two tests.

The third group of experiments is the accuracy validation test. The type of contact wire in this test is CTA120, and the rated diameter of the contact wire is 12.90 mm. Eight groups of data were collected via a dynamic test and ground measurement, respectively. The difference between the dynamic test data of residual thickness and the ground manual measurement data is listed in Table 2. The average difference value and the maximum difference value are 0.08 mm and 0.13 mm, respectively, which demonstrate that the method has good measurement accuracy. The measurement accuracy of contact wire wear in this study is basically consistent with that reported in the literature [8,9].

Table 2. Comparison between dynamic test results and ground measurement results.

Test Data (mm)	Ground Measurement Data (mm)	Difference (mm)
12.72	12.62	0.10
12.80	12.77	0.03
12.83	12.75	0.08
12.80	12.75	0.05
12.80	12.73	0.07
12.77	12.73	0.04
12.83	12.73	0.10
12.83	12.70	0.13

4.2. Discussion

Figure 17 shows the stagger effect on the measurement of the wear width of the contact wire. The black solid line with the zigzag shape indicates the stagger distribution along the line. The horizontal dash dot line represents the trajectory of the train roof center during the train's movement. Then, the measurement plane of the apparatus is perpendicular to the dash dot line. As a result, as shown in the partial enlarged drawing in Figure 17, w denotes the measured wear width of the contact wire, w_0 denotes the actual wear width, which is perpendicular to the direction of the contact wire, and there is a small angle θ between the measured dot line and the actual dot line.

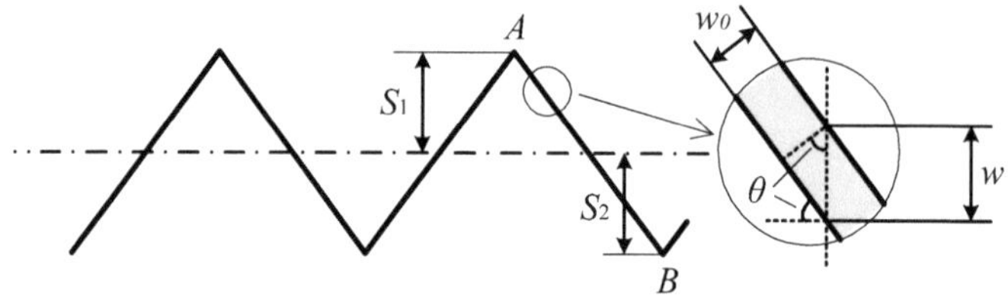

Figure 17. Stagger effect on the measurement of wear width of contact wire in a straight-line section.

For the straight-line section shown in Figure 17, points A and B denote two adjacent masts of the catenary with the design stagger of S_1 and S_2, respectively. The span of the two masts is L. Then, the angel, θ, can be calculated as follows:

$$\sin\theta = \frac{|S_1| + |S_2|}{L} \tag{11}$$

Then, the relationship between the measured wear width, w, and the actual wear width, w_0, is as follow:

$$w_0 = w\cos\theta \tag{12}$$

For a typical catenary design in a straight-line section in China, S_1 = 300 mm, S_2 = −300 mm, and L = 50 m. Then, we can calculate the value of $\sin\theta$; $\sin\theta$ = 0.0120, and $\cos\theta$ = 0.9999.

As shown in Figure 15, the stagger curve is arc-shaped in a curve-line section, and the angle, θ, varies from one mast to the midpoint of two neighboring masts. The maximum value of θ appears next to the mast, while the minimum value of θ locates at the midpoint of two neighboring masts and is approximately equal to 0. The maximum value of θ could be evaluated by calculating the change in stagger value per length along the catenary. For a typical curve line section, the maximum value of $\sin\theta$ is about 0.0200, and $\cos\theta$ = 0.9997.

It may seem that the stagger effect on the measurement of the wear width of the contact wire for a typical straight- or curved-line section is small. However, when the train passes through the turnout, the lateral position of the contact wire relative to the roof center changes rapidly, rather than the case of the normal curve-line section. Future work should be done by performing more experiments to further evaluate the stagger effect on the measurement of the wear width in the special section, such as turnout.

5. Conclusions

In order to guarantee the operational safety of the catenary infrastructure of railway lines, this paper develops an optical measurement method of contact wire wear using multi-view stereovision technology. The experimental results demonstrate that this method enables the accurate calculation of both the wear width and residual thickness of the contact wire, exhibiting excellent repeatability and precision. This research offers a promising avenue for future condition-based maintenance of contact wire wear, enhancing maintenance efficiency and safeguarding the integrity of catenary infrastructure. However, dynamic measurements on rolling stock present challenges, as the stagger arrangement can influence wear width measurements. Consequently, a further assessment of this stagger effect in regard to specific railway sections is crucial.

Author Contributions: Conceptualization, W.Z. and H.W.; methodology, W.Z. and Z.Q.; software, X.X.; validation, Z.Q., H.W. and W.Z.; formal analysis, Z.Q.; investigation, W.Z. and X.D.; resources, X.D. and H.L.; data curation, W.Z.; writing—original draft preparation, W.Z.; writing—review and editing, Z.Q. and X.D.; visualization, X.X. and H.W.; supervision, X.D.; project administration, H.L.; funding acquisition, H.L. All authors have read and agreed to the published version of the manuscript.

Funding: The work described in this paper was supported by Science and Technology Research and Development Plan of China National Railway Group Co., Ltd., Grant No. K2021T015.

Institutional Review Board Statement: Not applicable.

Informed Consent Statement: Not applicable.

Data Availability Statement: Data are contained within the article.

Conflicts of Interest: All Authors were employed by the company China Academy of Railway Sciences Corporation Limited, and the authors declare no conflict of interest.

References

1. Kießling, F.; Puschmann, R.; Schmieder, A. *Contact Lines For Electric Railways*, 1st ed.; China Railway Electrification Engineering Group Co. Ltd., Translator; China Electric Power Press: Beijing, China, 2004; pp. 369–371.
2. Wu, G.; Wu, J.; Wei, W.; Zhou, Y.; Yang, Z.; Gao, G. Characteristics of the sliding electric contact of pantograph/contact wire systems in electric railways. *Sensors* **2018**, *11*, 17. [CrossRef]
3. Seferi, Y.; Blair, S.; Mester, C.; Stewart, B. A novel arc detection method for DC railway systems. *Sensors* **2021**, *14*, 444. [CrossRef]
4. Fararroy, S.; Mair, C. *Review of Railway Overhead Wire Geometry Measurement Systems*; Current Collections for High Speed Trains Seminar: London, UK, 1998.
5. Liu, Z.; Liu, W.; Han, Z. A high-precision detection approach for catenary geometry parameters of electrical railway. *IEEE Trans. Instrum. Meas.* **2017**, *66*, 1798–1808. [CrossRef]
6. Borromeo, S.; Aparicio, J. Automatic systems for wear measurement of contact wire in railways. In Proceedings of the IEEE 2002 28th Annual Conference of the Industrial Electronics Society, IECON 02, Sevilla, Spain, 5–8 November 2002; Volume 4, pp. 2700–2705.
7. Borromeo, S.; Aparicio, J.; Martinez, P. MEDES: Contact wire wear measuring system used by the Spanish National Railway (RENFE). *Proc. Inst. Mech. Eng. Part F J. Rail Rapid Transit* **2003**, *217*, 167–175. [CrossRef]
8. Hidenori, S. Recent topics on power supply technology. *RTRI Rep.* **2021**, *62*, 79–82.
9. Chugui, Y.; Verkhogliad, A.; Bazin, V.; Kalichkin, S.; Kalikin, V.; Makarov, S.; Vykhristyuk, I. Optical remote dimensional inspection of live contact wire in train's electro-supply network. *Meas. Sci. Rev.* **2008**, *8*, 37–41. [CrossRef]
10. Nie, X. Research on Detection Algorithm of Contact Wire WEAR and Catenary Dynamic Geometric Parameters of Rail Transit. Master's Thesis, Southwest Jiaotong University, Chengdu, China, 2022.
11. Zhang, Z.; Yu, W. A study on image inspection of catenary wear. *J. China Railw. Soc.* **1995**, *17*, 118–122.
12. You, C. A study on dynamic measurement system of contact wire wear in electrified railway. *Procedia Comput. Sci.* **2019**, *154*, 210–220. [CrossRef]
13. Wan, S. Research on Dynamic Detection System of Wire Wear of Metro Rigid Catenary. Master's Thesis, Beijing Jiaotong University, Beijing, China, 2020.
14. Yu, L.; Gao, S.; Zhang, D.; Kang, G.; Zhan, D.; Roberts, C. A survey on automatic inspections of overhead contact lines by computer vision. *IEEE Trans. Intell. Transp. Syst.* **2022**, *23*, 10104–10125. [CrossRef]
15. Van Gigch, J.; Smorenburg, C.; Benschop, A. The contact wire thickness-measuring system (ATON) of The Netherlands railways. *Rail Int.* **1991**, *22*, 20–31.
16. Kusumi, S.; Nagasawa, H.; Mochizuka, H. Development of measuring apparatus for contact wire wear by sodium vapor lamps. *RTRI Rep.* **2000**, *41*, 117–119.
17. Zhou, D.; Wang, P.; Sun, C.; Dong, M.; Fu, L. Calibration method for trinocular stereovision system comprising projector and dual cameras. *Acta Opt. Sin.* **2021**, *41*, 1115001.
18. Shu, A.; Pei, H.; Duan, H. Trinocular stereo visual measurement method for spatial non-cooperative target. *Acta Opt. Sin.* **2021**, *41*, 0615001. [CrossRef]
19. Zhou, W.; Sun, Z.; Ren, S.; Zhang, W.; Wang, H.; Dai, P.; Wang, Y. Measurement method for geometric parameters of overhead contact line based on multi-view stereovision. *China Railw. Sci.* **2015**, *36*, 104–109.
20. Wang, W. *Dynamic Detection Technology of High-Speed Railway Infrastructure*, 1st ed.; Science Press: Beijing, China, 2017; pp. 108–115.
21. Shing, A.; Pascoschi, G. Contact wire wear measurement and data management. In Proceedings of the Institution of Engineering and Technology International Conference on Railway Condition Monitoring, Birmingham, UK, 29–30 November 2006; pp. 182–187.
22. Zhang, G. *Vision Measurement*, 1st ed.; Science Press: Beijing, China, 2008; pp. 23–41.

sensors

MDPI

Article

Near-Infrared Dual Greenhouse Gas Sensor Based on Hollow-Core Photonic Crystal Fiber for Gas-Cell In-Situ Applications

Jianing Wang [1], Bingqiang Li [1,2], Weiping Wu [1] and Guanyu Lin [1,*]

1 Changchun Institute of Optics, Fine Mechanics and Physics, Chinese Academy of Sciences, Changchun 130033, China; wangjn@ciomp.ac.cn (J.W.); libingqiang@ciomp.ac.cn (B.L.); rainciomp@sina.com (W.W.)
2 University of Chinese Academy of Sciences, Beijing 100049, China
* Correspondence: linguanyu@ciomp.ac.cn

Abstract: A greenhouse gas sensor has been developed to simultaneously detect multiple gas species within a hollow-core photonic bandgap fiber (HC-PBF) structure entirely composed of fibers. To enhance sensitivity, the gas cell consists of HC-PBF enclosed between two single-mode fibers fused with a reflective end surface to double the absorption length. The incorporation of side holes for gas diffusion allows for analysis of the relationship between gas diffusion speed, number of drilled side holes, and energy loss. As the number of drilled holes increases, the response time decreases to less than 3 min at the expense of energy loss. Gas experiments demonstrated detection limits of 0.1 ppm for methane and 2 ppm for carbon dioxide, with an average time of 50 s. In-situ testing conducted in rice fields validates the effectiveness of the developed gas detection system using HC-PBF cells, establishing all-fiber sensors with high sensitivity and rapid response.

Keywords: greenhouse gas; fiber sensor; infrared absorption

1. Introduction

The 2020 greenhouse gas bulletin from the World Meteorological Organization emphasizes carbon dioxide and methane as the primary components of greenhouse gases [1]. As per the Kyoto Protocol report, approximately 21% of greenhouse gases arise from agricultural activities [2]. Therefore, the development of a highly sensitive gas detection system tailored explicitly for agricultural applications becomes imperative. Ensuring accurate gas concentration measurements without environmental interference is essential.

Since 1996, HC-PBF has captured researchers' interest [3] due to its portable and flexible advantages [4–7]. Highly sensitive species detection, including liquid and gas, is carried out for qualitative and quantitative analysis using different technologies, such as Raman sensing and tunable diode laser absorption spectroscopy [8–10]. In gas detection applications, HC-PBFs are commonly used as gas cells. Considering light–gas interaction, HC-PBF maintains at least 95% optical power propagating within its central hollow cores [11]. The primary advantage of HC-PBF lies in its long absorption length despite its compact size, as demonstrated by numerous researchers [12]. However, the prolonged gas filling time into the hollow core hampers HC-PBF sensors from becoming a strong contender in gas sensing [13]. It is crucial to explore and uncover effective methods to enhance the response speed of HC-PBF while evaluating the potential impact of these methods.

HC-PBFs, functioning as gas cells, are commonly linked to single-mode fibers (SMFs) at one or both ends [14–17]. Using an open end of the HC-PBF represents the simplest approach for gas inlet and outlet [18]. In 2018, L. E. He et al. minimized mode interference effects by employing a free space coupling structure, achieving a methane detection limit of 4.35 ppm. However, due to the single open-ended structure, gas diffusion time extends to nearly 80 min [19]. Employing C-shaped rings spliced to PCF and SMF for connection can create pathways similar to the free space coupling method [17]. Besides gas diffusion driven by concentration gradients, actively exchanging gases can significantly enhance response speed. Vacuum chambers and extreme pressure applications are two common methods for actively increasing gas exchange [20,21]. Valiunas et al. effectively reduced response time, achieving sub-ppmv capability in detecting nitrous oxide gas [20]. However, employing active gas exchange methods inevitably disturbs external gas distribution and may be influenced by mechanical noise. Comparatively, employing multiple drilled side holes in HC-PCF allows for more rapid gas exchange without disturbing environmental gas distribution [22]. Besides the two possible reasons, strand thickness variation and surface roughness scattering, mentioned by T. Frosch, the energy loss from drilled holes should be considered [23]. Nevertheless, the impact of drilled HC-PBF on energy loss and response time requires testing to establish their relationship. Experiment data can provide a balanced trade-off between response time and signal-to-noise ratio.

A greenhouse gas sensor utilizing an all-fiber configuration based on wavelength modulation spectroscopy (WMS) and frequency-division multiplexing (FDM) has been developed for simultaneous multi-species detection. The diffusion gas cell, made from micron-sized diameter HC-PBF, is arranged in a compact setup with multiple side holes drilled into the cell to enhance gas exchange speed. In the experiment, methane and carbon dioxide, as the most typical greenhouse gases, were chosen as representative examples. By optimizing operational parameters, the sensor's performance, including detection limit, long-term stability, and the relationship among the number of holes, energy loss, and response time, was extensively investigated.

2. Materials and Methods

2.1. Sensor Configuration

A self-developed dual-gas sensor for methane (CH_4) and carbon dioxide (CO_2), based on HC-PBF, has been designed and implemented, with the sensor structure illustrated in Figure 1. The core sensing component is the gas cell formed by HC-PBF and SMFs. The HC-PBF, approximately 0.9 m in length, features an air core of nearly 10 μm, providing an absorption length of over 1.78 m in a reflection structure. Gas exchange occurs through drilled holes in the fiber, facilitating exchange between the interior and exterior gas. At one end of the gas cell, the SMF termination is coated with a reflective surface, reflecting over 95% of the energy back to the input end. The detected signal is then separated by a circulator and directed to the detector.

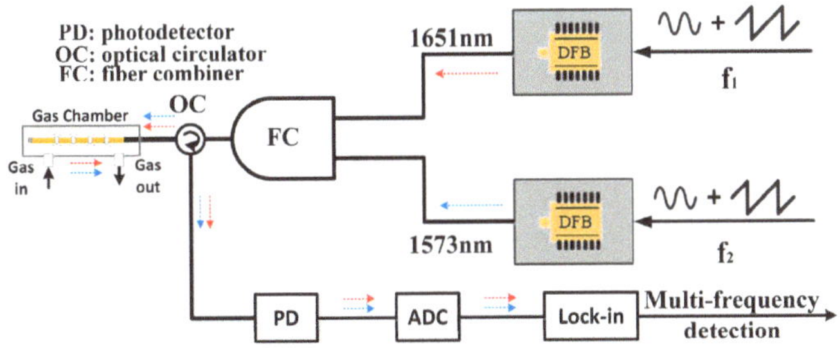

Figure 1. The implemented dual-gas detection sensor designed for in-situ greenhouse applications. PD refers to the photodiode; ADC stands for analog-to-digital converter. The red arrows and blue arrows represent CH_4 detection channel and CO_2 detection channel, respectively.

Two butterfly-packaged 14-pin distributed feedback (DFB) lasers, emitting wavelengths of 1653.7 nm and 1573 nm, are employed to scan absorption lines for methane and carbon dioxide detection, respectively. The gas cell comprises two main parts: an inner HC-PBF-based fiber cell and an outer protection chamber. The original fiber length is one meter. After fusion processing at both ends, the left length is reduced to 0.89 m, which is measured using the time domain reflection method. The laser beam enters the fiber cell through side-drilled holes and undergoes reflection at the opposite side, effectively doubling the absorption length to 1.78 m. The outer chamber serves to mechanically shield and maintain stable working conditions, ensuring consistent pressure and temperature. Analog electrical signals are converged through a sampling and impedance-matching circuit and further processed by a self-developed FPGA circuit. With this all-fiber optical structure, the developed sensor finds application in in-situ greenhouse monitoring, with the potential for optimization towards miniaturization and enhanced portability.

2.2. HC-PBF Fiber Characteristic and Fusion

The HC-PBF utilized in this design, HC-1550-02, is a mature product from NTK. This fiber has an outer diameter of 120 μm, and the central air core measures nearly 10 μm. The bandpass wavelength range is 1490–1680 nm, encompassing the required absorption wavelengths. Within this range, the transmission loss is below 20 dB/km, and the bending radius is approximately 10 cm. The gas cell development involves fusing SMFs with HC-PBF, as depicted in the scanning electron microscope photograph of the fusion joint in Figure 2a. According to finite element analysis (Figure 2b), over 95% of the beam energy is concentrated in the gas-filled hollow-core region, while the remaining energy disperses in the micro-structured cladding region. Energy experiments confirm an energy loss of approximately 0.6–0.8 dB per fusion end face, with fluctuations influenced notably by the process. In the capacity of a gas cell, multiple drilled side holes facilitate gas exchange. These holes have a diameter matching the air core, approximately 10 μm, and are designed as buried holes to mitigate noise and energy loss. The SEM image surface in Figure 2c displays the drilled holes, which measure between 11.2–14.8 μm, slightly larger than the intended size. Experimental energy loss registers between 0.17–0.42 dB with filled nitrogen in fiber, nearly 3.84–9.2% energy loss per hole, indicating a considerable impact of the drilling process on light beam propagation. When the number of drilled holes is increased to seven, the response time is less than 3 min and the energy loss is already higher than 33%.

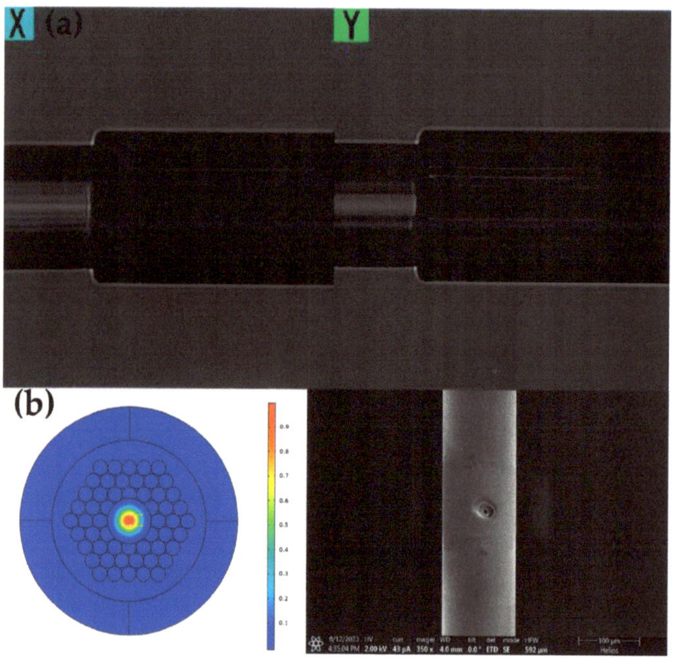

Figure 2. (**a**) SEM image showing the fusion end, (**b**) distribution of the fundamental mode in the HC-PBF, and (**c**) SEM image displaying the surface of the drilled hole.

2.3. Dual-Gas Detection Mechanism

In consideration of the HC-PBF fiber gas cell application, the absorption lines selected for CH_4 and CO_2 should be in the operation wavelength range from 1490 nm to 1680 nm. Therefore, the CH_4 absorption targeted line selected is around 1650.9 nm which is a typical absorption line for CH_4 detection. Within this absorption wavelength range, the common interference gases like CO_2 and H_2O are at least three times weaker than CH_4. When the absorption effect of interference gas is very low, this effect is able to be effectively suppressed by signal processing codes based on FPGA. Similarly, the selected absorption lines of CO_2 are around 1573 nm. The absorption strength ratio between CH_4 and CO_2 is nearly 80 which matches approximately the concentration in the atmosphere. In addition, before pumping into the gas chamber, the gas will undergo drying treatment to ensure that the water vapor concentration is not sufficient to affect the detection results. Upon the above design and analysis, the error caused by the interference gas has no effect on the detection precision of CH_4 and CO_2 in the selected absorption ranges. Detailed information on selected absorption wavelength ranges is shown in Figure 3.

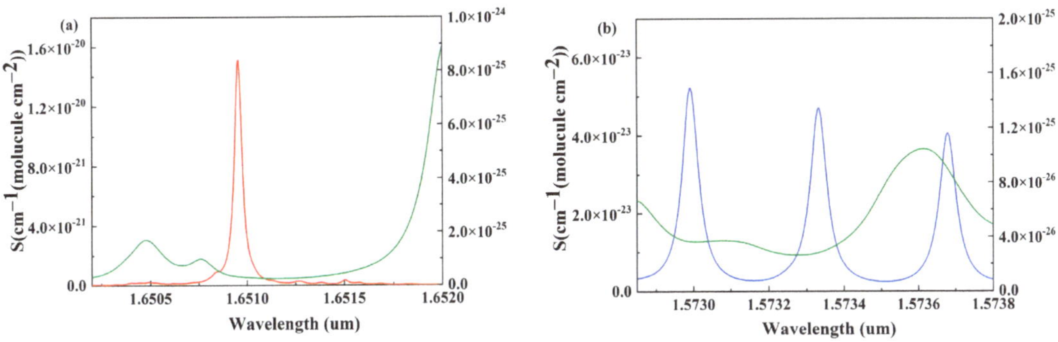

Figure 3. (**a**) Gas absorption spectrum of CH_4 at 1651 nm, the red and green lines represent CH_4 and H_2O absorption lines, respectively; (**b**) gas absorption spectrum of CO_2 at 1573 nm, the blue and green lines represent CO_2 and H_2O absorption lines, respectively.

2.4. Gas Cell Structure Design

The HC-PBF fiber is an exposed fiber with drilled holes. Due to the absence of protective armor, the operational state of the HC-PBF fiber is susceptible to external factors such as vibrations, temperature fluctuations, and airflow. These unpredictable interferences can significantly impact detection results. To mitigate this, a self-designed mechanical protection chamber has been developed and implemented. This chamber serves to shield the HC-PBF fiber from external influences and maintain a stable environmental condition conducive to reliable operation. The mechanical structure of the protection chamber is depicted in Figure 4. The chamber structure is primarily rectangular, with the HC-PBF fiber arranged along the inner wall. Figure 4a shows the design drawing of the mechanical structure, while Figure 4b presents an image of the chamber. The chamber's central region is filled to minimize internal gas volume, allowing for the pump in of external gases through air vents. Considering the sensitivity of the detection system to environmental temperature fluctuations, thermal insulation is applied to the chamber, as illustrated in Figure 4c.

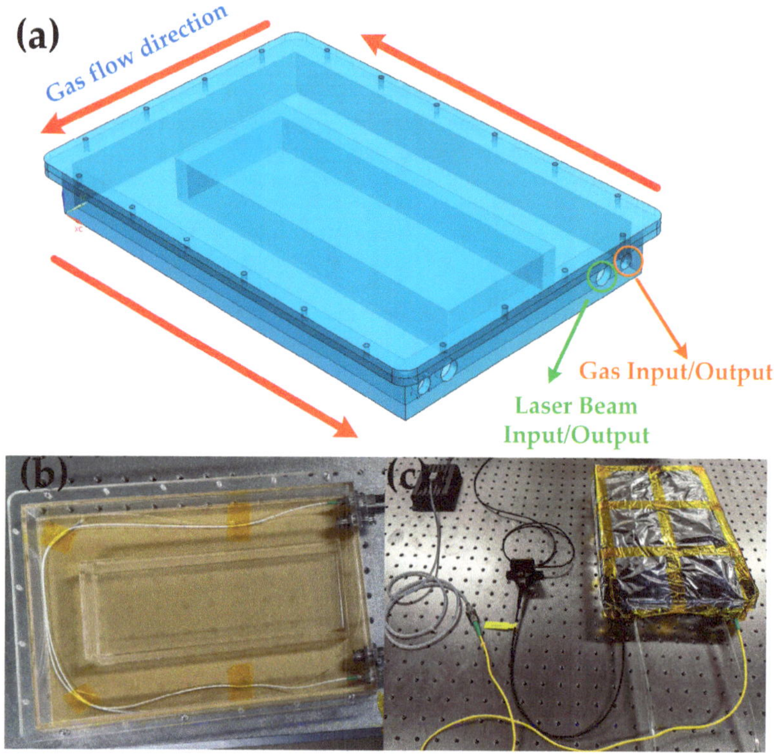

Figure 4. (**a**) Design drawing depicting the mechanical structure of the chamber, (**b**) an image showing the chamber, and (**c**) the gas cell with a thermal package.

2.5. Harmonic Detection Theory

Wavelength Modulation Spectroscopy (WMS) is commonly employed to enhance the signal-to-noise ratio (SNR) in detection. In the process, a low-frequency periodic saw-tooth signal and a high-frequency sine wave are utilized to scan the absorption lines and modulate the laser's injection current, respectively. Post modulation, the output light energy from the laser can be expressed as follows:

$$I(t) = I(t)[1 + u(t) + \text{nsin}(wt)], \tag{1}$$

where $u(t)$ represents the periodic saw-tooth signal, and n is the light intensity modulation parameter. According to the Beer–Lambert law, following the detection of gas absorption, the resulting light energy is

$$I'(t) = I_0[1 + u(t) + \text{nsin}(\omega t)] \exp(-\alpha(t)LC). \tag{2}$$

Therefore, the absorption coefficient can be obtained by using Lorentz linear fitting:

$$\alpha(t) = \frac{\alpha_0}{1 + (\frac{v_t - v_c}{v_{FWHM}})}, \tag{3}$$

where v_t is the wavelength of the emitted beam with modulation, v_c is the central wavelength of the absorption peak, a_0 is the absorption parameter at the absorption peak, and

υ_{FWHM} represents the full width at half maximum. Under typical conditions, the detected gas concentration, which is low, can be expressed as a function of

$$\alpha[\lambda(t)]LC << 1, \quad (4)$$

when the driven output laser wavelength is matched with the absorption peak, the output light signal can be sampled as

$$I'(t) = I_0[1 + \mathrm{n}\sin(wt) - \frac{\alpha_0 LC}{1 + m^2\sin^2(\omega t)}]. \quad (5)$$

Following the Fourier series expansion of the aforementioned function, various harmonics can be extracted, including the first harmonic (1*f*) and second harmonic (2*f*). In stable conditions, the output laser energy remains stable, allowing for the establishment of a relationship between the amplitude of the 2*f* signal and gas concentration. Considering environmental influences on the initial laser intensity, the ratio of amplitudes between 2*f* and 1*f* can be employed to mitigate unpredictable variations caused by changes in the initial intensity of the laser. This ratio function can be expressed as

$$\frac{A_{2f}}{A_{1f}} = \frac{-k\alpha_0 LC}{n}, \quad (6)$$

where k is the modulation parameter. This design establishes the relationship between $\frac{A_{2f}}{A_{1f}}$ and gas concentration.

In consideration of the previous research, the modulation depth in wavelength modulation should be nearly 2–2.5 times the half width at half maximum of the gas absorption lines. In this case, both the difficulty of hardware utilization of FPGA and the frequency domain distribution of environmental noise are considered. With the signal and noise test, to achieve a satisfying signal and noise ratio, the optimized modulation coefficient is set to 1.9 and 2.1, corresponding to CH_4 and CO_2, respectively.

3. Results

3.1. Waveform Measurements

Compared to previous reports, our research aims to develop an embedded detection system applied for on-situ applications rather than a testing prototype limited to laboratory applications. Therefore, in contrast to employing commercial products, our system is almost entirely self-developed and an implemented detection system, including signal sampling, actuation, and processing. The above-mentioned functions are realized in an embedded system based on an FPGA chip XC7A200T (Xilinx, San Jose, CA, USA), which is a Xilinx Artix-7 type chip. Compared with the detection system using PC, the challenge is how to use the limited hardware logic resources to achieve the expected detection performance. The system clock of this chip is set to 30 MHz with a maximum 120 KHz sample frequency. As a parallel process chip, the synchronization between laser driving and electrical signal demodulation is well ensured. As a price of parallel process, the greatest consumption of embedded logic resources results from iterative operations using multipliers and dividers. In the implementation of all functions, core iterative operations take place in the high-order low-pass finite impulse response (FIR) filter of the digital lock-in amplifier, with its order being directly proportional to the filtering performance. In consideration of the logic source utilization, a cascade integrator comb (CIC) filter and a lower order FIR are used to replace a high order FIR. After trial and error, the CIC order parameter is set to eight with an analog signal sampling of 20 KHz. The order of FIR is set to 246, and the 2*f* signal is extracted every 0.1 s. With a cascade filter, 32,600 logic elements and 110 embedded 18*18 multipliers were set as a standard resource cost to realize the lock-in amplification function.

To evaluate the sensor's performance, sample gas mixtures of known concentrations are generated by diluting 500 ppm CH_4, 2000 ppm CO_2, and pure N_2 using a commercial gas dilution instrument. With an absorption length of 1.78 m for the HC-PBF, the waveform is saved and observed using a self-developed CPU based on an FPGA chip. The observed original signal is depicted in Figure 5a. The processed 1*f* and 2*f* signals are displayed in Figures 5b and 5c, respectively.

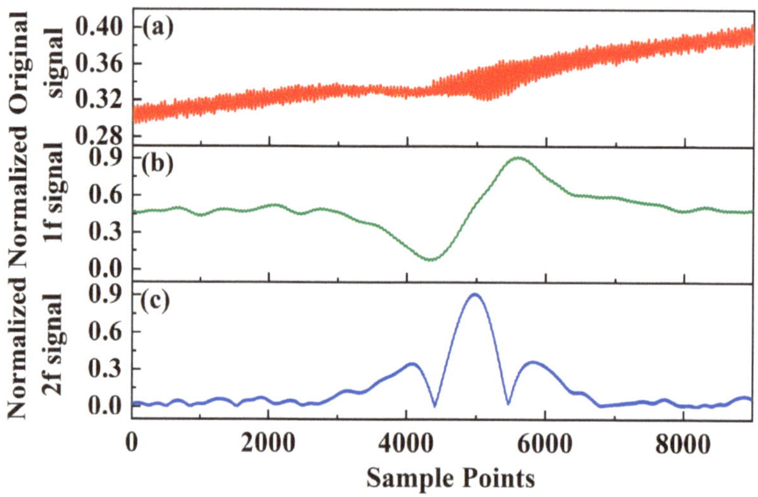

Figure 5. Normalized 2*f* amplitude plotted against modulation depth for optimization purposes. (**a**) the original detecting signal; (**b**) the demodulated 1*f* signal; (**c**) the demodulated 2*f* signal.

As a portable sensor, the power consumption of the CPU chip is a significant parameter. When the temperature of the laser is stable, the most power consumption comes from the embedded processor. During this experiment, power consumption was continuously monitored, and the average power consumption was about 1.71 W, including power consumption from the laser, DC-DC chip, FPGA, and other hardware components. It should be noted that power consumption in practical application environments should be re-evaluated because of the unpredictable changes in ambient temperature and variations in the detection period. In this case, the power consumption under the rice field application is lower than 2.12 W. A significant portion of the power consumption is attributed to laser temperature control.

3.2. 2f Signal Fitting

The signal-to-noise ratio of the demodulated 2*f* signal, without subsequent algorithm optimization, directly determines the system's detection performance. The standard deviation, approximately 0.327 mV (1σ), was determined within the non-absorption range. There is an inevitable fluctuation in achieving the peak value of the 2*f* signal. To obtain an accurately related 2*f* amplitude signal, the 2*f* signal is fitted before being recorded.

For comparing hardware resource utilization, the waveform of the 2*f* signal is separately fitted using Gaussian and sine wave fitting techniques. With the same hardware resources, the R-squared values for Gaussian fitting and sine wave fitting are 99.21% and 98.79%, respectively. Figure 6 shows the original demodulated 2*f* signal and the filtered 2*f* signal based on sine wave fitting. The fluctuation around the 2*f* peak signal is effectively suppressed.

3.3. Natural Diffusion Process of HC-PCF Cell

A specialized experiment was conducted to assess the optimal values for the number of holes and response time. With an increase in the drilled hole count, the corresponding energy loss and response time were recorded. During the experimental process, the protective chamber was purged with pure N_2 to prevent operational errors stemming from external pressure differences. Dynamic gas distribution was employed instead of static injection distribution for this purpose. In the response experiment, methane was used as the sample, and the amplitude of the 2*f* signal was recorded every second. The results for the response time are depicted in Figure 7a. When the number of drilled holes was increased to seven, the response time decreased to less than 3 min, meeting the requirements for many applications such as precision agriculture and industrial pollution gas detection. Specifically, the decrease in response time exhibited an almost exponential relationship with the number of drilled holes, as illustrated in Figure 7b. Overall, a noticeable reduction in response time resulting from an increased number of drilled holes aligns with some previous findings (e.g., ref. [19]). Discrepancies between our experimental results and reference papers may arise from factors like the total length of the microchannel, the drilled hole placement, and machining errors in their creation. Moreover, for the sake of faster response, each drilled hole consumes approximately 0.2–0.3 dB, implying a trade-off between energy loss and response speed.

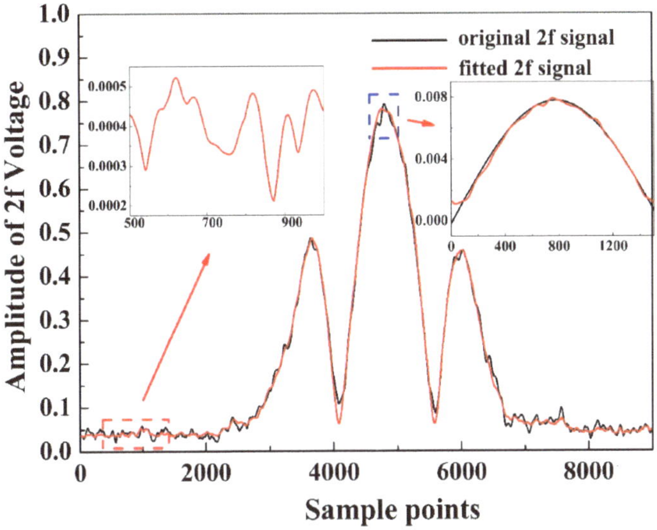

Figure 6. Normalized 2*f* signal flitting and standard deviation.

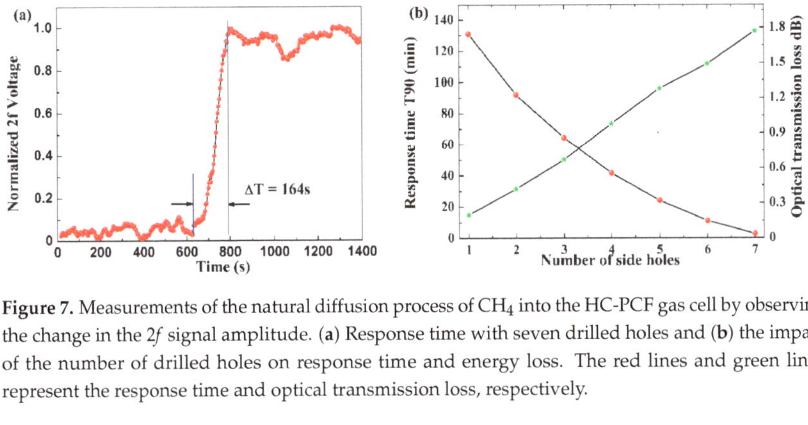

Figure 7. Measurements of the natural diffusion process of CH_4 into the HC-PCF gas cell by observing the change in the 2*f* signal amplitude. (**a**) Response time with seven drilled holes and (**b**) the impact of the number of drilled holes on response time and energy loss. The red lines and green lines represent the response time and optical transmission loss, respectively.

3.4. Calibration and Data-Fitting

Adjusting to match the gas concentration in the greenhouse, dual-gas calibrations are applied across various concentration ranges. Figure 8a and Figure 8b depicts the representative 2*f* signal amplitudes of the two species at different gas concentrations while maintaining a constant gas pressure, respectively. The data-fitting results demonstrate a linear relationship between 2*f* amplitude and concentration, with R-squared values of 0.996 and 0.993, respectively. The calibrated linear fitting results of CH_4 and CO_2 are shown as Equation (7) and Equation (8), respectively. The parameter C is the detecting gas concentration, and ΔU is the ratio of 2*f* signal and 1*f* signal.

$$C = 723.94 \times \Delta U - 1.121. \quad (7)$$

$$C = 1214.21 \times \Delta U - 17.62. \quad (8)$$

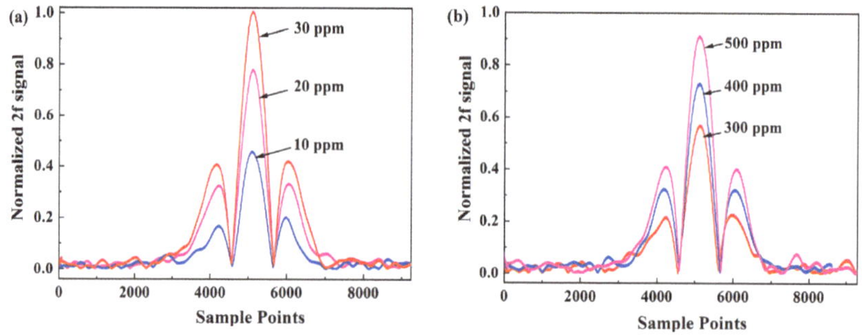

Figure 8. Dot plots for the extracted processed signal with different concentrations of (**a**) CH_4 and (**b**) CO_2.

3.5. Detection Limit

The Allan deviation was employed to determine the minimum detection limit of this system. With an optimal averaging time, the detection system can meet various application requirements across different application backgrounds. In the experiment, pure dry N_2 was used to flush the gas cell for 180 s before injecting the sample gas. The analysis is based on the Allan deviation. For CH_4 detection, as depicted in Figure 9a, the 1σ detection limit is approximately 1.12 ppmv for a 1s averaging time. Increasing the averaging time to 50 s improves the detection limit to 0.1 ppm, meeting the detection limit requirement under atmospheric conditions. Due to the weak absorption line, the detection limit for CO_2 is nearly 51.9 ppm for a 1s averaging time, as shown in Figure 9b. Extending the averaging time to 50 s enhances the detection limit to 2 ppm.

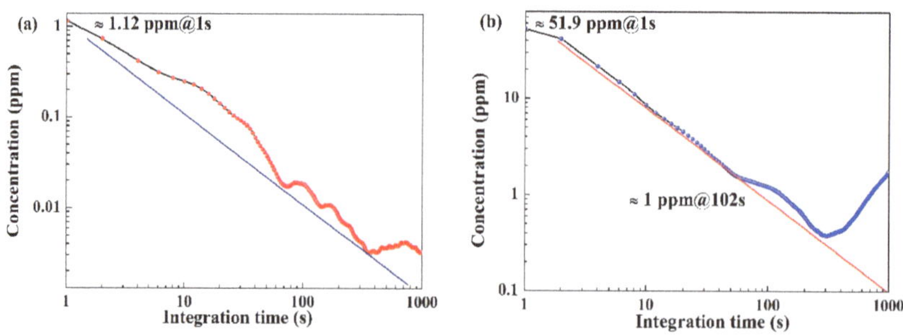

Figure 9. Allan deviation analysis of the sensor system based on one-hour measurements in a pure N_2 atmosphere concentration of (**a**) CH_4 and (**b**) CO_2.

3.6. Field Application

An on-situ test was conducted in a rice field located in Jilin Province, a city in northern China, a typical source and sink for greenhouse gases. Before the on-situ test, a comparative test with a G2401 was carried out to ensure the varying concentrations of water vapor would not influence our test results. Since the experiment took place in October, the selected rice plants were at a mature stage, reaching a height of over 1.2 m. The sampling location was near the roots of the rice field to monitor the greenhouse effect within the group of rice plants. As illustrated in Figure 10, the measured CH_4 concentration was slightly higher than the average atmospheric concentration, which was marginally lower than the anticipated concentration. The most probable reason is the decreased activity of methanogens in the mature-stage rice plants. Meanwhile, the cultivation method of mature rice is not a flooded mode, which significantly suppresses the generation of CH_4 from rice fields. In contrast, the concentration of CO_2 primarily aligned with photosynthesis, indicating satisfactory detection performance. Moreover, the rice field was an outdoor open structure, and the detected gas concentrations were influenced by environmental factors such as wind speed and human activities, contributing to the fluctuations and ripples in the detection results.

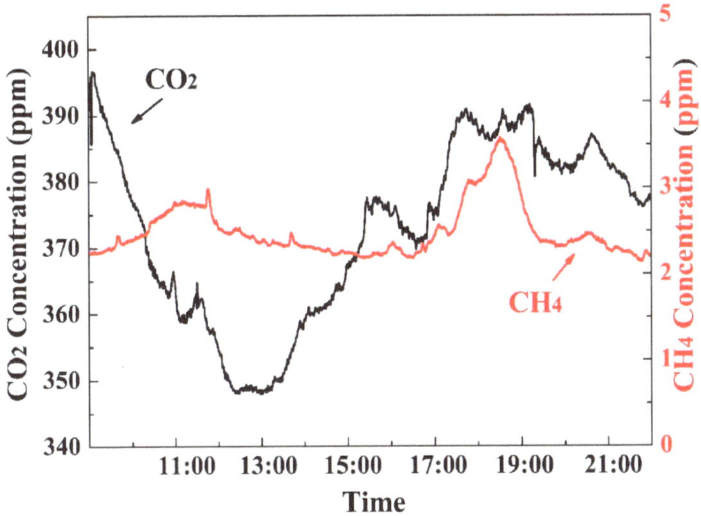

Figure 10. Continuous monitoring of CH_4 concentration and CO_2 concentration in a greenhouse on November 2023 in Jilin.

Compared to the indoor laboratory environments with nearly constant temperatures without wind, the on-situ application has a more stringent external environmental condition. The most significant influencing factor is the time consumed for temperature stabilization. When the sensor starts up, the time consumed is nearly 90 min, which is 30 times longer than the time consumed in the laboratory. Fortunately, the temperature change in ambient conditions is slow related to the sensor. After the warm-up process, the developed sensor will no longer require any adjustments.

4. Discussion

Compared to the other previously reported sensors using a hollow-core photonic crystal fiber, our research has a better performance in gas cell structure design and response time improvement. The single-end reflection structure doubles the absorption length without increasing the sensor volume. The self-developed mechanical protection structure ensures the applicability of the gas cell under in-situ environments, which is verified through testing in rice fields. What is more, the experiment data concerning drilled hole numbers, energy loss, and response time provides a reference for the trade-offs in further design.

5. Conclusions

The design and implementation of an in-situ trace gas sensor within an all-fiber structure were presented for the simultaneous detection of multiple greenhouse gas species. Gas experiments were conducted using a gas cell to determine the response time for gas diffusion within the core of an HC-PBF. The experimental results quantified the relationship between the number of drilled holes and response time. However, caution is required in increasing the number of drilled holes due to the associated increase in energy loss. Utilizing the self-developed gas cell, the detection limits for CH_4 and CO_2 were determined to be 1.12 ppm and 51.9 ppm, respectively, at 1 s. Performance was notably enhanced to 0.1 ppm and 2 ppm with an average time of 50 s, meeting atmospheric application requirements. An application test in a rice field confirmed the applicability of outdoor use. Future work will focus on improving stability by mitigating the effects of working conditions on the HC-PBF fiber cell. This includes controlling and limiting internal pressure and temperature within a narrower fluctuation range.

Author Contributions: Conceptualization, J.W.; Funding acquisition, G.L.; Methodology, J.W.; Project administration, J.W.; Software, B.L.; Supervision, G.L.; Writing—original draft, W.W.; Writing—review & editing, G.L. All authors have read and agreed to the published version of the manuscript.

Funding: National Natural Science Foundation of China (NO. 62005268), Key Research and Development Program of Jilin Province (NO. 20220203195SF), Youth Innovation Promotion Association CAS no. 2023229, and Open bidding for selecting the best candidates of ChangChun City (23JG06).

Institutional Review Board Statement: Not applicable.

Informed Consent Statement: Not applicable.

Data Availability Statement: Data are contained within the article.

Conflicts of Interest: The authors declare no conflict of interest.

References

1. World Meteorological Organization. *The State of Greenhouse Gases in the Atmosphere Based on Global Observation through 2019*; WMO: Geneva, Switzerland, 2019.
2. Wang, J.N.; Li, B.; Lin, G.Y.; Ma, Q.J.; Wang, S.R.; Piao, M.X. Near-infrared methane sensor based on a+ distributed feedback laser. *Spectrosc. Lett.* **2019**, *52*, 113–120. [CrossRef]
3. Knight, J.; Birks, T.; Russell, P.S.J.; Atkin, D. All-silica single-mode optical fiber with photonic crystal cladding. *Opt. Lett.* **1996**, *21*, 1547–1549. [CrossRef] [PubMed]
4. Bykov, D.; Schmidt, O.; Euser, T.; Russell, P.S.J. Flying particle sensors in hollow-core photonic crystal fibre. *Nat. Photonics* **2015**, *9*, 461–465. [CrossRef]
5. Jin, W.; Ho, H.; Cao, Y.; Ju, J.; Qi, L. Gas detection with micro-and nano-engineered optical fibers. *Opt. Fiber Technol.* **2013**, *19*, 741–759. [CrossRef]
6. Pang, M.; He, W.; Jiang, X.; Russell, P.S.J. All-optical bit storage in a fibre laser by optomechanically bound states of solitons. *Nat. Photonics* **2016**, *10*, 454. [CrossRef]
7. Amanzadeh, M.; Aminossadati, S.M. A new approach for drilling lateral microchannels in photonic crystal fibres. In Proceedings of the IEEE Photonics Conference (IPC), Waikoloa, HI, USA, 2–6 October 2016; pp. 779–780.
8. Yan, D.; Popp, J.; Pletz, M.W.; Frosch, T. Highly sensitive broadband Raman Sensing of Antibiotics in step-index hollo-core photonic crystal fibers. *ACS Photonics* **2017**, *4*, 138–145. [CrossRef]
9. Yan, D.; Popp, J.; Pletz, M.W.; Frosch, T. Fiber enhanced Raman sensing of levofloxacin by PCF bandgap-shifting into the visible range. *Anal. Methods* **2018**, *10*, 586–592. [CrossRef]
10. Yu, R.; Chen, Y.X.; Shui, L.L.; Xiao, L.M. Hollow-core photonic crystal fiber gas sensing. *Sensors* **2020**, *20*, 2996. [CrossRef]
11. Russell, P. Photonic crystal fibers. *Science* **2003**, *299*, 358–362. [CrossRef]
12. Yang, F.; Jin, W.; Cao, Y.; Ho, H.L.; Wang, Y. Towards high sensitivity gas detection with hollow-core photonic bandgap fibers. *Opt. Express* **2014**, *22*, 24894–24907. [CrossRef]
13. Lin, Y.; Liu, F.; He, X.; Jin, W.; Zhang, M.; Yang, F.; Ho, H.L.; Tan, Y.; Gu, L. Distributed gas sensing with optical fiber photothermal interferometry. *Opt. Express* **2017**, *25*, 31568–31585. [CrossRef]

14. Hu, H.-F.; Zhao, Y.; Zhang, Y.-N.; Yang, Y. Characterization of infrared gas sensors employing hollow-core photonic crystal fibers. *Instrum. Sci. Technol.* **2016**, *44*, 495–503. [CrossRef]
15. Cubillas, A.M.; Lazaro, J.M.; Silva-Lopez, M.; Conde, O.M.; Petrovich, M.N.; Lopez-Higuera, J.M. Methane sensing at 1300 nm band with hollow-core photonic bandgap fibre as gas cell. *Electron. Lett.* **2008**, *44*, 403–405. [CrossRef]
16. Cubillas, A.; Lazaro, J.; Conde, O.; Petrovich, M.; Lopez-Higuera, J. Gas sensor based on photonic crystal fibres in the 2v3 and v2+2v3 vibrational bands of methane. *Sensors* **2009**, *9*, 6261. [CrossRef]
17. Hu, L.E.; Zheng, C.T.; Yao, D.; Yu, D.; Liu, Z.; Zheng, J.; Wang, Y.; Tittel, F.K. A hollow-core photonic band-gap fiber based methane sensor system capable of reduced mode interference noise. *Infrared Phys. Tehchnol.* **2019**, *97*, 101–107. [CrossRef]
18. Kassani, S.H.; Park, J.; Jung, Y.; Kobelke, J.; Oh, K. Fast response in-line gas sensor using C-type fiber and Ge-doped ring defect photonic crystal fiber. *Opt. Express* **2013**, *21*, 14074–14083. [CrossRef] [PubMed]
19. Qixin, H.; Peipei, D.; Zhiwei, L.; Chuantao, Z.; Yiding, W. TDLAS-WMS based near-infrared methane sensor system using hollow-core photonic crystal fiber as gas-chamber. *Opt. Quant. Electron.* **2017**, *48*, 115.
20. Valiunas, J.K.; Tenuta, M.; Das, G. A gas cell based on hollow-core photonic crystal fiber (PCF) and its application for the detection of greenhouse gas (GHG): Nitrous oxide (N_2O). *J. Sens.* **2016**, *2016*, 7678315. [CrossRef]
21. Hoo, Y.L.; Liu, S.J.; Ho, H.L.; Jin, W. Fast Response microstructured optical fiber methane sensor with multiple side-openings. *IEEE Photonics Technol. Lett.* **2010**, *22*, 296–298. [CrossRef]
22. Hartung, A.; Kobelke, J.; Schwuchow, A.; Wondraczek, K.; Bierlich, J.; Popp, J.; Frosch, T.; Schmidt, M.A. Origins of modal loss of antiresonant hollow-core optical fibers in the ultraviolet. *Opt. Express* **2015**, *23*, 2557–2565. [CrossRef] [PubMed]
23. Kobelke, J.; Bierlich, J.; Schwuchow, A.; Lorenz, A.; Hartung, A. OH diffusion effects at preparation of antiresonant hollow core fibers. *Micro-Struct. Spec. Opt. Fibres* **2019**, *11029*, 1102904.

sensors

MDPI

Article

Modelling and Thermographic Measurements of LED Optical Power

Maria Strąkowska [1,*], Sebastian Urbaś [1], Mariusz Felczak [1], Błażej Torzyk [1], Iyad S. M. Shatarah [1], Rafał Kasikowski [1], Przemysław Tabaka [2] and Bogusław Więcek [1]

[1] Institute of Electronics, Lodz University of Technology, 90-924 Lodz, Poland; sebastian.urbas@dokt.p.lodz.pl (S.U.); mariusz.felczak@p.lodz.pl (M.F.); iyad.shatarah@p.lodz.pl (I.S.M.S.); rafal.kasikowski@p.lodz.pl (R.K.); boguslaw.wiecek@p.lodz.pl (B.W.)

[2] Institute of Electrical Power Engineering, Lodz University of Technology, 90-924 Lodz, Poland; przemyslaw.tabaka@p.lodz.pl

* Correspondence: maria.strakowska@p.lodz.pl

Abstract: This paper presents a simple engineering method for evaluating the optical power emitted by light-emitting diodes (LEDs) using infrared thermography. The method is based on the simultaneous measurement of the electrical power and temperature of an LED and a heat source (resistor) that are enclosed in the same plastic packaging under the same cooling conditions. This ensures the calculation of the optical power emitted by the LED regardless of the value of the heat transfer coefficient. The obtained result was confirmed by comparing it with the standard direct measurement method using an integrated sphere. The values of the estimated optical power using the proposed method and the integrated sphere equipped with a spectrometer were consistent with each other. The tested LED exhibited a high optical energy efficiency, reaching approximately $\eta \approx 30\%$. In addition, an uncertainty analysis of the obtained results was performed. Compact modelling based on a thermal resistor network (R_{th}) and a 3D-FEM analysis were performed to confirm the experimental results.

Keywords: IR thermography; LED; optical power; thermal modelling

1. Introduction

The optical power of light sources is usually measured with Ulbricht integrating spheres [1], which are equipped with calibrated and certified spectrometers for precise spectral measurements. This means that optical power measurements can only be performed in laboratories equipped with advanced measurement systems. In the general case, this ensures the high accuracy of optical power measurements. There is another approach to such measurements using goniometers, which is more difficult to apply in practice. It requires a dark room, and the entire measurement is performed step by step for each emission direction. It is characterized by a long measurement time and is obviously not suitable for use in the case of diodes operating in a lighting system in the field.

In many practical cases, there is a need to measure the optical power of LEDs using simple technical methods. Typically, LEDs emit light in one or more narrow bands. In addition, due to production tolerances, manufacturers indicate the light intensity or luminous flux with a large scatter, which causes difficulties in selecting the appropriate element for new solutions where an exact value of radiant power is required.

Nowadays, highly efficient LEDs are available, in which a significant part of the electrical power is converted into radiation [2–5]. On the other hand, the high optical power emitted involves a large amount of thermal power dissipated, which increases the temperature of the device. Knowing the radiant power helps to develop effective heat management for LED light sources. The radiation characteristics of LEDs are expressed in terms of either photometric or radiometric parameters. The photometric parameters of light sources are calculated and measured in relation to green color radiation at a wavelength of λ = 555 nm. Typically, manufacturers provide the photometric parameters, while for energy analyses, radiometric data are needed [6].

Measuring the emitted infrared radiation and the electrical power consumption allows one to obtain the optical power and efficiency of an LED.

Thermography, as a noninvasive, cost-effective, and fast method, could be applied in many fields of engineering practice, including electronics [7,8]. In electronics, it is often used for estimating the heat loss of components and entire systems. In many cases, it can replace much more expensive and sophisticated measurement equipment. The Joule heating of an LED can be compared with the Joule heating of a resistor, both being encapsulated in identical housings. Measuring the emitted infrared radiation and the electrical power consumption allows one to obtain the optical power and efficiency of an LED. The results of the thermographic method and the radiometric measurement of the luminous flux using an integrating sphere and calibrated spectrometer are presented in this paper. This paper considers both thermal and radiation problems and hence both research domains are taken into account. Heat transfer analysis can support the investigations of optical phenomena. In the literature, there are numerous thermal analyses of LED structures based on different modelling approaches, such as 3D-FEM (finite element method), analytical conjugate analyses, and compact simplified models [4,9–13]. In this research, a heat transfer analysis using compact 3D-FEM modelling was performed to validate the obtained experimental results.

2. Materials and Methods

2.1. Thermal Modelling

High-efficiency LEDs were sealed in a variety of plastic packages. An example cross-section of a 5 × 5 mm^2 package used in this research is shown in Figure 1. It is an RGB module with semiconductor diodes placed in a row in the middle of the substrate. Metal connectors/electrodes play a very important role in heat dissipation. Above the semiconductors is a semi-transparent diffusive material that acts as an optical lens and provides an emission angle of 110°. The rated current is 20 mA. For the red diode, the forward voltage is about 2.0 V and the typical luminous intensity is 1090 mcd. The central wavelength is 620 nm and the full width at half maximum (FWHM) wavelength is about 20 nm.

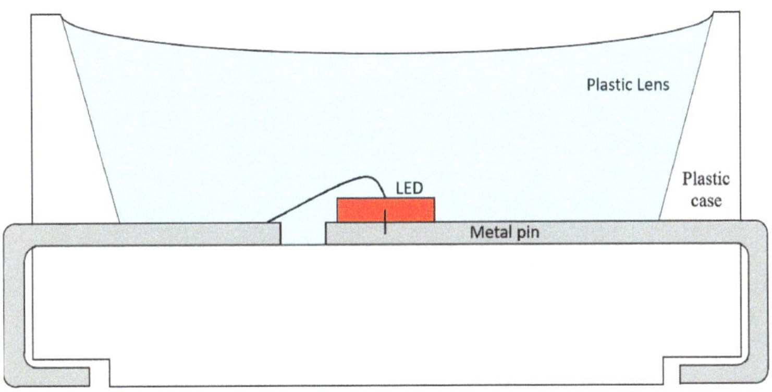

Figure 1. Cross-section of a high-efficiency LED.

In this research, an R-LED was considered. This light source was attached to the substrate closer to one side of the housing. This research began with heat transfer modelling to estimate the temperature rise above the ambient temperature. Two models were developed: a simple, compact one for rapidly estimating the steady-state LED temperature using Matlab R2023b software, and a 3D model implemented in the ANSYS R19.1 environment using the FEM to solve heat transfer differential equations. For the external walls of the model, convective boundary conditions were set. The model assumptions were comparable with R_{th} network compact thermal modelling. In the compact R_{th} network model, the power cables were modelled as infinite wires, while in the FEM, the boundary conditions were modified to consider heat dissipation by the powering connections. In this way, heat transfer through the wires was simulated by increasing the values of the heat transfer coefficients. The simulation was carried out for power of $P = 27.72$ mW dissipated in the diode, corresponding to the thermal power obtained in measurements.

2.1.1. R_{th} Network—Compact Thermal LED Model

A simple thermal compact LED model consisting of a network of thermal resistances R_{th} is shown in Figure 2. The heat source is inside the plastic housing. A part of the generated thermal power P_{TD} is dissipated to the environment in all directions: vertically to the top and bottom surfaces, represented by the thermal resistance R_{1e}, and horizontally to all four side surfaces, represented by R_{2e1}, R_{2e2}, and R_{3e}. A significant part of the dissipated power is transferred through the electrical connectors soldered to the LED pins (R_{pin}). Finally, all power is transferred to the environment by free convection, modelled using the heat transfer coefficients represented by thermal resistances R_{h1}, R_{h2}, R_{h3}, and R_{h4}.

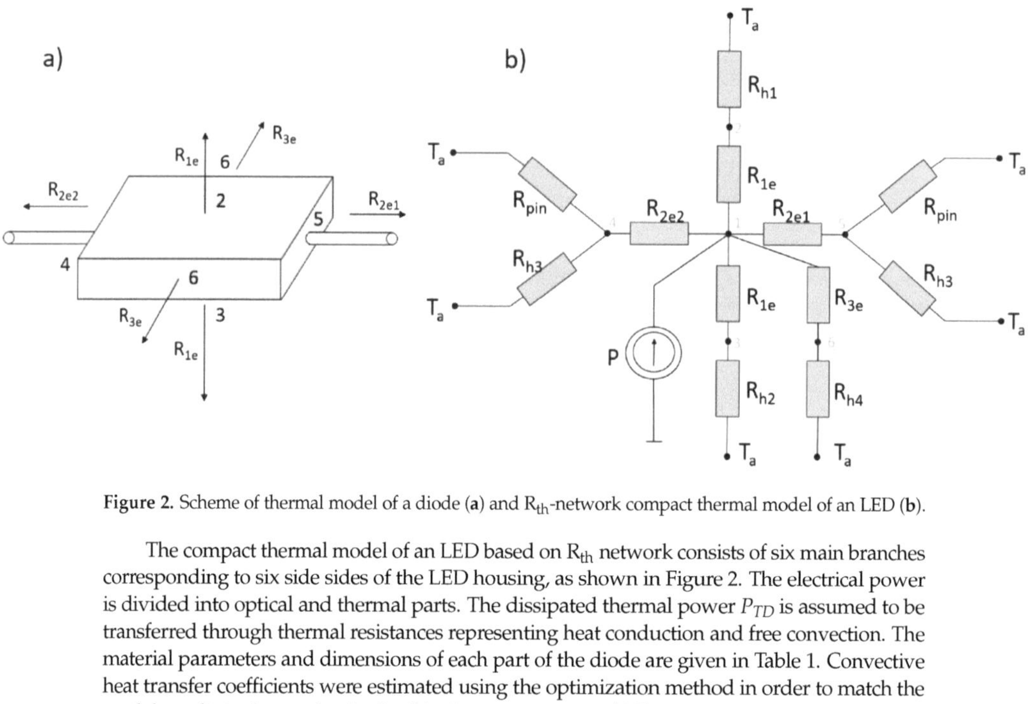

Figure 2. Scheme of thermal model of a diode (**a**) and R_{th}-network compact thermal model of an LED (**b**).

The compact thermal model of an LED based on R_{th} network consists of six main branches corresponding to six side sides of the LED housing, as shown in Figure 2. The electrical power is divided into optical and thermal parts. The dissipated thermal power P_{TD} is assumed to be transferred through thermal resistances representing heat conduction and free convection. The material parameters and dimensions of each part of the diode are given in Table 1. Convective heat transfer coefficients were estimated using the optimization method in order to match the model results to the results obtained in the measurements [4,8].

The resistance R_{1e} associated with heat conduction to the top and bottom surfaces of the diode takes the form (1):

$$R_{1e} = \frac{d_1}{k_{epoxy} S_{1e}} \tag{1}$$

The resistance R_{h1} corresponding to convection cooling is expressed as (2):

$$R_{h1} = \frac{1}{h_1 S_d} \tag{2}$$

Table 1. Parameters of the compact thermal model of the LED.

Parameter	Value	Description
k_{epoxy}	2 W/(m·K)	Thermal conductivity of epoxy resin
k_{Cu}	300 W/(m·K)	Thermal conductivity of metal contacts containing mainly copper (wire)
S_d	$4.92 \times 4.92 \times 10{-}^{6}$ m^2	Diode top and bottom surface
S_{1e}	$3 \times 3 \times 10^{-6}$ m^2	Epoxy resin replaced cross-section surface
S_{2e}	$4.92 \times 2.7 \times 10^{-6}$ m^2	Left and right diode surface
S_{3e}	$4.92 \times 3 \times 10^{-6}$ m^2	Side surfaces of the diode
r	300×10^{-6} m	Connecting wire radius
d_1	1.5×10^{-3} m	Half the thickness of the epoxy layer
d_2	2×10^{-3} m	The length of the inner connector to the outer conductor
d_{Cu}	2.5×10^{-3} m	The length of the copper pad on which the diode is placed
h_1	10.61 W/m^{2}K	Heat transfer coefficient for the top surface
h_2	5.96 W/m^{2}K	Heat transfer coefficient for the bottom surface
h_3	13.33 W/m^{2}K	Heat transfer coefficient for the side's surfaces
h	12.09 W/m^{2}K	Heat transfer coefficient for the connecting wire

Temperature measurements with the IR camera take place on the upper surface of the diode (node 2). On both sides (left—node 4, right—node 5), there are connecting wires that transfer a large part of the heat to the environment—R_{pin}. The value of R_{pin} is calculated assuming an infinite length of the wire in Figure 3.

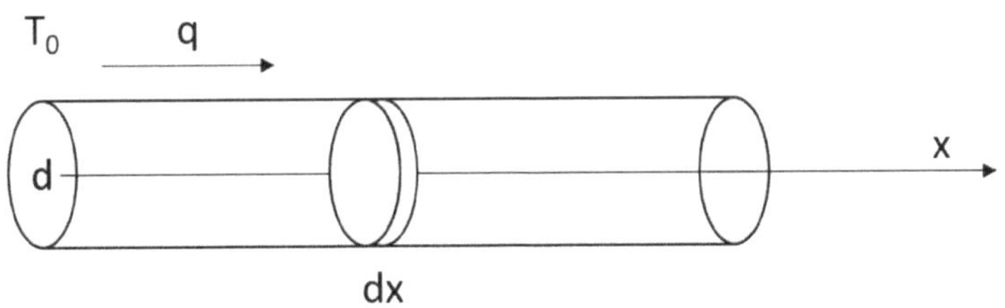

Figure 3. Connection wire model.

The one-dimensional thermal model of the wire is represented by Equations (3)–(5).

$$-dq * \pi r^2 = 2\pi r * dx * hT \tag{3}$$

$$q = -k\frac{dT}{dx} \tag{4}$$

$$\frac{d^2T}{dx^2} - \frac{2h}{rk}T = 0 \tag{5}$$

The temperature T in all equations is the temperature difference between the thermal object and the environment. In other words, we assumed that the ambient temperature is equal to 0. In practice, this means that in order to find the real object temperature when modelling, the actual ambient temperature must be added to the temperature obtained by the model.

The general solution of (5) takes the form:

$$T = T_0 e^{-\frac{x}{L}} \tag{6}$$

where the diffusion length L is expressed as:

$$L = \sqrt{\frac{rk}{2h}} \tag{7}$$

The boundary condition for $x = 0$ allows us to obtain a special solution of (5).

$$\left.\frac{dT}{dx}\right|_{x=0} = -\frac{T_0}{L} \tag{8}$$

$$q = -\frac{kT_0}{L} = -\frac{P}{\pi r^2} \tag{9}$$

Finally, the thermal resistance of the wire R_{pin} can be presented as (10).

$$R_{pin} = \frac{T_0}{P} = \frac{L}{k\pi r^2} = \frac{1}{\pi r}\sqrt{\frac{1}{2hkr}} \tag{10}$$

On the left side of the diode, there is a thin wire leading to the diode's left electrical contact—Figure 1. The thermal resistance here, R_{2e2}, is slightly different to the resistance on the right side, R_{2e1}, where the diode lies directly on the thermal pad. The equations of thermal resistances for these branches are given by (11)–(13):

$$R_{2e2} = \frac{d_2}{k_{Cu} * S_{2e}} \tag{11}$$

$$R_{2e1} = \frac{d_{Cu}}{k_{Cu} * S_{2e}} \tag{12}$$

$$R_{h3} = \frac{1}{h_3 * S_{3e}} \tag{13}$$

The two other branches (back and forward) are the same, so they are analyzed together as node 6. The corresponding resistances are given by (14) and (15).

$$R_{3e} = \frac{d_2}{k_{epoxy} * 2 * S_{2e}} \tag{14}$$

$$R_{h4} = \frac{1}{h_3 \cdot 2 \cdot S_{2e}} \tag{15}$$

2.1.2. Three-Dimensional FEM Model

A 3D FEM model was developed in the ANSYS environment (version R19.1). For geometry preparation, Space Claim software was used (part of the ANSYS R19.1 software package). The dimensions and material properties correspond to the R_{th} network compact thermal model as well as the real diode and package sizes.

In this case, steady-state simulation was performed. For this purpose, Workbench software (part of the ANSYS R19.1 software package) was used. The simulation type was Steady-State Thermal.

2.2. Thermographic Method of LED Optical Power Evaluation

The proposed method consists of the simultaneous measurement of the temperature of the tested LED, T_D, the heating element (resistor—R), T_R, and the power cables on both sides of the elements: T_{1D}, T_{2D}, T_{1R}, and T_{2R}. All measurements were performed using a thermal imaging camera, as shown in Figure 4. The emissivity values of the measured elements were estimated at the value $\varepsilon = 0.92$.

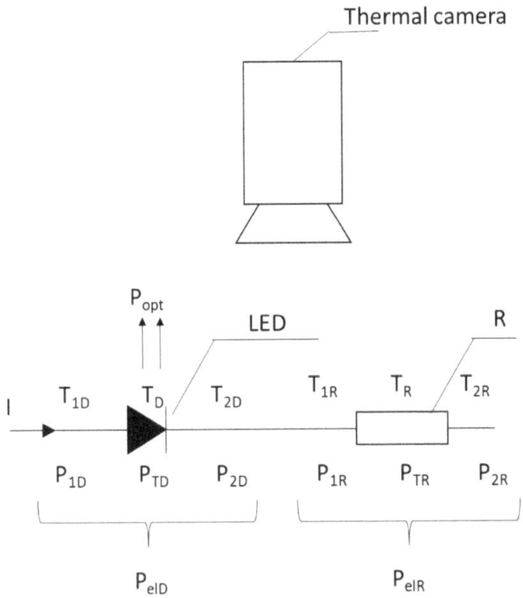

Figure 4. The concept of thermovision measurements of the optical power of LEDs.

The flowchart illustrating the method and experiment is shown in Figure 5.

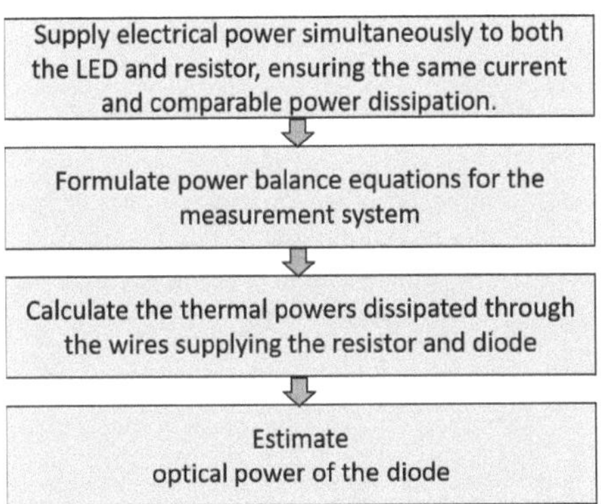

Figure 5. Block diagram of the proposed measurement method for LED optical power.

Electrical power, P_{elD} and P_{elR}, is supplied to both elements connected in series simultaneously. This allows us to obtain the same current and comparable power dissipated in them. The key issue of the proposed method is to ensure equal convection cooling conditions of both elements: the LED diode and resistor R. This can be achieved by using the same shape and geometric dimensions for both elements. This will ensure equal conditions for heat dissipation from them to the environment. In addition, there should be equal and stable thermal cooling conditions around the LED and the resistor R during the measurement. This makes it possible to obtain identical values of the averaged convective and radiation heat transfer coefficient h for both elements.

As a result, power balance equations can be obtained for the measurement system shown in Figure 4:

$$P_{elD} = P_{TD} + P_{opt} = hT_D + P_{12D} + P_{opt} \tag{16}$$

$$P_{elR} = P_{TR} = hT_R + P_{12R} \quad (17)$$

where P_{elR} and P_{elD}—the electrical power supply of the resistor and the diode, respectively, P_{TR} and P_{TD}—thermal powers dissipated to the environment by the resistor and the diode, respectively, P_{12R} and P_{12D}—thermal powers dissipated to the environment through the supply wires of the resistor and the diode, respectively, P_{opt}—radiant power of the diode, T_R and T_D—temperature differences between the resistor and diode and the ambient temperature, respectively, and h—heat transfer coefficient.

The thermal power dissipated to the environment through the resistor and diode connecting wires is calculated by determining the temperature gradient along the power cables on both sides of each element, $P_{1D(R)}$ and $P_{2D(R)}$, with the knowledge of the cable thermal conductivity coefficient k and its diameter $2r$—Figure 6.

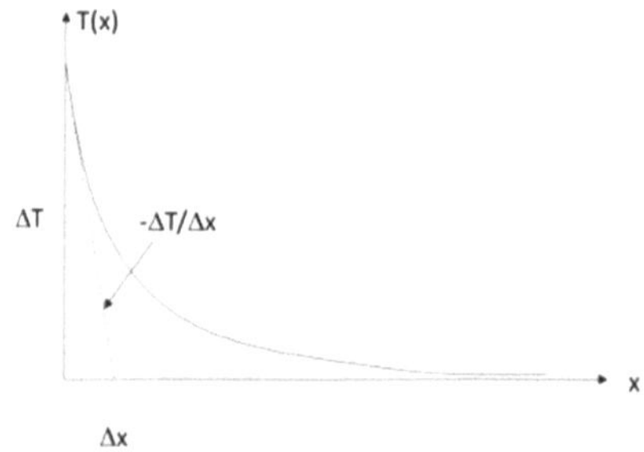

Figure 6. Determination of the thermal power dissipated to the environment through the wires supplying the resistor and the diode.

The thermal power transferred to the environment through the wires supplying the resistor and the diode can be determined by:

$$P_{12D(R)} = P_{1D(R)} + P_{2D(R)} = -k\frac{\Delta T_{1D(R)}}{\Delta x}\pi r^2 - k\frac{\Delta T_{2D(R)}}{\Delta x}\pi r^2 \quad (18)$$

where T_{1D} and T_{2D}—temperature of the diode supply cable on side 1 and 2, respectively, and T_{1R} and T_{2R}—temperature of the cable supplying the resistor on side 1 and 2, respectively.

From Equations (16) and (17), the optical power P_{opt} of the diode is finally expressed as:

$$P_{opt} = P_{elD} - P_{12D} - \frac{T_D}{T_R}(P_{elR} - P_{12R}) \quad (19)$$

2.3. Measurement Setup

A diagram of the measuring system is presented in Figure 7. The system consists of a resistor (1) and an LED (2). Both elements, the LED and resistor R, are connected in series and supplied from the current source (4) with an adjustable current in the range of 10–30 mA. Both measuring elements (LED and R) were placed in a trough (3) with dimensions of L = 25 cm, W = 10 cm, and H = 5 cm. In order to ensure the lack of thermal impact of both heat sources, the distance between them should not be less than 15 cm. A support (6) placed in the middle of the trough separates and insulates both elements. The tested elements were suspended in the trough by means of power supply wires at a height of approx. 1/2 of the height H from the bottom of the trough to ensure stable thermal conditions during the measurement. The trough was closed except for the upper part, over which the thermal imaging camera was placed. The trough with the support was made via 3D printing. Similarly, using the same method, a thermal element (1) was made, in which a heat source in the form of a small resistor with dimensions similar to the dimensions of the semiconductor structure of the LED diode was embedded.

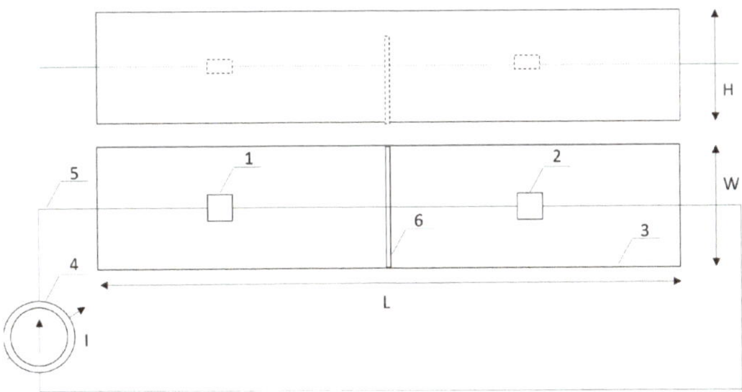

Figure 7. Experimental setup and the tray with the measured elements: resistor R (1) and an LED (2).

A special measurement stand equipped with an IR microbolometer camera, a precise, adjustable current source, and current/voltage meters was prepared. Measurements were performed under laboratory conditions inside a closed chamber where the free convection is stable and homogenous. Airflow was reduced to a minimum throughout the experiment. The IR camera was places on top of the measurement section and it could monitor the temperature of the objects through a dedicated speculum made for an IR camera in the measurement section of the stand—Figure 8.

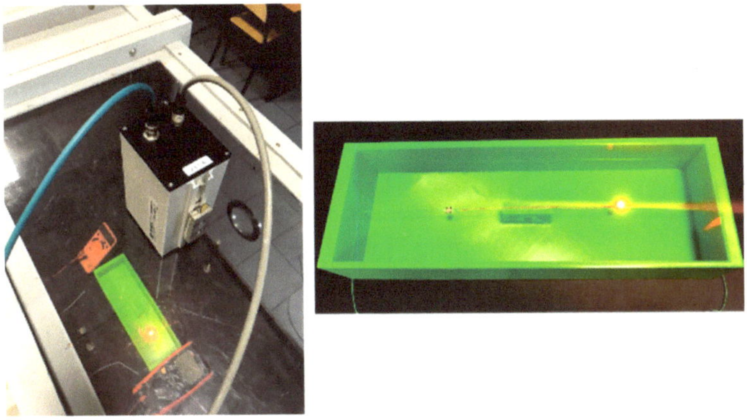

Figure 8. Scheme of the measurement system.

In order to prepare a non-radiant heat source, an SM-0603 surface-mounted resistor was inserted into the LED housing using 3D printing.

3. Results and Discussion

In order to compare the obtained results, the optical power emitted by the LED was measured directly using an integrating sphere—Figure 9. Using an integrated sphere is the most reliable and standard method for measuring radiant power, commonly used by light source producers, including LED manufacturers. This direct measurement is the standard method in this field. The used integrated sphere was calibrated in order to obtain proper values of emitted radiant power. A sphere with a diameter of 500 mm was equipped with a spectrometer, enabling the measurement of luminous flux and radiant power in the spectral range of $\lambda \in (340–1700)$ nm.

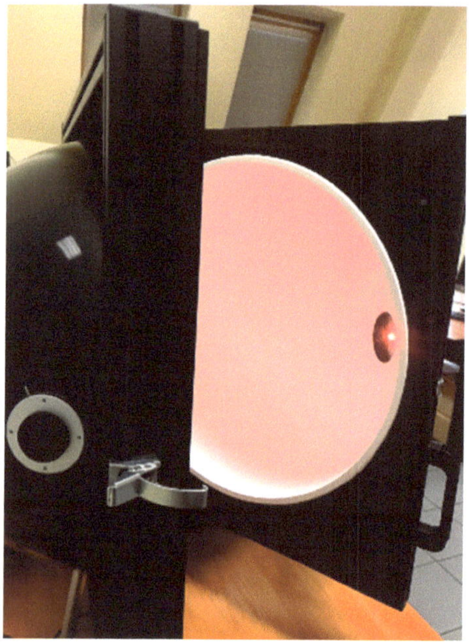

Figure 9. Integrating sphere with a mounted LED.

In order to power the analyzed elements, a precise laboratory power supply was applied. For measuring the voltage and current of the diode and the resistor, two digital multimeters were used: one as a voltmeter and the other as an ampere meter. A *DIAS 640Lc* thermal imaging camera with *Pyrosoft* Compact software ver. 3.4.1.1 was used for the tests. This thermal imaging camera allows for precise non-contact temperature measurements from −20 °C to 500 °C. The IR camera has an uncooled microbolometer array with 640 × 480 pixels operating in the spectral range of 8–14 µm. The measurement frequency of this IR camera was 50 frames per second and the *NETD* (*Noise Equivalent Temperature Difference*) was less than 0.08 K.

3.1. Modelling Results

The result of the simulation of the R_{th} network thermal model presented in Section 2 gives the values of temperatures in all nodes (Table 2).

Table 2. Temperature in all nodes obtained by the R_{th} network compact model.

Node No	1	2	3	4	5	6
ΔT, °C	7.07	6.93	6.99	7.07	7.07	6.98

As can be seen, the temperature inside the diode housing is almost uniform. The temperature rise in node 2 corresponds to the temperature value on the upper surface of the diode, which was measured during the experiment and estimated to be 6.93 °C for a thermal power equal to P_{TD} = 27.72 mW.

An example of the simulation results obtained using the 3D-FEM model is shown in Figure 10.

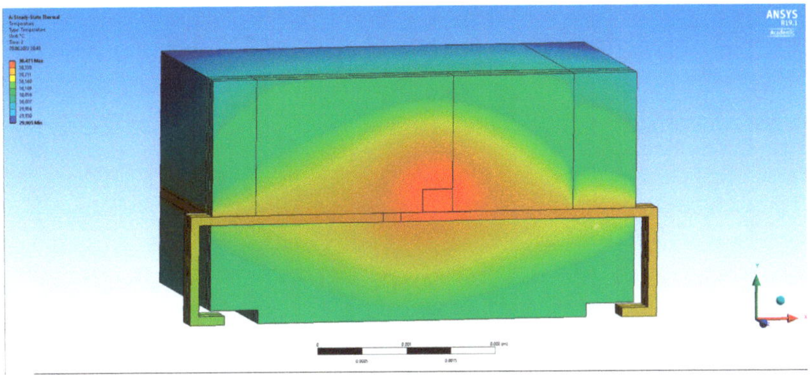

Figure 10. Temperature distribution inside the LED enclosure in a steady state.

The initial temperature as well as the boundary air temperature was set to 25 °C. The results of the modelling, presented as the values of the temperature at the specified measurement points, are shown in Table 3.

Table 3. Temperature obtained by the 3D FEM thermal model.

Measuring Point	Heat Source	Top Surface	Bottom Surface	Electrical Contact 1	Electrical Contact 2	Side Surfaces
ΔT°C	7.46	6.95	7.02	7.26	7.24	7.11

The geometry is not very complex; thus, the 3D model could be used without incurring a very long computation time. It took about 10 h using a desktop computer with an AMD EPYC 7301 16-Core Processor, 2.20 GHz, and 128 GB of RAM, which is not much for a 3D model.

Simulations of developed models were performed based on the shape and the structure of the measured diode and including the conditions of the measurement setup. In the FEM analysis in ANSYS, the mesh was hexahedral and almost regular in the whole simulated structure. Its side dimensions were equal to about 5×10^{-5} m, which balances the computation time and results accuracy.

The simulation was carried out for a thermal power equal to 27.72 mW. The temperature difference between the diode and the environment was 6.93 °C in the measurement, the same as the R_{th} model (node 2), and 6.95 °C in the 3D FEM analysis. Moreover, the temperature distributions in the corresponding nodes of the R_{th} compact model and surfaces from the 3D FEM are comparable. Both thermal models gave similar results—Tables 2 and 3. The difference between the results could be reduced by using a denser mesh and a more accurate geometry.

3.2. Optical Power Evaluation

Optical power measurements were separated into six experimental sessions that lasted a few hours in one day. Each session consisted of six measurements to average the results. The maximum temperature of both the LED and the resistor was taken for the calculations. The average temperature could also be considered, but the maximum temperature was selected for further calculations due to the difficulties in precisely determining the LED area for averaging. The experiments were carried out with RGB LEDs in 5 × 5 mm housings and a non-radiating resistive heat source with resistance $R = 220\ \Omega$. During this experiment, the thermal resistance of the LED package inside the measuring stand was estimated at approximately $R_{th} \approx 255$ K/W. Exemplary results of temperature measurements using the IR camera are shown in Figure 11 and Table 4.

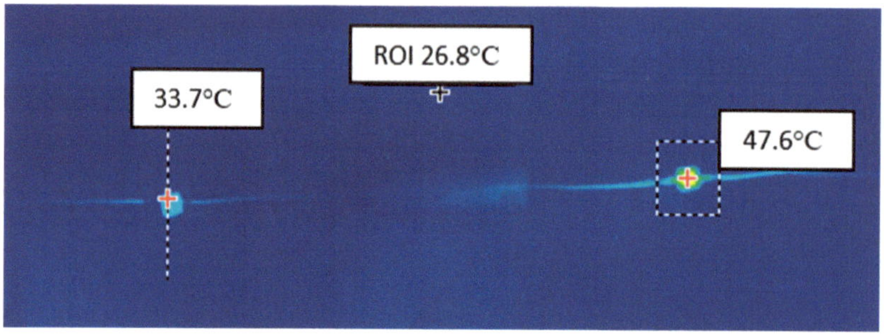

Figure 11. Thermogram of a series connection of the diode (on the left) and the resistor (on the right).

Table 4. Measurement results of temperature T_D and T_R and optical power P_{opt} for a supplying current of $I = 20$ mA for six measurement sessions lasting a few hours. The last row presents mean values.

P_{elD} mW	P_{elR} mW	T_{amb} °C	T_d °C	T_r °C	$\Delta T_d = T_D$ °C	$\Delta T_r = T_R$ °C	P_{12D} mW	P_{12R} mW	P_{rad} mW	η %
38.7	87.3	26.3	33.8	47.7	7.05	20.95	1.5	11	11.55	29.80
38.7	87.3	26.7	33.77	47.6	7.02	20.85	1.6	9	10.77	27.79
38.8	87.6	27.3	33.73	47.72	6.98	20.97	1.5	9	11.15	28.77
38.7	87.3	26.9	33.67	47.70	6.92	20.95	1	9	11.88	30.66
38.7	87.2	26.8	33.47	47.52	6.72	20.77	3	10	10.72	27.66
38.8	87.5	26.5	33.62	47.7	6.87	20.95	3	10	10.39	26.82
38.8	**87.4**	**26.75**	**33.68**	**47.66**	**6.93**	**20.91**	**1.93**	**9.67**	**11.08**	**28.58**

All temperature results are presented as the excess of ambient temperature. The temperature of the LED is much lower due to the significant amount of optical energy emitted. The main achieved results confirm the high optical energy efficiency of the tested LED, reaching $\eta \approx 30\%$.

Using the integrating sphere, it was possible to measure both the spectrum and the power density of the emitted radiation within the 350–750 nm wavelength range [1]. The spectrum of emitted LED radiation at a current of $I = 20$ mA is shown in Figure 12. It confirms the narrow band of the radiation and the high optical power effectiveness of the tested LED. The optical power measured with the integrating sphere was $P_{opt} = 11.3$ mW and agrees with the thermographic measurement very well. The optical efficiency of the tested LED was high and estimated at $\eta \approx 30\%$. This result agrees with the manufacturer's data.

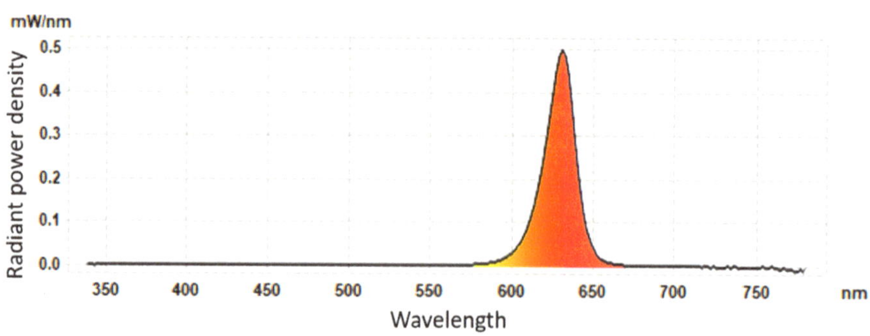

Figure 12. Power density of emitted LED radiation vs. wavelength for $I = 20$ mA.

The measurements carried out in the integrating sphere for different currents flowing through the diode confirm the linear relationship between the optical power generated and the current of the diode—Figure 13.

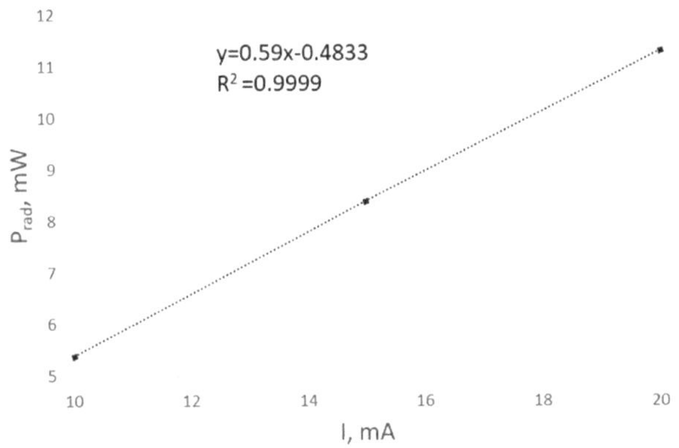

Figure 13. Optical power of a diode as a function of the current.

3.3. Uncertainty Analysis

The measurement uncertainty was calculated according to the specifications given in the literature [14–17]. Optical power (19) depends on the temperature and both the electrical power supplied to the diode and the resistor and the thermal power dissipated in them. It was assumed that the measurement uncertainty of radiant power depends on the uncertainty of measurements of T_R and T_D and the uncertainty of power dissipated in the connecting wires—Equation (20). The uncertainty of electrical power is negligible because this power was measured using professional electrical quantity meters with a high accuracy and a high resolution.

In order to obtain reliable results from thermographic measurements, it is necessary to determine the maximum permissible range of variability in the measured quantity with a given probability, called the confidence level. This value refers to the expanded uncertainty, denoted as U. Expanded uncertainty provides the range of potential variation in measurement results due to random fluctuations in the measured data and systematic errors of the apparatus used. It uses a coverage factor to increase uncertainty as the number of measurements decreases.

The expanded uncertainty of the optical power measurement can be expressed as:

$$U_{P_{opt}} = k_p u_c(P_{opt} = f(T_D, T_R, P_{12D}, P_{12R})) \quad (20)$$

where k_p is the coverage factor depending on the number of measurements performed and u_c is the combined uncertainty.

The combined uncertainty can be expressed as:

$$u_c(P_{opt}) = \sqrt{\left(\frac{\partial P_{opt}}{\partial T_D}u_c(T_D)\right)^2 + \left(\frac{\partial P_{opt}}{\partial T_R}u_c(T_R)\right)^2 + \left(\frac{\partial P_{opt}}{\partial P_{12D}}u_c(P_{12D})\right)^2 + \left(\frac{\partial P_{opt}}{\partial P_{12R}}u_c(P_{12R})\right)^2} \quad (21)$$

The combined uncertainty consists of A- and B-type uncertainties, u_A and u_B, relating to the stochastic nature of the measurement and the maximum absolute error of the measurement devices.

$$u_c = \sqrt{u_A{}^2 + u_B{}^2} \quad (22)$$

Both uncertainties u_A and u_B were calculated independently for both temperature values of the diode T_D and the resistor T_R. The entire measurement process consisted of six sessions in which six measurements were taken in order to average the results of the measured quantities. The entire measurement lasted for several hours at a variable ambient temperature. A variable ambient temperature is one of the main sources of increasing uncertainty in temperature measurements with bolometric cameras. Averaging the measurement results allowed us to reduce the type A uncertainty of the temperature measurement significantly.

$$u_A(x) = \sqrt{\frac{\sum_{i=1}^{N}(x_i - \overline{x})^2}{N(N-1)}} \tag{23}$$

where x is the measured value (temperature of the diode and resistor and power of the connecting wires) for each measurement and $\overline{x}$ is its average value. The symbol x refers to the temperature of the diode, T_D, and the resistor, T_R, and the power, P_{12R} and P_{12D}, and $N = 6$ means the number of measurements in each session.

Note that all temperature values presented in this work refer to the excess in temperature over the ambient temperature. Type A uncertainty results are shown in Table 5.

Table 5. Uncertainties u_A, u_B, and u_c for measurements of temperature, T_D and T_R, as well as for power dissipated in the connecting wires, P_{12D} and P_{12R}.

Parameter	Mean Value	u_A	u_B	u_c
T_D (°C)	6.93	0.05	0.08	0.09
T_R (°C)	20.91	0.03	0.24	0.24
P_{12D} (mW)	1.93	0.35	0.00	0.35
P_{12R} (mW)	9.67	0.33	0.00	0.33

The systematic error component of uncertainty $u_B(T)$—type B uncertainty—is determined by the maximum absolute error ΔT_{max} of a single measurement from the thermal imaging camera. It was assumed that the source of this measurement uncertainty is the IR camera itself [16,17]. The accuracy of the thermal camera used in this work is ΔT_{max} = 2% of the measured value.

$$u_B(T) = \frac{\Delta T_{max}}{\sqrt{3}} = \frac{0.02T}{\sqrt{3}} \tag{24}$$

For the power dissipated in the connecting cables, the uncertainty u_B is scaled by the factor $kS/\Delta x$, where k is the thermal conductivity of the cable, $S = \pi r^2$ is the cross-sectional area, and Δx is the length of the cable for which the gradient is calculated.

$$u_B(P) = \frac{kS}{\Delta x}\frac{\Delta T_{max}}{\sqrt{3}} = \frac{kS}{\Delta x}\frac{0.02T}{\sqrt{3}} \tag{25}$$

where k = 200 W/(m·K), Δx = 1.26 cm, $S = \pi r^2$, and r = 0.1 mm.

According to uncertainty standards [17,18], if the uncertainty of u_B dominates over u_A, then the random variable has a uniform probability distribution and the value of the coefficient k_p (20) is equal to $\sqrt{3}$ for a 100% confidence level.

Based on (19) and (21), the combined uncertainty of the optical power measurement u_c is estimated as:

$$u_c(P_{opt}) = \sqrt{\left(\frac{T_D(P_{elR} - P_{12R})}{{T_R}^2}u_c(T_R)\right)^2 + \left(\frac{(P_{elR} - P_{12R})}{T_R}u_c(T_D)\right)^2 + u_c^2(P_{12D}) + \left(\frac{T_D}{T_R}u_c(P_{12R})\right)^2} \tag{26}$$

Using actual measurement values, U = 1.17 mW was estimated, so the final result of estimating the LED optical power is P_{opt} = 11.08 ± 1.17 mW for a thermal camera with a 2% accuracy. When this accuracy is lowered to 1%, the uncertainty U = 0.92 mW. In order to further reduce the uncertainty, it is necessary to increase the number of measurements that are averaged.

4. Conclusions

This article presents a simple engineering method for evaluating the optical power of LEDs by measuring the temperature using infrared thermography. The presented results seem promising. In addition, during this research, a compact thermal model in the form of a network of thermal resistances was constructed. Advanced 3D-FEM thermal modelling was performed in the ANSYS environment to compare the results with the measurements. In order to verify the obtained results, the optical power of the diode was measured using an integrating sphere. The results were very close to each other. In order to show the practical potential of the proposed new thermographic method of measuring the optical power of LEDs, an uncertainty analysis was carried out. This analysis shows how the accuracy of such measurements can be improved. First, the accuracy of the IR camera must be high, at least 1%. The temperature then needs to be averaged over both time and space. It is recommended to choose an area of interest containing at least a dozen pixels of the thermal image to average the temperature.

The proposed method of measuring the optical power of LEDs is an engineering alternative to using an integration sphere and spectrometer that can only be used in the laboratory, as it cannot be used remotely, but, on the other hand, it is very accurate. It needs expensive equipment, requires calibration, and cannot be used while diodes are operating in a system.

Author Contributions: Conceptualization, B.W.; Software, M.S. and M.F.; Formal analysis, B.W.; Investigation, M.S. and S.U.; Resources, B.T., I.S.M.S. and P.T.; Writing—original draft, M.S., S.U. and M.F.; Writing—review & editing, R.K.; Supervision, B.W. All authors have read and agreed to the published version of the manuscript.

Funding: This research received no external funding.

Data Availability Statement: Dataset available on request from the authors.

Conflicts of Interest: The authors declare no conflicts of interest.

References

1. Mujahid, M.A.; Panatarani, C.; Maulana, D.W.; Wibawa, B.M.; Joni, I. Integrating sphere design for characterization of LED efficacy. *AIP Conf. Proc.* **2016**, *1712*, 030010.
2. Kong, D.-J.; Kang, C.-M.; Lee, J.-Y.; Kim, J.; Lee, D.-S. Color tunable monolithic InGaN/GaN LED having a multi-junction structure. *Opt. Express* **2016**, *24*, A667–A673. [CrossRef] [PubMed]
3. Horng, R.-H.; Shen, K.-C.; Kuo, Y.-W.; Wuu, D.-S. GaN light emitting diodes with wing-type imbedded contacts. *Optics Express* **2013**, *21*, A1–A6. [CrossRef] [PubMed]
4. Janicki, M.; Ptak, P.; Torzewicz, T.; Górecki, K. Compact Thermal Modeling of Color LEDs—A Comparative Study. *IEEE Trans. Electron Devices* **2020**, *67*, 3186–3190. [CrossRef]
5. Górecki, K.; Ptak, P. Compact Modelling of Electrical, Optical and Thermal Properties of Multi-Colour Power LEDs Operating on a Common PCB. *Energies* **2021**, *14*, 1286. [CrossRef]
6. Zwinkels, J.C.; Ikonen, E.; Fox, N.P.; Ulm, G.; Rastello, M.L. Photometry, radiometry and 'the candela': Evolution in the classical and quantum world. *Metrologia* **2010**, *47*, 5. [CrossRef]
7. Kasikowski, R.; Więcek, B. Fringing-Effect Losses in Inductors by Thermal Modeling and Thermographic Measurements. *IEEE Trans. Power Electron.* **2021**, *36*, 9772–9786. [CrossRef]
8. Kopeć, M.; Więcek, B. Low-cost IR system for thermal characterization of electronic devices. *Meas. Automat. Monitor.* **2018**, *64*, 103–107.
9. Poppe, A.A. Multi-domain compact modeling of LEDs: An overview of models and experimental data. *Microelectron. J.* **2015**, *46*, 1138–1151. [CrossRef]
10. Ptak, P.; Górecki, K.; Heleniak, J.; Orlikowski, M. Investigations of Electrical and Optical Parameters of Some LED Luminaires—A Study Case. *Energies* **2021**, *14*, 1612. [CrossRef]
11. Kim, Y.-Y.; Joo, J.-Y.; Kim, J.-M.; Lee, S.-K. Compact Measurement of the Optical Power in High-Power LED Using a Light-Absorbent Thermal Sensor. *Sensors* **2021**, *21*, 4690. [CrossRef] [PubMed]
12. Rongier, C.; Gilblas, R.; Le Maoult, Y.; Belkessam, S.; Schmidt, F. Infrared thermography applied to the validation of thermal simulation of high luminance LED used in automotive front lighting. *Infrared Phys. Technol.* **2022**, *120*, 103980. [CrossRef]
13. Chang, C.-H.; Teng, T.-P.; Teng, T.-C. Influence of Ambient Temperature on Optical Characteristics and Power Consumption of LED Lamp for Automotive Headlamp. *Appl. Sci.* **2022**, *12*, 11443. [CrossRef]

14. Joint Committee for Guides in Metrology. *JCGM 100: Evaluation of Measurement Data—Guide to the Expression of Uncertainty in Measurement, Technical Report*; Joint Committee for Guides in Metrology: Paris, France, 2008.
15. Minkina, W.; Dudzik, S. *Infrared Thermography: Errors and Uncertainties*; John Wiley & Sons: Hoboken, NJ, USA, 2009; ISBN 978-0-470-74718-6.
16. Minkina, W.; Dudzik, S. Simulation analysis of uncertainty of infrared camera measurement and processing path. *Measurement* **2006**, *39*, 758–763. [CrossRef]
17. EFSA Scientific Committee. Guidance on Uncertainty Analysis in Scientific Assessments. *EFSA J.* **2018**, *16*, 1.
18. Fotowicz, P. A method of approximation of the coverage factor in calibration. *Measurement* **2004**, *35*, 251–256. [CrossRef]

sensors

MDPI

Article

A Novel Combined Method for Measuring the Three-Dimensional Rotational Angle of a Spherical Joint

Qianyun Yang, Kai Ouyang, Long Yang, Rao Fu and Penghao Hu *

Anhui Province Key Laboratory of Measuring Theory and Precision Instrument, School of Instrument Science and Opto-Electronic Engineering, Hefei University of Technology, Hefei 230009, China; 2021170006@mail.hfut.edu.cn (Q.Y.); 2020010001@mail.hfut.edu.cn (K.O.); yangl@mail.hfut.edu.cn (L.Y.); 2023110008@mail.hfut.edu.cn (R.F.)
* Correspondence: hupenghao@hfut.edu.cn

Abstract: To improve the measurement accuracy of the three-dimensional rotation angle of a spherical joint, a novel approach is proposed in this study, which combines magnetic detection by a Hall sensor and surface feature identification by an eddy current sensor. Firstly, a permanent magnet is embedded in the ball head of a spherical joint, and Hall sensors are set and distributed in the ball socket to measure the variation in the magnetic flux density when the spherical joint rotates, which are related to the 3D rotation angle. In order to further improve the measurement accuracy and robustness, we also set grooves on the ball head and use eddy current sensors to synchronously identify the rotation angle of the ball head. After the combination of two signals is performed, a measurement model is established using the RBF neural network by training, and the real-time measurement of the 3D rotation angle of the spherical joint is realized. The feasibility and superiority of this method are validated through experiments. The experimental results indicate that the measurement accuracy is substantially promoted compared to the preliminary measurement scheme based on spherical coding; the average measurement error of the single axis is reduced by $9'9''$. The root mean square errors for the measurements of the 3D rotation angles in this proposed method are as follows: pitch angle α has an error of $1'8''$, yaw angle β has an error of $2'15''$, and roll angle γ has an error of $29'6''$.

Keywords: spherical joint; 3D rotation angle; RBF neural network

1. Introduction

The spherical joint is a mechanism that can present three rotary degrees of freedom [1–3]. It is compact in structure and flexible in motion, and is widely used in parallel mechanisms, machine tools, measuring instruments, medical devices, optical devices, and various other equipment [4–6]. As a purely passive component, a spherical joint cannot determine its own rotation direction and rotation angle value [7,8]. If an embedded precise measurement method for its rotation angle can be obtained, the spherical joint becomes an intelligent device with broader application prospects, which is beneficial for improving the motion accuracy and facilitating the control of equipment for which spherical joints are used [9,10].

In recent decades, the identification of spherical rotation direction and angle measurement techniques have been researched and developed [11,12]. Several measurement methods based on different principles have emerged, which mainly include optical [13,14], magnetic [15], and inertial fields [16,17].

For example, Min Li utilized embedded sensors to simultaneously measure the magnetic flux density and back electromotive force of a spherical motor [18]. The data from these sensors were input into a sensor fusion system based on Kalman filtering to estimate the three degrees of freedom of angular displacement and the angle in real time. The idea for the implementation of this system is to use embedded non-contact sensors to measure the magnetic flux density of the spherical motor and the back electromotive force generated by the stator. Then, these two quantities are taken as two sets of decoupled inputs for the sensor fusion system based on the Kalman filter. This system includes an artificial neural network for estimating the rotor position and a mathematical model for calculating the angular velocity. Experimental results show that, within a certain measurement range, the average measurement error of the system in a single axis is 0.08°.

Jae-Hyeok Kim et al. proposed using a precision mechanical sensor called the attitude and heading reference system (AHRS), comprising a gyroscope, accelerometer, and magnetometer, to measure the tilt angle of a spherical motor [19]. The AHRS sensor is arranged on the ball joint. When the motor tilts, the sensor outputs a corresponding signal. By processing the signal, the tilt angle can be obtained. The measurement accuracy values of the two axes are 0.27° and 0.83°, but this method cannot measure the self-rotation angle of the motor, and the accuracy still needs to be improved.

Wang Q et al. also proposed a sensorless rotor attitude detection method based on the mutual inductance voltage of a stator coil [20]. In the online detection process of the rotor position, the three-dimensional angle of the rotor was inversely calculated based on the real-time collected mutual inductance voltage information, using an intelligent optimization algorithm, combined with the distribution law of the mutual inductance voltage and the constraints of the rotor structure. This detection method has a good online detection effect, with a standard deviation of the group within 1.8°; but, the accuracy is not high.

Yang S et al. proposed a two-degree-of-freedom angle displacement measurement method using a spherical capacitive sensor to measure a spherical pair [21]. The capacitance sensor proposed in this method had a four-quadrant differential electrode configuration. Compared with other angle measurement detection methods, it has an integrated structure, occupies a small space, and is convenient to integrate into the sphere. However, this method was unable to calculate measurements around the rotation axis.

Under the support of the National Natural Science Foundation of China, our team proposes a measurement scheme for three-dimensional rotational angles in the spherical joint space based on eddy current sensors and pseudo-random coding [22]. As shown in Figure 1a, a sensor array consisting of multiple sensors is used to identify spherical encoding. Two-dimensional spherical grooves are employed. The groove width is generated using pseudo-random encoding, and the groove depth is generated using an arithmetic sequence, ensuring the uniqueness of the spherical head's three-dimensional encoding. An artificial neural network is employed to establish a measurement model between the output voltage of the eddy current sensors and the spatial three-dimensional (3D) rotational angles. Consequently, the measurement of 3D rotational angles in the spherical joint space is realized. In this scheme, the pitch angle α and the twist angle β are within the range of $-10°$ to $10°$, with root mean square errors of $22'32''$ and $25'58''$, respectively. The rotational angle, γ, of the spherical joint along the axis of the spherical joint rod is within the range of 0° to 120°, with a root mean square error of $30'17''$. Figure 1b presents an alternative approach proposed by our team for measuring the two-dimensional rotational angle of a ball joint [23]. In this approach, a cylindrical permanent magnet is embedded at the bottom of the ball head, and a Hall sensor installed in the ball socket is used to measure the rotational angle of the ball head in any direction in space. Finally, the measured values are decomposed into rotational angle components α and β around the X- and Y-axes, respectively. This approach has two modeling methods: one is to establish an equivalent magnetic charge model, and the other is to establish a neural network model. The experimental results demonstrate that the neural network model has higher accuracy, a simpler structure, faster data processing speed, and the highest single-axis angle measurement accuracy can reach $4'$. However, this method has a lower accuracy for measuring the rotational angle, γ, of the spherical joint along the axis of the spherical joint rod; thus, it can be considered unsuitable for measuring the γ angle.

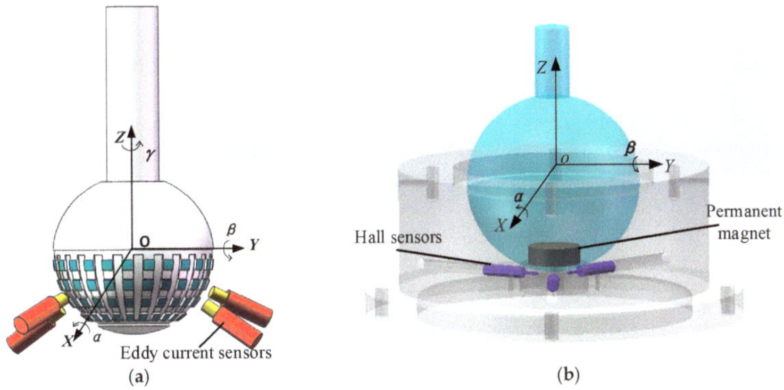

Figure 1. Two measurement schemes (**a**,**b**) for the rotation angle of a spherical joint.

The accuracy level of the measurement scheme for the three-dimensional rotation angle of the spherical joint space based on eddy current sensors and pseudo-random codes is not able to meet the needs of the precision engineering field. Therefore, in order to solve the problem that the measurement system of the spherical hinge based on the magnetic effect method cannot effectively measure the rotational angle, γ, of the spherical joint along the axis of the spherical joint rod, this article combines the advantages of the magnetic effect and spherical coding methods based on previous research and constructs a new combined measurement scheme.

2. Measurement Plan Design

This scheme embeds a permanent magnet into the bottom of the ball head and uses three Hall sensors to detect the magnetism of the ball head. When the spherical joint rotates, the Hall sensors perceive the 3D rotation angle of the ball hinge through the change in the magnetic field. However, this scheme has a low measurement accuracy for the roll angle, γ. Therefore, a one-dimensional groove was machined on the surface of the metal ball head to improve the measurement accuracy of the self-rotation angle, γ, using the distance measuring principle of eddy current sensors, and also to enhance the measurement accuracy of the pitch angle, α, and yaw angle, β. The overall design is shown in Figure 2. This measurement scheme limits the three translational degrees of freedom along the X-, Y-, and Z-axes of the ball hinge, and does not limit the three rotational degrees of freedom around the X-, Y-, and Z-axes.

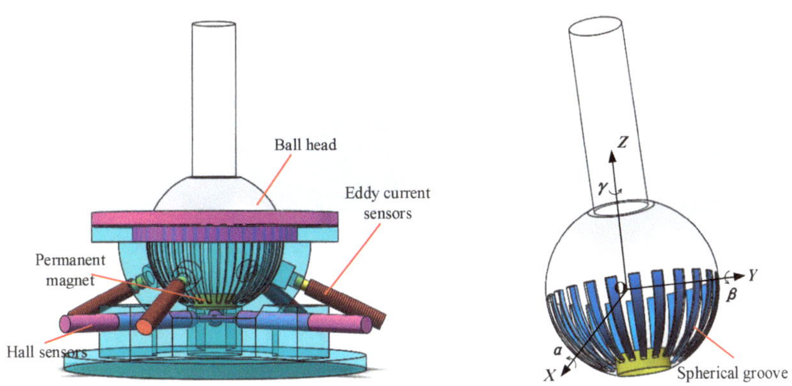

Figure 2. Schematic diagrams of measurement scheme.

2.1. Sensor Placement Location

First, three Hall sensors were horizontally placed at the bottom of the ball socket. Then, the sensor positioning fixture inside the ball socket was used to fix the three Hall sensors in the same plane. Two different sensor placement schemes are used here for comparison. Scheme 1: as shown in Figure 3a—sensor S1 is located on the *X*-axis and senses the magnetic field in the *X*-axis direction, sensor S2 is located on the *Y*-axis and senses the magnetic field in the *Y*-axis direction, and sensor S3 is located on the axis that is 135°counterclockwise from the *Y*-axis, sensing the magnetic field in the axial direction. Scheme 2: as shown in Figure 3b—sensors S1 and S2 are rotated by 90°, respectively. S1 measures the magnetic field component in the *Y*-axis direction and S2 measures the magnetic field component in the *X*-axis direction. The position of S3 remains unchanged. The coordinates in Figure 3 represent the position of the sensor in a plane coordinate system.

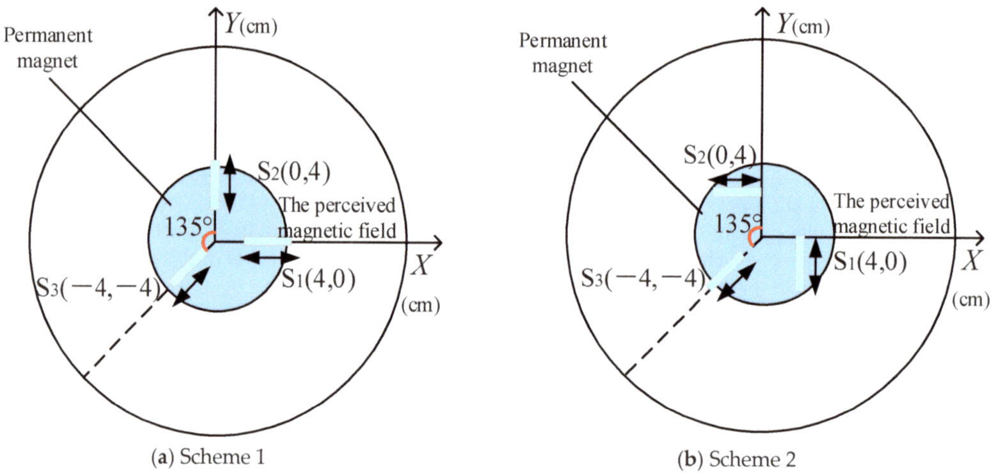

Figure 3. Hall sensor-placement angle design.

Using COMSOL and MATLAB, simulations were conducted for two different schemes. During the finite element simulation process in COMSOL, the γ angle remained constant, while the α and β angles varied within a range of $\pm 20°$. An RBF (radial basis function) neural network model was constructed in MATLAB, and the simulated α and β angles were inputted to fit the error results, as shown in Figure 4. It can be seen that scheme 2 has higher measurement accuracy results for the α and β angles, so scheme 2 is adapted to place the Hall sensor. The design of the ball socket is shown in Figure 5. Based on the spherical coding scheme described above, the eddy current sensor was installed at any four asymmetric positions on the surface of the spherical socket, which could improve the measurement accuracy of the three-dimensional rotation angle of the spherical joint to a certain extent [22].

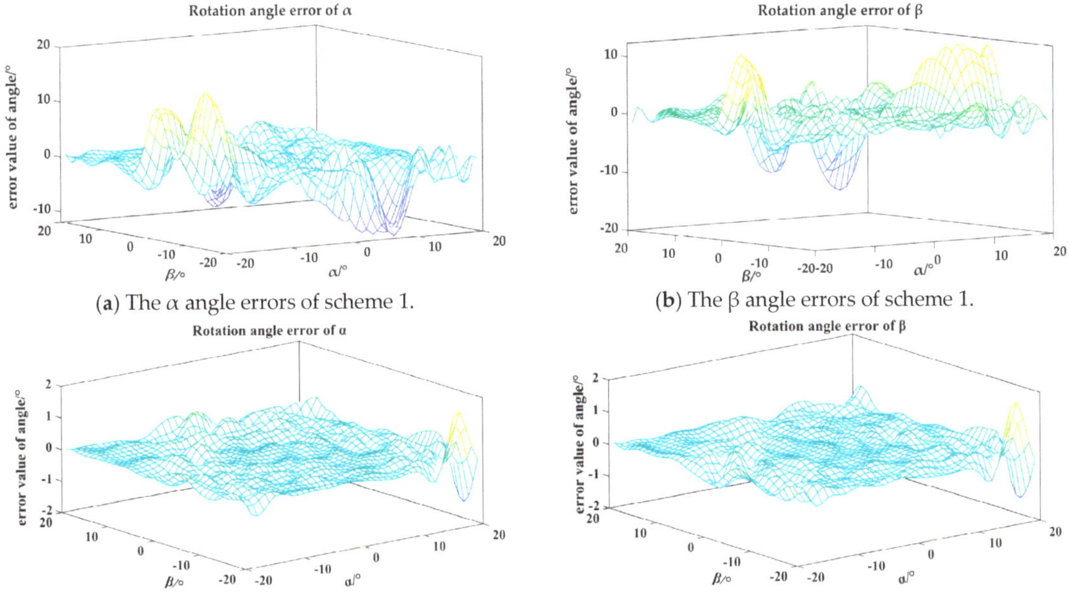

(**a**) The α angle errors of scheme 1. (**b**) The β angle errors of scheme 1.

(**c**) The α angle errors of scheme 2. (**d**) The β angle errors of scheme 2.

Figure 4. The α and β angle errors of the neural network fitting process.

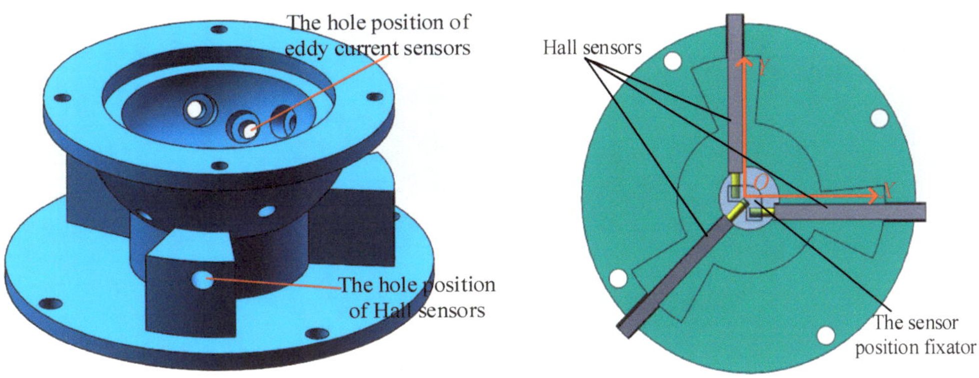

(**a**) Three-dimensional view of ball socket. (**b**) Bottom of ball socket.

Figure 5. The design scheme of the ball socket.

2.2. Spherical Groove Design

Using the AD/DC module of the COMSOL physics field simulation tool, a three-dimensional finite element model was established, as shown in Figure 6b. The colored graphics on the sensor represent the magnitude of the magnetic induction, the direction of the arrows represents the direction of the current, and the color of the arrows represents the magnitude of the current. From the simulation results, it can also be seen that the eddy current field is mainly distributed around the sensor, and the apparent range of the eddy current field does not exceed the size of the sensor. This also indicates that the width of the grooves and protrusions should not exceed the size of the sensor; otherwise, there is a "blind spot" in the measurement.

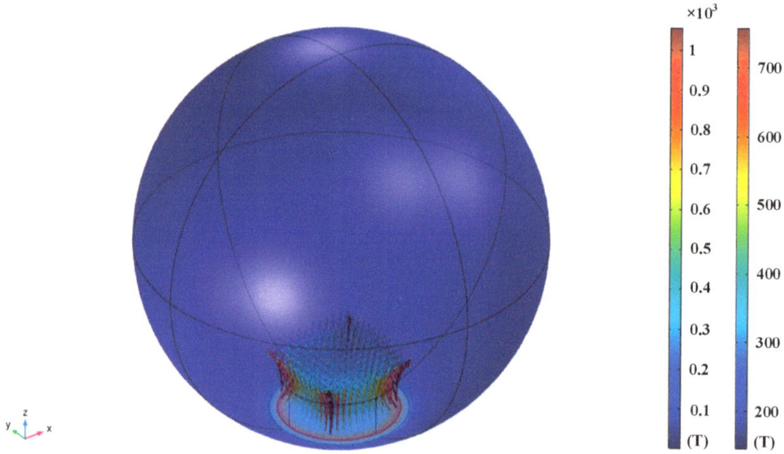

(**a**) Simulation results of permanent magnets.

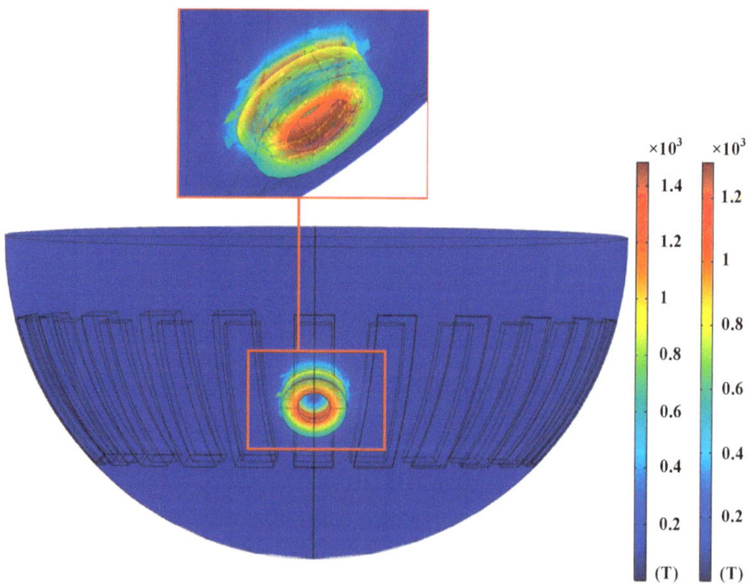

(**b**) Simulation results of eddy current sensor.

Figure 6. Simulation results based on COMSOL.

At the same time, the simulation also shows that the direction of the eddy current is opposite to the direction of the sensor current. From Figure 6a, it can be seen that the magnetic field generated by the magnet is distributed around the magnet, and the farther away from the magnet, the weaker the magnetic field. The eddy current sensor can only sense the magnetic field around the sensor, and the influence on the eddy current sensor is minor when the magnet is far away from the eddy current sensor. Moreover, since the magnetic field around the magnet is a uniform demagnetized field, this fixed influence is beneficial to the measurement of the angle when we need the eddy current sensor to generate specific data to identify the angle.

In addition, by simulating and analyzing the grooves with different parameters on a metal block, the parameter settings of the grooves on the surface of the spherical head were determined. Based on the research results for the output characteristics of the eddy current sensors, it is known that the inductance of the sensor will undergo significant changes when scanning grooves with different parameters, especially when the sensor is located at the center of the groove, where the difference value is the maximum [24]. For grooves with the same width, the larger the groove depth, the greater the change in inductance; for grooves with the same depth, the larger the groove width, the greater the change in inductance. Additionally, the sensor is more sensitive to the groove depth than groove width.

Therefore, in this study, only the parameter setting of the groove depth was modified. Finally, through a large number of combination schemes, the optimal solution was determined as follows: the spherical head was divided into 24 groups, with each group consisting of a 15° interval, and the groove width occupied 7°l (where l was the arc length when the central angle of the equatorial plane of the sphere was 1°). Among these groups, 12 groups had groove depths starting from 0.1 mm and increased by 0.1 mm each time, while the other 12 groups ad groove depths starting from 1.25 mm and decreased by 0.1 mm each time. Based on the design scheme described above, the physical structures of the ball head and ball socket are shown in Figure 7. The ball head was made of aluminum alloy and the ball socket was manufactured using 3D printing with nylon material, which does not affect the sensor measurements.

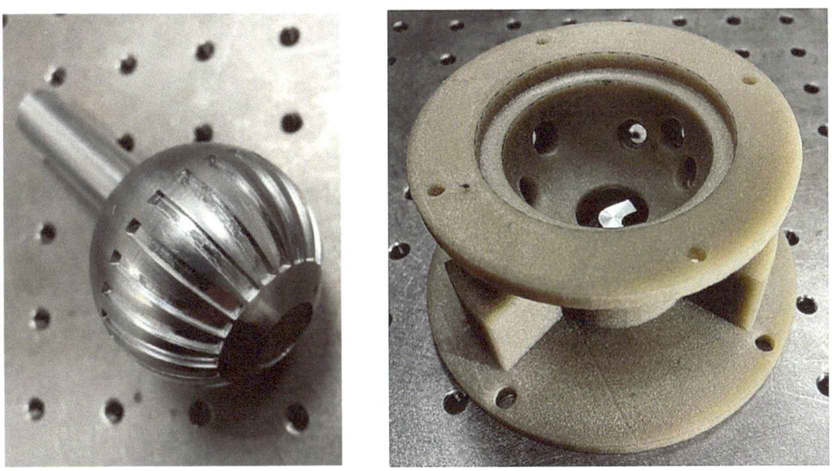

(**a**) Ball head. (**b**) Ball socket.

Figure 7. Physical diagrams of the ball head and socket.

3. RBF Neural Network

Based on the team's previous experience of using neural network modeling [22], establishing a measurement model for the rotational angle of the spherical joint space based on artificial neural networks can simplify the algorithm model, eliminate the complex and lengthy model derivation process, and the high robustness of the neural network can also compensate for the defects in prototype structural parameters, installation errors, and gap errors during ball head movements. Among them, the RBF neural network can approximate any nonlinear function with arbitrary precision and has a good generalization ability. When the network parameters are determined, the output of the network is the linear weighted sum of the hidden layer node outputs, so various linear optimization algorithms can be used to solve the network weights, speed up the learning speed, and avoid local minimum value problems [25–27].

The RBF neural network structure, as shown in Figure 8, is composed of an input layer, hidden layer, and output layer. The transformation from the input layer space to the hidden layer space is non-linear, while the transformation from the hidden layer space to the output layer space is linear.

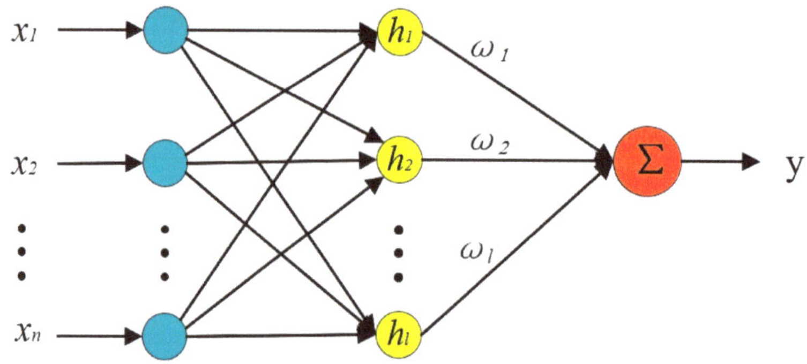

Figure 8. Structure of the RBF neural network.

The network input is the output of the Hall sensor and eddy current sensor, denoted as $X = [x_1, x_2, \ldots, x_7]^T$. The network output is the predicted values of three rotational angles, denoted as $Y = [y_1, y_2, y_3]^T$. The expression of the output layer of the RBF network is:

$$y(x_i) = \sum_{i=1}^{l} \omega_i h_i(x) \tag{1}$$

$$h(x) = \exp\left(-\frac{||x - c_i||^2}{\sigma_i^2}\right), i = 1, 2, \ldots, l \tag{2}$$

where ω_i is the ith output weight vector and l is the number of nodes in the hidden layer; $h_i(x)$ is the activation function and the Gaussian function is the most commonly used radial basis function; and c_i represents the center parameter of the kernel function for the ith hidden layer neuron, while σ_i is the expansion constant for the ith hidden node.

Due to the fact that the center point selection of the RBF algorithm uses the K-means algorithm, the training process adjusts the weights of the network using either gradient descent or least squares method, which often leads to overfitting and reduces the model's generalization ability. While the RBF neural network optimizes the centers of the hidden layer (c_i), expansion constants (σ_i), and output weights (ω_i) as particles in the particle swarm algorithm, this approach effectively avoids overfitting and other problems that may arise during model training [28–30]. Therefore, this paper used the PSO algorithm to optimize the RBF neural network, improved its robustness and generalization ability, and enhanced the accuracy of the measurement system.

The optimization process of the PSO (particle swarm optimization) algorithm for RBF neural networks can be roughly divided into the following steps. First, the particle swarm is generated by determining the structure of the RBF neural network. The particles are then mapped to the RBF neural network, establishing the RBF neural network model. Then, the expansion constant (σ_i) is calculated using Formula (3). Based on the expansion constant (σ_i) values, the global best particle and individual best particle are updated, and then it is determined whether the threshold targets are met. If the requirements are not met, the velocity and position of the particles are updated and the particle swarm is regenerated. If the requirements are met, the optimal particle is outputted and the particles are mapped to the RBF neural network for testing.

The calculation formula of the extended constant is:

$$\sigma = \frac{d_{max}}{\sqrt{2n}} \tag{3}$$

where d_{max} represents the maximum distance between the selected centers. n represents the number of samples.

The optimization goal of the PSO algorithm is to minimize the error function value between the actual output and the expected output of the RBF neural network [31]. The fitness function is set as the objective:

$$\mathrm{F} = \sum_{i=1}^{n} \sum_{j=1}^{e} (d_{ij} - o_{ij}) \tag{4}$$

where n represents the number of samples, e represents the number of outputs of the neural network, d_{ij} represents the j expected output of the i sample of the RBF neural network, and o_{ij} represents the j actual output of the i sample of the RBF neural network.

The RBF and PSO optimized RBF neural network models were established. The data of sensors S1, S2, and S3 obtained from the simulation of scheme 2 in Section 2.1 were used as inputs, and the corresponding alpha and beta angles were used as outputs. The RBF neural network and the improved PSO optimized RBF neural network were tested. In order to reduce the impact of randomness on the network performance, 50 simulation tests were conducted.

The test results are shown in Figure 9. From the graph, it can be observed that, for each simulation, the fitting effect of the PSO optimized RBF neural network model is superior to the unoptimized RBF neural network. The PSO optimized RBF neural network model is capable of controlling the root mean square errors of the α and β angles to around 0.09°, whereas the unoptimized RBF neural network model yields an RMSE of the α and β angles of around 0.16°. The reason why the unoptimized results in the graph appear as a straight line is because this study utilizes MATLAB's newrb function for the RBF neural network. Under the condition of a consistent training set and unchanged sample order, the results of multiple training using this function are consistent and not affected by randomness.

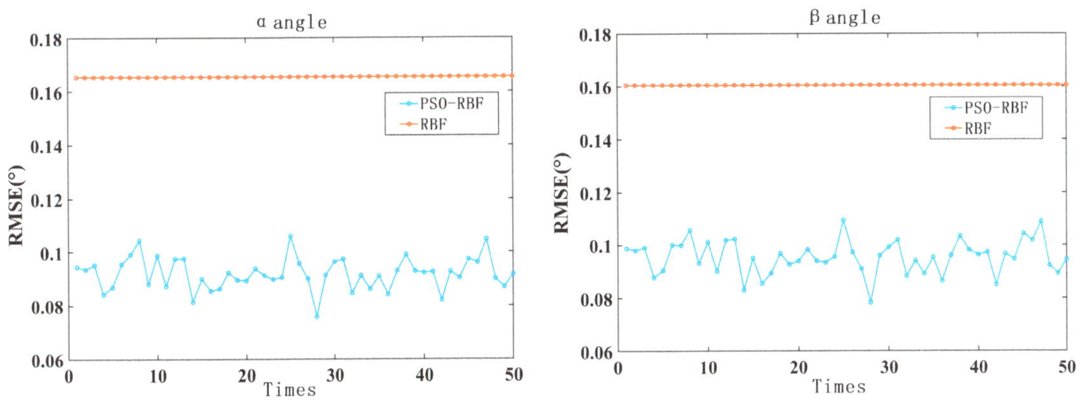

(**a**) α angle with PSO-RBF and RBF neural networks. (**b**) β angle with PSO-RBF and RBF neural networks.

Figure 9. The comparison of test results for α and β angles with the PSO optimized RBF neural network and unoptimized RBF neural network.

Through the validation of the γ angle with the COMSOL simulation data in Section 2.2, it can be seen in Figure 10 that the error distribution of the optimized RBF neural network is similar to that of the unoptimized RBF neural network, with a slight improvement in the accuracy. Through the data analysis, the improved PSO algorithm optimized the maximum error of the γ angle tested by the RBF neural network model to be 1.68°, with a root mean square error of 0.52°, while the unoptimized RBF neural network model tested a maximum error of 1.73° for the γ angle, with a root mean square error of 0.59°.

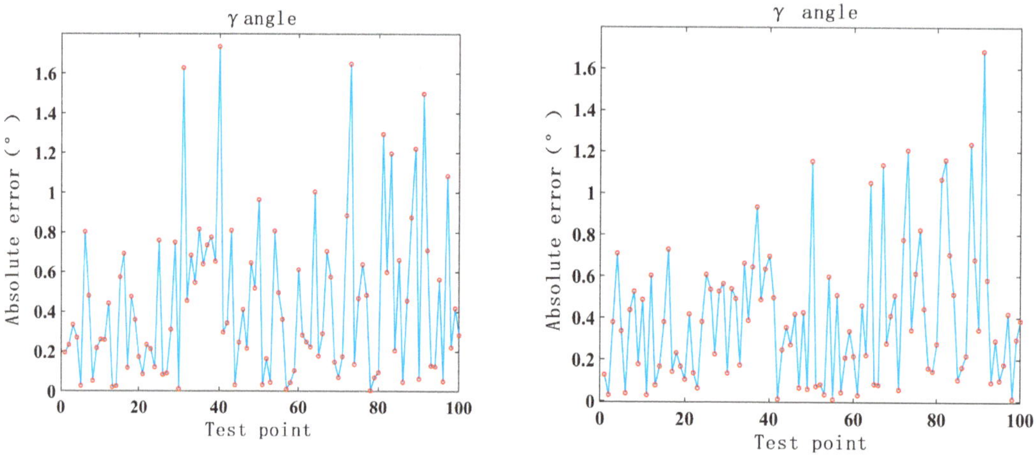

(**a**) Absolute error of angle for unoptimized γ. (**b**) Absolute error of angle for optimized γ.

Figure 10. The comparison of test results for γ with the PSO optimized RBF neural network and unoptimized RBF neural network.

4. Experiment

4.1. Experimental Equipment

The three-dimensional diagram of the experimental setup, as shown in Figure 11, includes three calibrated rotary stages for rotation around the *X*-, *Y*-, and *Z*-axes, namely, the RPI (rotary precision instrument), LS (Lian Sheng), and PI (Physik Instrumente) stages, with respective accuracies of $\pm 1''$, $\pm 4''$, and $\pm 2''$. The ball head was composed of an aluminum alloy material with a diameter of 50 mm. The permanent magnet used was a cylindrical sintered neodymium iron boron magnet, with a residual magnetic parameter of 1.2 mT and magnetization direction along the axis, with a diameter of 15 mm and height of 5 mm. The Hall sensor had a measurement range of 0~3 T and a resolution of 10–5 mT. The eddy current sensor had a measurement range of 1.5 mm and a resolution of 0.15 um.

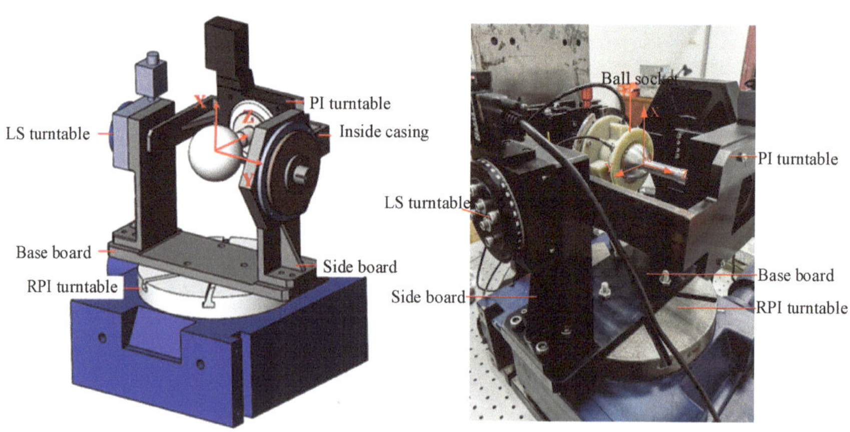

(**a**) Three-dimensional diagram of the calibration device.

Figure 11. *Cont.*

(b) Vertical view of the experimental platform.

Figure 11. Experimental installation.

The experimental process was carried out according to the rotation sequence of Euler angle x–y–z, first rotating the RPI table (α angle), then rotating the LS table (β angle), and finally rotating the PI table (γ angle).

4.2. Experimental Data Analysis

A total of 21*21*181 sets of the data were collected in the experiment, with a measurement range of $''-10^\circ \leq''\ \alpha \leq 10^\circ$, $''-10^\circ \leq''\ \beta \leq 10^\circ$, $''-90^\circ \leq''\ \gamma \leq 90^\circ$ and a sampling interval of 1°. Due to the amount of data, inputting all the data into the network resulted in excessively long training times, so the data were partitioned. Based on γ, each 20° was taken as an independent region, with a total of 9 regions. Table 1 shows the test results for each region, ME represents the mean error, and RMSE represents the root mean square error. It can be seen that the results between regions are not significantly different. The average difference in the RMSE value of the three rotation angles is less than 1′.

Table 1. Comparison of test results across all regions.

	α		β		γ	
	ME	**RMSE**	**ME**	**RMSE**	**ME**	**RMSE**
Q1	50.4″	1′22.8″	1′19.2″	2′9.6″	22′15.6″	28′44.4″
Q2	61.2″	1′19.2″	1′22.8″	2′20.4″	23′31.2″	29′38.4″
Q3	50.4″	1′8.4″	1′26.4″	2′34.8″	23′24″	29′49.2″
Q4	46.8″	1′4.8″	1′26.4″	2′31.2″	23′34.8″	29′52.8″
Q5	46.8″	1′1.2″	1′26.4″	2′24″	23′38.4″	30′7.2″
Q6	43.2″	57.6″	1′19.2″	2′20.4″	22′44.4″	28′37.2″
Q7	54″	1′12″	1′15.6″	1′48″	23′13.2″	29′24″
Q8	54″	1′8.4″	1′8.4″	1′40.8″	21′46.8″	27′32.4″
Q9	54″	1′12″	1′15.6″	1′58.8″	22′26.4″	28′8.4″

By integrating the measurement results from nine regions, the RMSE values of the rotation angles α, β, and γ are 1′8″, 2′15″, and 29′6″, respectively. The test results of all regions are shown in Figure 12, where the color represents the magnitude of the error. Due to the large number of test points, only points with significant errors are shown in this figure. Figure 12a shows the test error of the α angle in all regional test sets. For the α angle, most of the errors are below 0.1°. The errors in each region are relatively balanced, and the points with large errors are concentrated near 0° for α, β, and γ. Figure 12b shows the test errors for the beta angle of the test sets in all regions. For the beta angle, most of the errors are below 0.4°. As with the alpha angle, the larger errors are concentrated near 0° for α, β, and γ. Figure 12c shows the test errors of the γ angle in all regional test sets. Most of the errors of the γ angle are below 1°, which is equivalent to α and β angles with large errors and uniform error distributions.

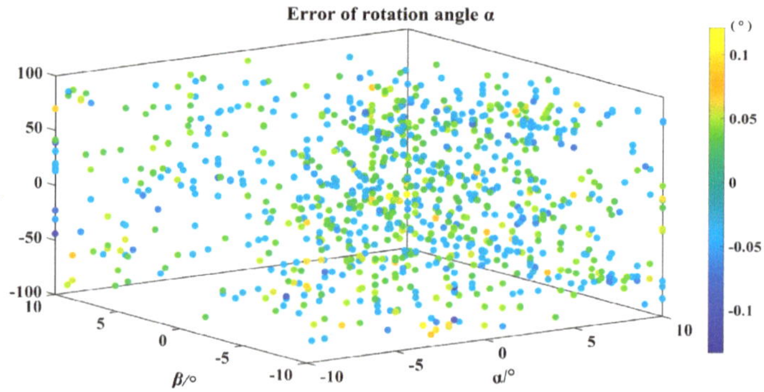

(**a**) α angle error plots for the test set in all areas.

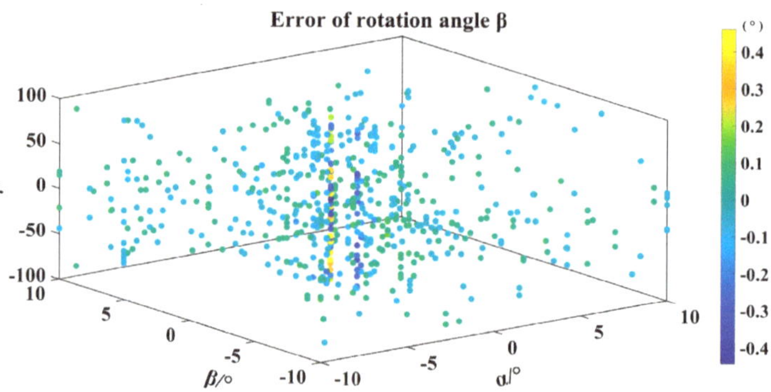

(**b**) β angle error plots for the test set in all areas.

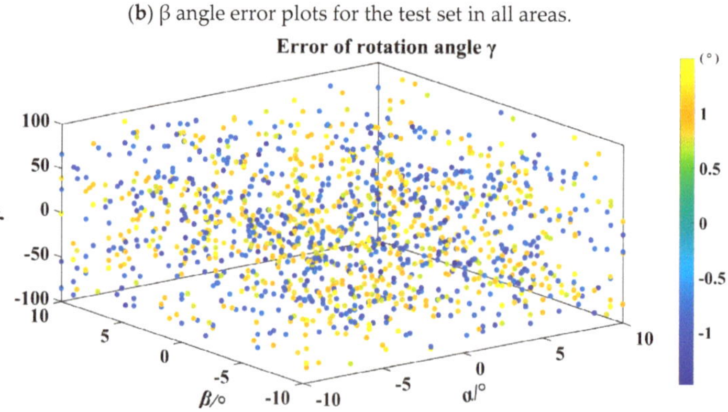

(**c**) γ angle error plots for the test set in all areas.

Figure 12. Three rotation angle error plots for the test set in all areas.

5. Analysis of Uncertainty

From the perspective of research progress and experimental experience, the main error factors affecting the measurement accuracy of the three-dimensional rotation of the spherical joint included the error of the permanent magnet structure parameters and magnetic field eccentricity, the eccentricity error of the ball head, the rotational error of the turntable, the error of repeated measurements, the stability error of the sensor, and the fitting calculation error of the neural network mode. Based on these sources of error, several types of uncertainty sources can currently be observed and verified through calculations.

5.1. Uncertainty Introduced by Sensor Measurement Repeatability

Under the same experimental conditions, measurements were repeated 10 times at the same spherical position (5°, 5°, 5°). Based on the results of the 10 repeated measurements, the uncertainty components introduced by the measurement repeatability error were evaluated, as shown in Table 2.

Table 2. Repeatability measurement data for (5°, 5°, 5°).

	1	2	3	4	5	6	7	8	9	10
$\alpha/°$	4.9264	4.9364	4.9344	4.9264	4.9625	4.9731	4.9311	4.9645	4.9564	4.9820
$\beta/°$	4.9689	4.9785	4.9805	4.9689	4.9536	4.9566	4.9720	4.9431	4.9513	4.9852
$\gamma/°$	4.3970	4.5035	4.5538	4.3970	4.557	4.6025	4.4078	4.5124	4.6031	4.3395

The standard uncertainty calculation formula for repetitive errors is as follows [32]:

$$u_R = \sqrt{\frac{\sum_{i=1}^{n}(a_i - \overline{a})^2}{10 \times 9}} \tag{5}$$

where u_R is the uncertainty introduced by sensor measurement repeatability, a_i is the respective angular value for the i measurement, and $\overline{a}$ is the average value of the angles. The calculation results are shown in Table 3.

Table 3. Standard uncertainty components introduced by repeatability.

Rotation Angle	$\overline{a}/°$	$u_R/°$	v_R
α	4.94932	0.006549	9
β	4.96586	0.004435	
γ	4.48736	0.030254	

5.2. Uncertainty Introduced by Drift

With the continuous changes in the experimental environment (temperature, vibration, etc.), the measurement system experienced a drift in the parameters of its components or mechanisms during the operation, which affected the accuracy of the measurement results. To assess the uncertainty introduced by drift to the system, the measurement system was kept stationary in a constant-temperature laboratory for a while, and the system's output values were recorded in real time. The uncertainty was then evaluated by calculating the range difference of the drift data during this period.

Under the assumption that the drift error follows a uniform distribution, the formula for calculating the standard uncertainty is as follows [32]:

$$u_D = \frac{a}{\sqrt{3}} \tag{6}$$

where u_D is the uncertainty introduced by drift and a is the maximum value of the error of the measured value. The probability that the measurement error falls within the interval $(x - a, x + a)$ is 1. The calculation and results are shown in Table 4.

Table 4. Standard uncertainty components introduced by drift.

Rotation Angle	$a/^\circ$	$u_D/^\circ$	v_D
α	0.0038	0.00219	5305
β	0.0133	0.00768	
γ	0.0394	0.02275	

Due to the fact that the two uncertainties are caused by different errors, it can be considered that they are independent of each other. Therefore, the formula for the combined standard uncertainty is [32]:

$$u_C = \sqrt{\left(u_R^2 + u_D^2\right)} \tag{7}$$

The synthetic degrees of freedom is [32]:

$$v_C = \frac{u_C^4}{\frac{u_R^4}{v_R} + \frac{u_D^4}{v_D}} \tag{8}$$

The combined uncertainty results are shown in Table 5.

Table 5. Combined uncertainty.

Rotation Angle	$u_C/^\circ$	v_C
α	0.006905	11
β	0.008868	141
γ	0.037853	22

The confidence probability was set to $p = 0.95$, we determined the inclusion factor, k, by checking the t-distribution table through the degrees of freedom, and calculated the spreading uncertainty. The results are:

$$(\alpha, \beta, \gamma) = (4.94932, 4.96586, 4.48796) \pm (0.015191, 0.017381, 0.078356) \tag{9}$$

6. Conclusions

This article presented a new method for the precise measurement of the 3D rotation angles of a spherical joint based on Hall and eddy current sensors. The spatial position matching of the permanent magnet and Hall sensor was optimized. The relationship between the output of the eddy current sensor and groove parameters was explored to determine the appropriate groove scheme for the ball head. A measurement model was established using the PSO-RBF neural network algorithm, and the feasibility of the method was verified through experiments. The three-dimensional rotation angles of the spherical joint were measured, with the root mean square errors of rotation angles α, β, and γ being $1'8''$, $2'15''$, and $29'6''$, respectively, and the mean errors being $51''$, $1'20''$, and $22'57''$. Compared with the spherical encoding three-dimensional rotation angle measurement scheme with root mean square errors of $22'32''$, $25'58''$, and $30'17''$, the accuracy was significantly improved.

The innovation of this method lies in the combination of the eddy current and Hall sensors, which solves the problem that the two-dimensional rotation angle scheme of ball hinge based on a magnetic effect cannot effectively measure rotation angle γ on the rotation axis. This scheme enabled the measurement of the three-dimensional rotation angle and provided a new method for the measurement of the three-degree-of-freedom rotation angle. On this basis, the PSO-RBF neural network algorithm was used to improve the precision of the spherical hinge measurement system. This scheme presents benefits for industrial precision manufacturing applications because it can work in harsh environments, such as situations where grating measurements are difficult to perform. It has excellent environmental adaptability, high accuracy, and fast response results, among other advantages. When implementing this scheme, there are no limitations on the required dimensions. As long as it is within the processable range of the spherical head, this scheme can be successfully used, and it has a wide range of applications. Compared to previous sensor modeling methods, this scheme simplifies the establishment of measurement models and reduces the impact of data processing on measurement accuracy.

Author Contributions: Conceptualization, K.O.; methodology, L.Y.; software, Q.Y.; validation, K.O., L.Y. and R.F.; formal analysis, P.H.; investigation, R.F.; resources, P.H.; data curation, L.Y.; writing—original draft preparation, Q.Y.; writing—review and editing, P.H.; visualization, Q.Y.; supervision, K.O.; project administration, K.O.; funding acquisition, L.Y. All authors have read and agreed to the published version of the manuscript.

Funding: This research was funded by the National Natural Science Foundation of China, grant number 52175505.

Institutional Review Board Statement: Not applicable.

Informed Consent Statement: Not applicable.

Data Availability Statement: Data are contained within the article.

Conflicts of Interest: The authors declare no conflict of interest.

References

1. Palpacelli, M.C.; Carbonari, L.; Palmieri, G. Details on the Design of a Lockable Spherical Joint for Robotic Applications. *J. Intell. Robot. Syst.* **2016**, *81*, 169–179. [CrossRef]
2. Guckert, M.L.; Naish, M.D. A Compact 3 Degree of Freedom Spherical Joint. *ASME J. Mech. Robot. August* **2011**, *3*, 264–278. [CrossRef]
3. Jäntsch, M.; Schmaler, C.; Wittmeier, S.; Dalamagkidis, K.; Knoll, A. A Scalable Joint-Space Controller for Musculoskeletal Robots with Spherical Joints. In Proceedings of the IEEE International Conference on Robotics & Biomimetics, Karon Beach, Thailand, 7–11 December 2011; pp. 2211–2216.
4. Gao, X.; Zhang, S.; Deng, J.; Liu, Y. Development of a Small Two-Dimensional Robotic Spherical Joint Using a Bonded-Type Piezoelectric Actuator. *IEEE Trans. Ind. Electron.* **2021**, *68*, 724–733. [CrossRef]
5. Folgheraiter, M.; Yessirkepov, S.; Yessaly, A. An Actuated Spherical Joint for Humanoid Robotics Applications. In Proceedings of the IEEE International Conference on Cybernetics and Intelligent Systems, IEEE Conference on Robotics, Automation and Mechatronics, Bangkok, Thailand, 18–20 November 2019; pp. 571–576.
6. Wang, W.; Zhang, M.; Zhu, Y.; Tang, C. A review on the measurement methods of multi-dimensional turning Angle of ball hinge. *J. Electron. Meas. Instrum.* **2017**, *31*, 1–8.
7. Khalid, A.; Mekid, S. Intelligent spherical joint based tri-actuated spatial parallel manipulator for precision applications. *Robot. Comput. Integr. Manuf.* **2018**, *54*, 173–184. [CrossRef]
8. Yun, H.; Jeon, H.; Yang, S.; Jun, M.B. Calibration of Industrial Robots with Spherical Joint using Single Wire Encoder. *Manuf. Lett.* **2022**, *33*, 46–50. [CrossRef]
9. Wright, S.E.; Mahoney, A.W.; Popek, K.M.; Abbott, J.J. The Spherical-Actuator-Magnet Manipulator: A Permanent-Magnet Robotic End-Effector. *IEEE Trans. Robot.* **2017**, *33*, 1013–1024. [CrossRef]
10. Bai, S.; Li, X.; Angeles, J. A review of spherical motion generation using either spherical parallel manipulators or spherical motors. *Mech. Mach. Theory* **2019**, *140*, 377–388. [CrossRef]
11. Mashimo, T.; Awaga, K.; Toyama, S. Development of a Spherical Ultrasonic Motor with an Attitude Sensing System using Optical Fibers. In Proceedings of the IEEE International Conference on Robotics and Automation, Rome, Italy, 10–14 April 2007; pp. 4466–4471.
12. Bai, K.; Li, K.; Lu, J. A magnetic flux model based method for detecting multi-DOF motion of a permanent magnet spherical motor. *Mechatronics* **2016**, *39*, 217–225. [CrossRef]
13. Li, Z.; Zhu, Y.; Xie, B.; Wang, Y.; Guo, X.; Sun, H. Position Detection Method of Piezoelectric Driven Spherical Motor Based on Laser Detection. *Micromachines* **2022**, *13*, 662. [CrossRef]

14. Wu, T.; Guo, X.; Wang, Q.; Li, J.; Gong, N. Real-time attitude detection method of permanent magnet spherical motor based on optical sensor. *Sens. Microsyst.* **2021**, *40*, 5.
15. Gofuku, A.; Yokomitsu, N.; Yano, T.; Kasashima, N. A Rotor Posture Measurement System by Analyzing Sensed Magnetic Field from Arrayed Hall Sensors. In Proceedings of the 2019 12th International Symposium on Linear Drives for Industry Applications, Neuchatel, Switzerland, 1–3 July 2019; pp. 1–5.
16. Rong, Y.; Wang, Q.; Lu, S.; Li, G.; Lu, Y.; Xu, J. Improving attitude detection performance for spherical motors using a MEMS inertial measurement sensor. *IET Electr. Power Appl.* **2019**, *25*, 198–205. [CrossRef]
17. Wang, D.; Zhang, C.; Mao, Y.; Bai, K. A transfer alignment method for inertial attitude correction of spherical motor on moving base. *Meas. Sci. Technol.* **2021**, *32*, 105023.
18. Li, M.; Kok-Meng, L.; Hanson, E. Sensor Fusion Based on Embedded Measurements for Real-Time Three-DOF Orientation Motion Estimation of a Weight-Compensated Spherical Motor. *IEEE Trans. Instrum. Meas.* **2022**, *71*, 9508009. [CrossRef]
19. Kim, J.H.; Kim, M.J.; Moon, J.H.; Kang, D.W. A Study on AHRS Position Sensing Accuracy for Control of the Traction Spherical Motor. *Trans. Korean Inst. Electr. Eng.* **2020**, *69*, 225–228. [CrossRef]
20. Xu, J.; Wang, Q.; Li, G.; Zhou, R.; Wen, Y.; Ju, L.; Zhou, S. Sensorless Posture Detection of Reluctance Spherical Motor Based on Mutual Inductance Voltage. *Appl. Sci.* **2021**, *11*, 3515. [CrossRef]
21. Yang, S.; Xu, Y.; Xu, Y.; Ma, T.; Wang, H.; Hou, J.; Liu, D.; Shen, H. A Novel Method for Detecting the Two-Degrees-of-Freedom Angular Displacement of a Spherical Pair, Based on a Capacitive Sensor. *Sensors* **2022**, *22*, 3437. [CrossRef]
22. Yang, L.; Hu, P.; Ma, K.; Zhang, J.; Dang, X.; Liu, S. A new method for measuring 3D rotation angle of spherical joint. *Measurement* **2022**, *190*, 110661–110675. [CrossRef]
23. Hu, P.; Cheng, S.; Liao, P.; Liu, W.; Zhu, Q. Intelligent ball hinge optimization design and accuracy improvement. *Chin. J. Sci. Instrum.* **2018**, *39*, 132–140.
24. Ma, K.; Yang, Q.; Zhang, J.; Dang, X.; Hu, P. A New 2D displacement measurement method based on an eddy current sensor and absolute encoding. *Machines* **2022**, *10*, 1077. [CrossRef]
25. Leong, T.K.; Saratchandran, P.; Sundararajan, N. Real-time performance evaluation of the minimal radial basis function network for identification of time varying nonlinear systems. *Comput. Electr. Eng.* **2002**, *28*, 103–117. [CrossRef]
26. Abiodun, O.I.; Jantan, A.; Omolara, A.E.; Dada, K.V.; Mohamed, N.A.; Arshad, H. State-of-the-art in artificial neural network applications: A survey. *Heliyon* **2018**, *10*, 112–153. [CrossRef] [PubMed]
27. Agatonovic-Kustrin, S.; Beresford, R. Basic concepts of artificial neural network (ANN) modeling and its application in pharmaceutical research. *J. Pharm. Biomed. Anal.* **2000**, *22*, 717–727. [CrossRef] [PubMed]
28. Pradhan, A.; Bisoy, S.K.; Das, A. A Survey on PSO Based Meta-Heuristic Scheduling Mechanism in Cloud Computing Environment. *J. King Saud Univ. Comput. Inf. Sci.* **2021**, *34*, 4888–4901. [CrossRef]
29. Rezaee-Jordehi, A. Enhanced leader PSO (ELPSO): A new PSO variant for solving global optimisation problems. *Appl. Soft Comput. J.* **2015**, *26*, 401–417. [CrossRef]
30. García-Gonzalo, E.; Fernández-Martínez, J.L. A Brief Historical Review of Particle Swarm Optimization (PSO). *J. Bioinform. Intell. Control.* **2012**, *1*, 3–16. [CrossRef]
31. Imran, M.; Hashim, R.; Khalid, N.E. An Overview of Particle Swarm Optimization Variants. *Procedia Eng.* **2013**, *53*, 491–496. [CrossRef]
32. Fei, Y.; Chen, X.; Qin, L.; Song, M.; Xv, L.; Huang, Q. *Error Theory and Data Analysis*, 7th ed.; China Machine Press: Beijing, China, 2015; pp. 83–89.

sensors

Article

Maximum Acceptable Tilt Angle for Point Autofocus Microscopy

Huixu Song *, Qingwei Li and Zhaoyao Shi

Beijing Engineering Research Center of Precision Measurement Technology and Instruments, Beijing University of Technology (BJUT), Beijing 100124, China; liqwei@emails.bjut.edu.cn (Q.L.); shizhaoyao@bjut.edu.cn (Z.S.)
* Correspondence: huixu_song@bjut.edu.cn

Abstract: The complete and accurate acquisition of geometric information forms the bedrock of maintaining high-end instrument performance and monitoring product quality. It is also a prerequisite for achieving the 'precision' and 'intelligence' that the manufacturing industry aspires to achieve. Industrial microscopes, known for their high accuracy and resolution, have become invaluable tools in the precision measurement of small components. However, these industrial microscopes often struggle to demonstrate their advantages when dealing with complex shapes or large tilt angles. This paper introduces a ray-tracing model for point autofocus microscopy, and it provides the quantified relationship formula between the maximum acceptable tilt angle and the beam offset accepted in point autofocus microscopy, then analyzing the maximum acceptable tilt angle of the objects being measured. This novel approach uses the geometric features of a high-precision reference sphere to simulate the tilt angle and displacement of the surface under investigation. The research findings show that the maximum acceptable tilt angles of a point autofocus microscope vary across different measured directions. Additionally, the extent to which the maximum acceptable tilt angles are affected by the distances of the beam offset also varies. Finally, the difference between the experiment results and the theoretical results is less than 0.5°.

Keywords: point autofocus microscopy; small components; complex components; maximum acceptable tilt angle; reference sphere

1. Introduction

In the era of artificial intelligence, complex and tiny components have seen wide application, especially in strategic sectors [1–3] such as next-generation communications, service robots, automotive electronics, intelligent sensors, defence equipment and aerospace. Both the geometric accuracy and product quality of these components are crucial for ensuring stable and accurate system operations. Accurate evaluation of product characteristics and quality relies on obtaining sufficient information about their condition, which in turn depends on the continuous improvement of measurement theory and technological devices. Therefore, high-precision measurement methods are indispensable tools for guiding the high-precision processing of complex and small components, thereby improving product quality. Complex small components, despite their overall diminutive size, often possess geometric features with large depth-to-width or length-to-diameter ratios [4]. Traditional contact measurement methods struggle to effectively obtain comprehensive information about complex small surfaces [5]. Furthermore, for components with different sizes but equivalent accuracy levels, the tolerance requirements for smaller ones are stricter than those for larger ones [6,7]. This makes it challenging for non-contact measurement methods such as structured light and machine vision to meet the required measurement accuracy. In this context, point autofocus microscopy technology provides a better solution for measuring the geometric errors of complex small components [8,9].

Point autofocus microscopy can be used to measure various complex surfaces, especially for small machine elements, such as micro gears and micro splines. Similarly to other microscopic measurement systems, point autofocus microscopy has limitations, including a narrow measurement field of view and a small maximum acceptable tilt angle [10]. To effectively gather complete surface information from a complex and small component, multiple measurements and data stitching are typically required to keep the tested region within the microscopy's maximum acceptable tilt angle [11,12]. However, this approach not only decreases the measurement efficiency but also amplifies potential errors [13–15]. Increasing the maximum acceptable tilt angle of the microscopic measurement systems can effectively reduce the need for data stitching and expand the measurement range. Simultaneously, it can improve the ability of point autofocus microscopy to measure the geometric information of complex surfaces. Nikolaev et al., 2016 conducted a study on the maximum acceptable tilt angle for a focus variation microscope, primarily focusing on the impact of different tilt angles of the measured surface on surface roughness measurements [16]. Thomas et al., 2021 investigated the maximum acceptable tilt angle for a coherence scanning interferometer by modeling a coherence scanning interferometer [17]. Gao et al., 2023 analyzed the maximum measurable tilt angle accepted by a confocal microscope under different numerical aperture (NA) objectives [18]. The maximum acceptable tilt angle of point autofocus microscopy is related to the offset distance and direction of the incident laser beam when the reflected laser beam from the workpiece remains within the receiving range of the objective [19]. Therefore, quantifying the relationship between the offset of the incident laser beam and the maximum acceptable tilt angle is crucial for expanding the measurement range of point autofocus microscopy.

This article expounds the principles of point autofocus microscopy and quantifies the relationship between the laser beam offset and the maximum acceptable tilt angle in the second part. In the third part, a maximum acceptable tilt angle measurement method based on a reference sphere and the parameters of several crucial components are given. In the fourth part, via theoretical calculations and experimental verification, we determine the maximum acceptable tilt angle of a point autofocus microscope with a sub-micrometre measurement accuracy. Finally, in the fifth part, the contents of this article are summarized and a reasonable point autofocus microscopy measurement scheme is given.

2. Basic Principle and Optical Model

2.1. Measurement Principle of Point Autofocus Microscopy

Point autofocus microscopy is a non-contact surface measuring tool that can automatically focus a collimated laser beam onto a target surface. This, in conjunction with the motion of a two-dimensional motion stage (2D motion stage), allows for dynamic scanning measurements of a surface [20]. Using an Olympus 100× objective (NA = 0.8), the focused spot can reach a minimum diameter of 1 μm. This data then enable a more in-depth analysis of the geometric accuracy of the measured surface.

Figure 1 illustrates the main structure of point autofocus microscopy, which comprises multiple components, including a laser source, LED source, beam splitter, objective, 2D motion stage, tube lens, focusing lens, PSD sensor, CCD camera (CCD, Charge Coupled Device), objective scanner (PZT, Piezoelectric Transducer) and motion control system. These components collectively form two systems: the laser measurement system and the white light imaging system. In the laser measurement system, the laser beam is reflected by the beam splitter and transmitted through the objective at a certain offset distance from the optical axis of the objective. The focused laser beam forms a micron-sized spot on the workpiece surface. After reflecting off the workpiece, the laser beam passes through the objective again and travels through the beam splitter as collimated light. It is then focused by the focusing lens onto the centre of the PSD sensor. When the 2D motion stage moves to a new position, the laser spot on the workpiece no longer aligns with the focal plane of the objective. Therefore, the laser beams passing through the objective, beam splitter and focusing lens deviate from their original positions, and the final laser beam is no longer focused on the centre of the PSD sensor. This shift in the focused position of the laser spot generates the corresponding photoelectric signals in the PSD sensor. These signals control the objective scanner, moving it in the direction of the optical axis with nanometre precision until the spot returns to the centre of the PSD sensor again. The distance that the objective scanner moves along the optical axis represents the difference in distance between the two measured positions on the workpiece towards the optical axis. This, combined with the motion of the 2D motion stage, facilitates the measurement of the 3D surface. However, the white light imaging system works entirely differently. Here, the broadband light emitted by the LED source is reflected by the beam splitter and transmitted through the objective without any offset. The white light reflected from the workpiece is projected onto the imaging plane of the CCD camera after transmission through the objective and the tube lens. When the surface of the workpiece aligns with the focal plane of the objective, the clearest image of the surface appears on the CCD imaging plane. In other words, when the workpiece surface aligns with the focal plane of the objective, the laser beam is focused on the centre of the PSD sensor, presenting the clearest image of the surface on the CCD camera.

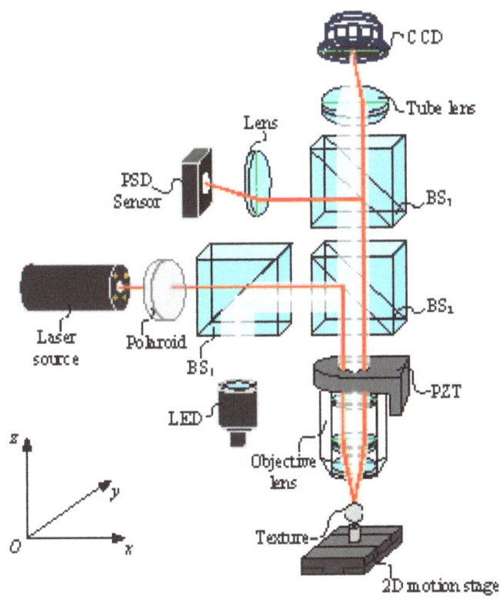

Figure 1. The schematic diagram of point autofocus microscopy.

2.2. The Relationship between Laser Beam Offset and Maximum Acceptable Tilt Angles

The maximum acceptable tilt angles of the point autofocus microscope are dependent on the direction and distance of the laser beam offset. As depicted in Figure 2, when the laser beam is positioned in the negative x-axis direction with an offset of Δx, A_1 and A_2 represent the maximum acceptable tilt angles for the clockwise and anti-clockwise rotation of the measured surface around the y-axis, respectively. Conversely, for both clockwise and anti-clockwise rotations of the surface measured around the x-axis, the maximum acceptable tilt angle is depicted at A_3; however, A_3 will change according to the direction and distance of the laser beam offset.

To quantify the maximum acceptable tilt angles under different directions and distances of the laser beam offset, an optical model of point autofocus microscopy is established, as shown in Figure 3. This model incorporates several key parameters. For the objective, these include a refractive index of n_1, a working distance of WD_1, a centre thickness of t_1 and a curvature radius of R_1. The focusing lens has a refractive index of n_2, a working distance of WD_2, a centre thickness of t_2 and a curvature radius of R_2. The distance between the objective and the centre of the BS_3 is l_1, while l_2 represents the distance between the focusing lens and the centre of the BS_3. In this model, the laser beam enters the optical system with the pose of $[d_{in}, \theta_{in}]$, and exits with the pose of $[d_{out}, \theta_{out}]$. Here, d_{in} and θ_{in} represent the offset displacement and deflection angle of the laser beam reflected by the measured surface when it enters the objective with respect to the optical axis of the objective. Similarly, d_{out} and θ_{out} represent the offset displacement and deflection angle of the laser beam when it passes through the focusing lens and enters the PSD sensor relative to the optical axis of the focusing lens. Applying the principle of paraxial ray tracing (see (1)), it is possible to calculate the position of the spot on the PSD sensor [21,22].

$$\begin{aligned} d_{out} = & \begin{bmatrix} 1 & f_2 \end{bmatrix} \cdot \begin{bmatrix} 1 & 0 \\ 0 & n_2 \end{bmatrix} \cdot \begin{bmatrix} 1 & t_2 \\ 0 & 1 \end{bmatrix} \cdot \begin{bmatrix} 1 & 0 \\ \frac{1-n_2}{n_2 R_2} & \frac{1}{n_2} \end{bmatrix} \cdot \begin{bmatrix} 1 & l_1 \\ 0 & 1 \end{bmatrix} \\ & \cdot \begin{bmatrix} 1 & 0 \\ \frac{n_1-1}{R_1} & n_1 \end{bmatrix} \cdot \begin{bmatrix} 1 & t_1 \\ 0 & 1 \end{bmatrix} \cdot \begin{bmatrix} 1 & 0 \\ 0 & \frac{1}{n_1} \end{bmatrix} \cdot \begin{bmatrix} d_{in} \\ \theta_{in} \end{bmatrix} \end{aligned} \quad (1)$$

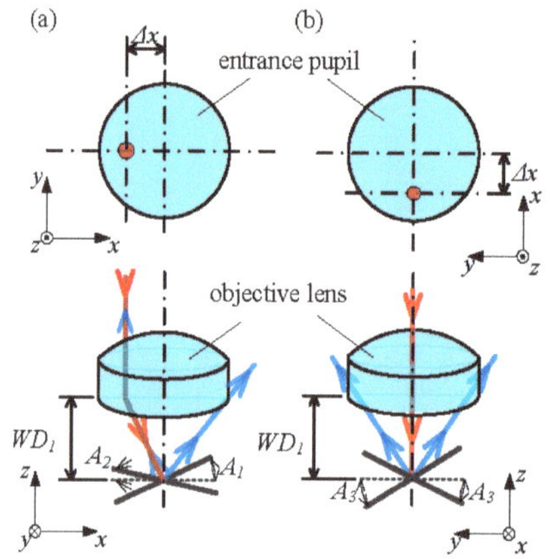

Figure 2. The relationship between laser beam offset and maximum acceptable tilt angles. (**a**) The measured surface tilts around the y-axis; (**b**) The measured surface tilts around the x-axis. The red arrow represents incident laser beam, and the blue arrow represents reflected laser beam.

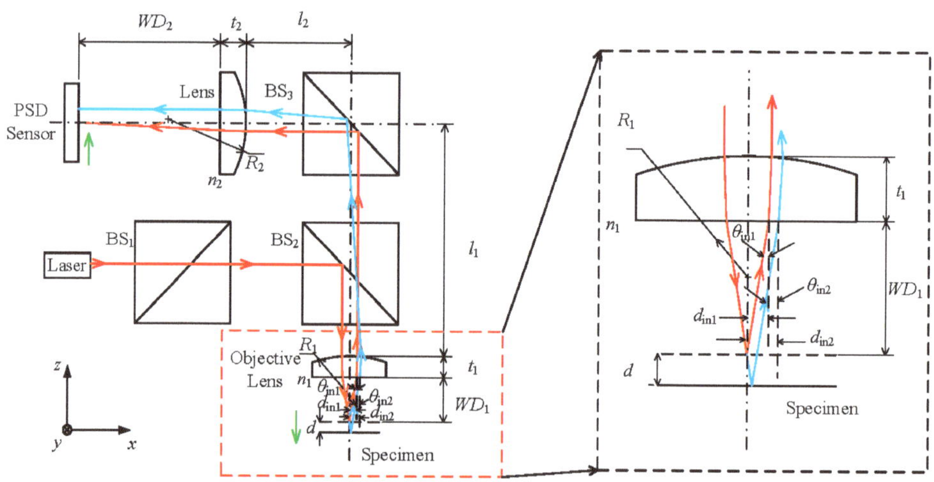

Figure 3. Optical model of point autofocus microscopy.

Figure 3 illustrates two propagation paths of the laser beam depending on the location of the measured surface. The solid red line represents the propagation path of the laser beam when the measured surface is located at the focal plane of the objective. Reflected by the measured surface, the laser beam enters the objective with the pose of $[d_{\mathrm{in1}}, \theta_{\mathrm{in1}}]$. The reflected laser beam is focused onto the PSD sensor surface with the pose of $[d_{\mathrm{out1}}, \theta_{\mathrm{out1}}]$.

Alternatively, the solid blue line in Figure 3 represents the propagation path of the laser beam when the measured surface is moved a certain distance d_{in} in the negative z-axis direction. Reflected by the measured surface, the laser beam enters the objective with the pose of $[d_{\text{in2}}, \theta_{\text{in2}}]$. The reflected laser beam is focused onto the PSD sensor surface with the pose of $[d_{\text{out2}}, \theta_{\text{out2}}]$. Formula (2) can be used to determine the position of the laser spot on the PSD sensor.

$$d_{out2} = \begin{bmatrix} A & B \end{bmatrix} \cdot \begin{bmatrix} d_{in1} \\ \theta_{in1} \end{bmatrix} = A \cdot d_{in2} + B \cdot \theta_{in2} \tag{2}$$

Here,

$$\begin{aligned} \begin{bmatrix} A & B \end{bmatrix} = \begin{bmatrix} 1 & f_2 \end{bmatrix} \cdot \begin{bmatrix} 1 & 0 \\ 0 & n_2 \end{bmatrix} \cdot \begin{bmatrix} 1 & t_2 \\ 0 & 1 \end{bmatrix} \cdot \begin{bmatrix} 1 & 0 \\ \frac{1-n_2}{n_2 R_2} & \frac{1}{n_2} \end{bmatrix} \cdot \begin{bmatrix} 1 & l_1 \\ 0 & 1 \end{bmatrix} \\ \cdot \begin{bmatrix} 1 & 0 \\ \frac{n_1-1}{R_1} & n_1 \end{bmatrix} \cdot \begin{bmatrix} 1 & t_1 \\ 0 & 1 \end{bmatrix} \cdot \begin{bmatrix} 1 & 0 \\ 0 & \frac{1}{n_1} \end{bmatrix} \end{aligned} \tag{3}$$

According to the geometric characteristics of the optical model, θ_{in1} equals θ_{in2}. Therefore, changes in the position of the laser spot on the PSD sensor can be determined using Formula (4).

$$d_{out2} - d_{out1} = A \cdot (d_{in2} - d_{in1}) = A \cdot \Delta d_{in} \tag{4}$$

Step 1: Figure 2a illustrates a scenario where the measured surface tilts anti-clockwise around the y-axis. As shown in Figure 4, d_{in} of the reflected laser beam increases as the surface moves away when the tilt angle $\theta < \theta_{\text{lim}}$, resulting in $\Delta d_{\text{in}} > 0$. When the tilt angle θ continues to increase to θ_{lim}, the reflected laser beam aligns with the incident laser beam and d_{in} remains constant irrespective of surface movements; hence, $\Delta d_{\text{in}} = 0$. At this inclination angle, the PSD sensor fails to accurately determine the displacement and direction of the objective. However, if the tilt angle θ continues to increase, d_{in} of the reflected laser beam decreases as the surface moves away, resulting in $\Delta d_{\text{in}} < 0$. The position where the reflected laser beam aligns with the incident laser beam determines the maximum acceptable tilt angle A_1, which can be calculated using Formula (5). In this formula, Δx represents the offset displacement of the incident laser beam on the objective.

$$A_1 = \arctan\frac{\Delta x}{WD_1} \tag{5}$$

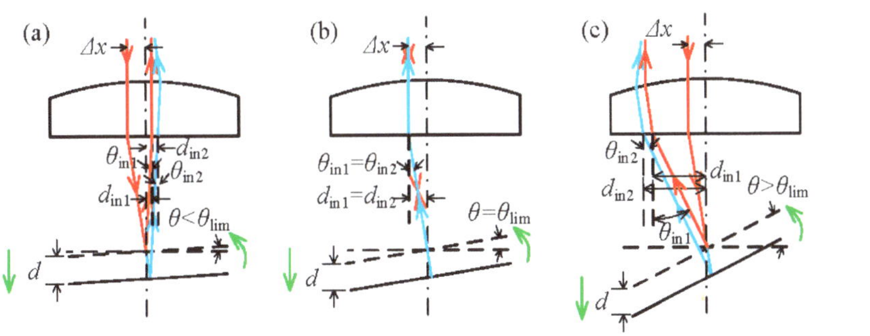

Figure 4. The relative position for the incident and reflected laser beam when the measured surface tilts anti-clockwise around the y-axis. (**a**) The tilt angle $\theta < \theta_{\text{lim}}$; (**b**) The tilt angle $\theta = \theta_{\text{lim}}$; (**c**) The tilt angle $\theta > \theta_{\text{lim}}$. The red arrow represents the laser beam before the tilt of the measured surface, and the blue arrow represents the laser beam after the tilt of the measured surface.

Step 2: Figure 2a illustrates a scenario where the measured surface tilts clockwise around the y-axis. As shown in Figure 5, d_{in} of the reflected laser beam increases as the surface moves away and decreases as the surface moves closer. This indicates a consistent correlation between the motion direction of the laser spot on the PSD sensor and the motion direction of the measured surface. This is necessary to ensure that the laser beam reflected by the measured surface does not exceed the effective diameter range of the objective. The maximum acceptable tilt angle, A_2, can be determined using Formula (6), where α represents the aperture angle of the objective.

$$A_2 = \frac{\frac{\alpha}{2} - \arctan\frac{\Delta x}{WD_1}}{2} \tag{6}$$

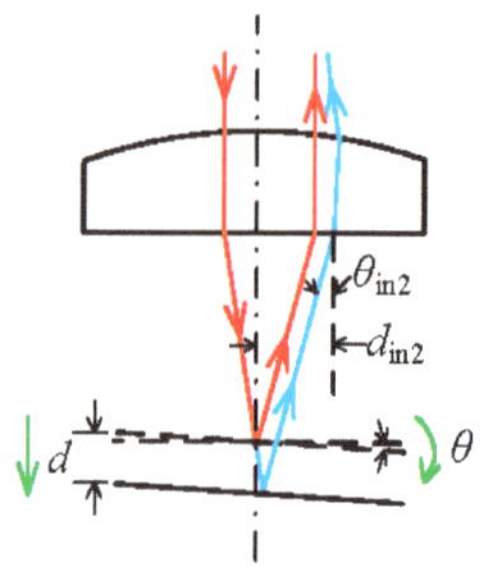

Figure 5. The relative position for the incident and reflected laser beam when the measured surface tilts clockwise around the y-axis. The red arrow represents the laser beam before the tilt of the measured surface, and the blue arrow represents the laser beam after the tilt of the measured surface.

Step 3: Figure 2b illustrates a scenario where the measured surface tilts clockwise and anti-clockwise around the x-axis. As shown in Figure 6, $|d_{in}|$ of the reflected laser beam increases as the surface moves away and decreases as it moves closer. This demonstrates a distinct and fixed correlation between the motion direction of the laser spot on the PSD sensor and the motion direction of the measured surface. It is important to ensure that the laser beam reflected by the measured surface does not exceed the effective diameter range of the objective. The maximum acceptable tilt angle, A_3, can be calculated using Formula (7). Here, D represents the effective diameter of the objective.

$$A_3 = \frac{\arctan\frac{\sqrt{\left(\frac{D}{2}\right)^2 - (\Delta x)^2}}{WD_1}}{2} \tag{7}$$

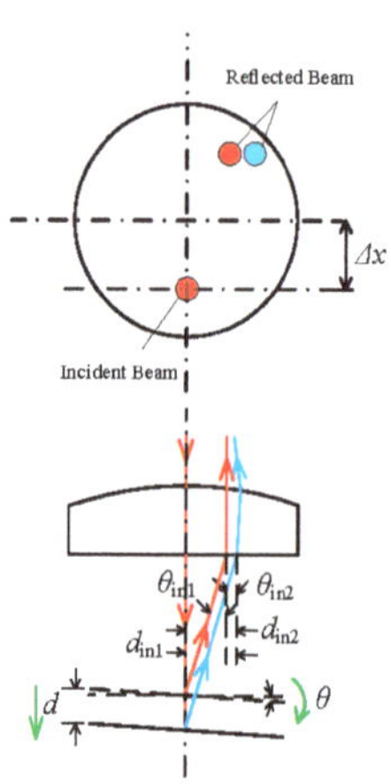

Figure 6. The relative position for the incident and reflected laser beam when measured surface tilts clockwise and anti-clockwise around the x-axis. The red arrow represents the laser beam before the tilt of the measured surface, and the blue arrow represents the laser beam after the tilt of the measured surface.

3. Experimental Validation

3.1. Experimental Method

A maximum acceptable tilt angle is crucial in point autofocus microscopy, which measures its accuracy on curved or inclined surfaces. A reference sphere is used in the experiment (as shown in Figure 7) to simulate the displacement and tilt angle changes of the workpiece.

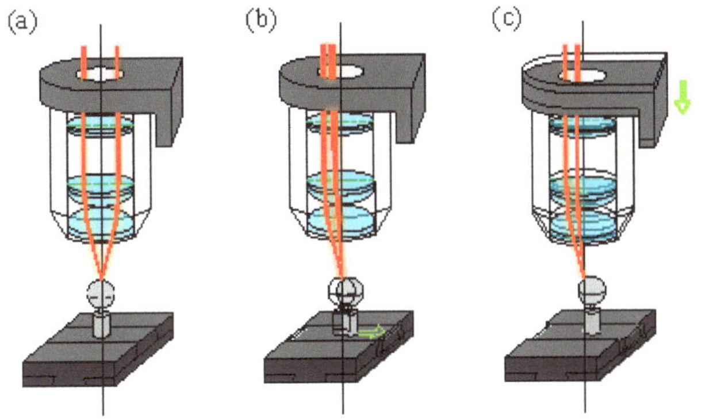

Figure 7. Schematic of the experimental process. (**a**) The initial status; (**b**) The motion of the reference sphere during measurement; (**c**) The motion process of the objective scanner during the measurement. The green arrow indicates the direction of component movement.

Figure 7a shows the initial status, where the centre of the reference sphere is located on the extension line of the optical axis of the objective. By adjusting the objective scanner, the apex of the reference sphere is located on the focal plane of the objective. Figure 7b shows the motion of the reference sphere during measurement. The reference sphere moves in the positive direction of the x-axis in incremental steps by controlling the 2D motion stage. This movement causes the measured point on the reference sphere to shift away from the objective, resulting in the deviation of the laser beam by the reference sphere. Consequently, it causes the spot on the PSD sensor to deviate from the centre. Figure 7c shows the motion process of the objective scanner during the measurement. The PSD sensor generates a deviation signal, processed and used to prompt the objective scanner to move in the negative direction of the z-axis. The objective approaches the reference sphere until the measured point on the sphere aligns with the focal plane of the objective. The distance moved by the objective scanner indicates the coordinate difference z_{m} (also known as the measuring value) between the two points measured on the reference sphere along the optical axis of the objective. Using the geometric parameters and horizontal motion distance of the reference sphere, we can accurately calculate the coordinate difference z_{t} (also known as theoretical value) between the two points measured along the optical axis of the objective (see (8)).

$$z_t = R - \sqrt{R^2 - x^2} \tag{8}$$

Simultaneously, we can also calculate the tilt angle at each measurement position of the reference sphere. R represents the radius of the reference sphere, and x signifies the displacement of the reference sphere along the x-axis. In the experiment, the 2D motion stage carries the reference sphere in 50 μm steps, moving in both the positive and negative directions of the x-axis, as well as the positive direction of the y-axis. A high-precision incremental-length gauge performs the precise displacement of the reference sphere. Given the sub-micrometre measurement accuracy of the system, the measurement error of the point autofocus microscopy is maintained within 1 μm. Ultimately, the tilt angles corresponding to the extreme error position are obtained and defined as the measured value A_{m} of the maximum acceptable tilt angles (see (9)).

$$A_{\mathrm{m}} = \arcsin\frac{x}{R} \tag{9}$$

3.2. Experimental System

This article presents an experimental method for quantifying the maximum acceptable tilt angles of point autofocus microscopy and constructs an experimental system, as shown in Figure 8. The accuracy of the entire experimental system relies on several crucial components. These include the objective, reference sphere, objective scanner, PSD sensor and high-precision incremental-length gauges. First, our objective is the LMPlanFL N 50 from Olympus, Tokyo, Japan, and its main technical parameters are listed in Table 1.

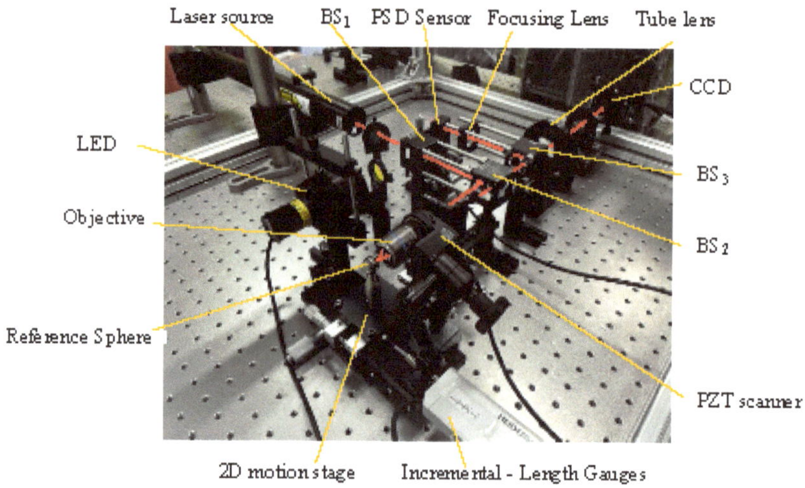

Figure 8. The protype of the experimental system.

Table 1. The technical parameters of objectives.

Olympus LMPlanFL N 50 × Object Lens		Units
NA	0.5	
Working Distance	10.6	mm
Focal Length	3.6	mm
Resolution	0.67	μm

The reference sphere is an STL Precision Ball from Hexgon, Stockholm, Sweden, and its main technical parameters are listed in Table 2.

Table 2. The technical parameters of reference sphere.

Precision Balls		Units
Diameter	15.8756	mm
Roundness	0.06	μm

The objective scanner is P73.Z200S from COREMORROW, Harbin, China, and its main technical parameters are listed in Table 3.

Table 3. The technical parameters of objective scanner.

P73.Z200S		Units
Travel	200	μm
Resolution	5.5	nm
Positioning Error	±0.6	μm
Repeatability	±0.5	μm

The PSD sensor is the PDP90A from Thorlabs, Newton, NJ, USA, and its main technical parameters are listed in Table 4.

Table 4. The technical parameters of PSD sensor.

PDP90A		Units
Saturation Power	100	μw
Minimum Power	20	μw
Resolution	0.75	μm
Sensor Size	9 × 9	mm

The high-precision incremental-length gauge is MT 2500 from Heidenhain, Traunreut, Germany, and its main technical parameters are listed in Table 5.

Table 5. The technical parameters of high-precision incremental-length gauge.

MT 2500		Units
Measurement Range	25	mm
Position Error	0.2	μm
Repeatability	0.02	μm

The experimental protype is shown in Figure 8.

4. Results and Discussion

Owing to the restriction of the objective's entrance pupil and the diameter of the laser beam, the maximum offset of the laser beam is 4 mm. To verify the quantitative relationship between the offset and the maximum acceptable tilt angles, an experiment has been designed to measure the maximum acceptable tilt angles A_{m1}, A_{m2} and A_{m3} at offsets of 1 mm, 2 mm, 3 mm and 4 mm, respectively. Each offset distance will undergo five repeated measurements to ensure data accuracy. The standard deviation σ of the measurement data and the expanded uncertainty u (k = 2) can be calculated using Formulas (10) and (11), respectively. n represents the number of repeated measurements, and e_i represents the measurement error for the ith (1, 2, 3, 4, 5) group at each measurement position. In order to achieve a sub-micrometre measurement accuracy, the expanded uncertainty u should be less than 1 μm.

$$\sigma = \sqrt{\frac{\sum_{i=1}^{n}(e_i - \overline{e})^2}{n-1}} \tag{10}$$

$$u = k \cdot \frac{\sigma}{\sqrt{n}} \tag{11}$$

Figure 9 shows the measurement error when the measured surface tilts anti-clockwise around the y-axis with the laser beam offset at 1 mm, 2 mm, 3 mm and 4 mm. In Figures 9–11, T1, T2, T3, T4 and T5, respectively, represent the experimental data of the five repeated measurements.

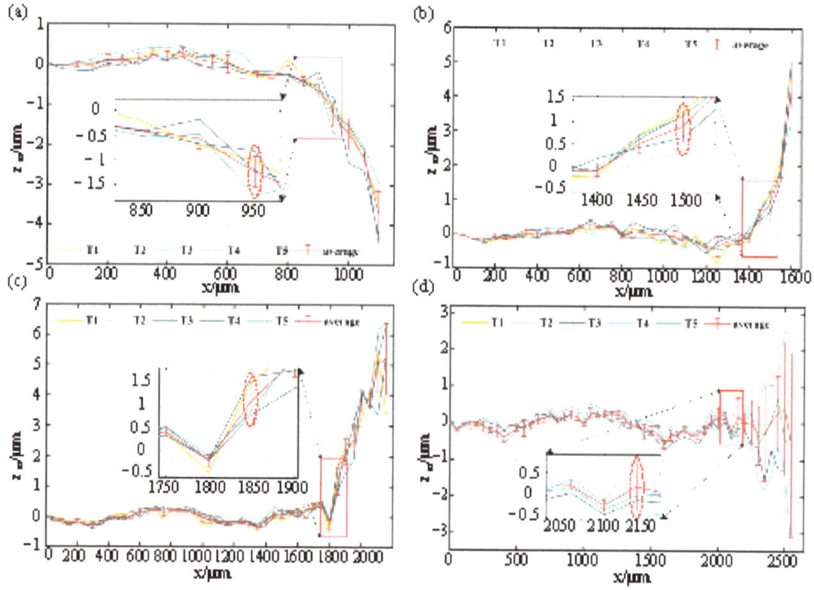

Figure 9. The measurement error when the measured surface tilts anti-clockwise around the y-axis. (**a**) The laser beam offset at 1 mm; (**b**) The laser beam offset at 2 mm; (**c**) The laser beam offset at 3 mm; (**d**) The laser beam offset at 4 mm.

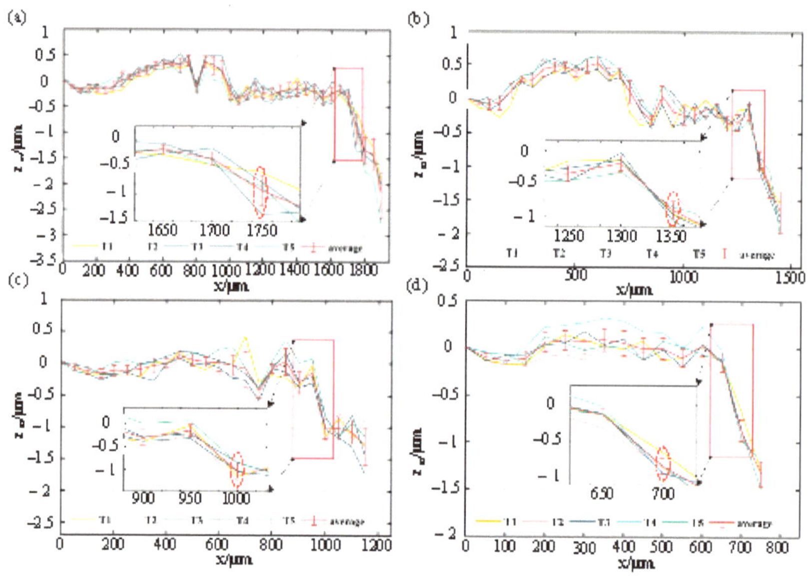

Figure 10. The measurement error when the measured surface tilts clockwise around the y-axis. (**a**) The laser beam offset at 1 mm; (**b**) The laser beam offset at 2 mm; (**c**) The laser beam offset at 3 mm; (**d**) The laser beam offset at 4 mm.

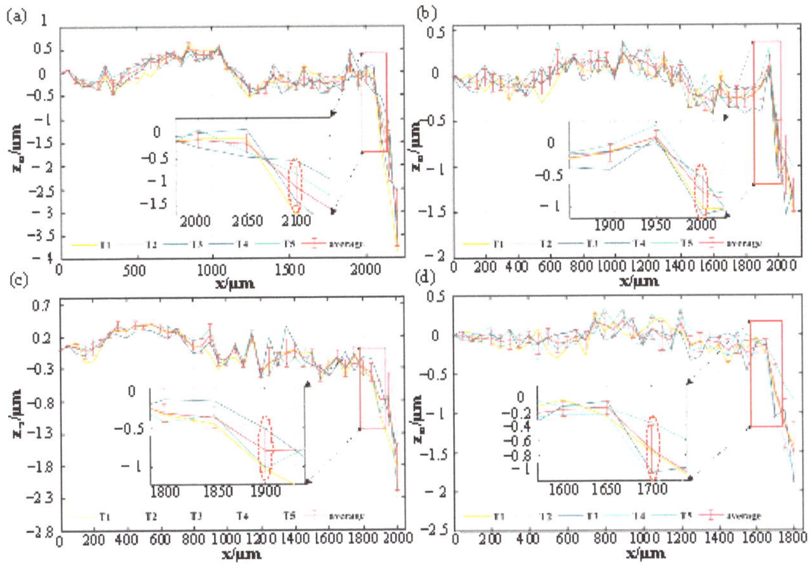

Figure 11. The measurement error when the measured surface tilts clockwise and anti-clockwise around the x-axis. (**a**) The laser beam offset at 1 mm; (**b**) The laser beam offset at 2 mm; (**c**) The laser beam offset at 3 mm; (**d**) The laser beam offset at 4 mm.

The maximum offset distance of the reference sphere within sub-micrometre measurement accuracy is circled with a red ellipse. Applying Formula (5), we can determine the theoretical maximum acceptable tilt angles A_1 for the four offset distances, which are 5.3°, 10.7°, 15.5° and 20.3°. Similarly, Formula (9) allows us to calculate the measured maximum acceptable tilt angles A_{m1} with for these four offset distances, yielding values of 5.3°, 10.5°, 15.3° and 20.7°. The analysis in step 1 of Section 2.2 shows that the maximum acceptable tilt angles are not directly affected by the objective lens when the measured surface tilts anti-clockwise around the y-axis, but are determined by the laser incident angle. Under this tilt direction, the maximum acceptable tilt angles can reach 20.3°.

Figure 10 shows the measurement error when the measured surface tilts clockwise around the y-axis for the same offset distances.

Applying Formula (6), we calculate the theoretical maximum acceptable tilt angles A_2 for the four offset distances. The results are 12.3°, 9.7°, 7.1° and 4.7°. Using Formula (9), we determine the measured maximum acceptable tilt angles A_{m2} for these offsets as 12.4°, 9.4°, 6.9° and 5.0°. Due to the laser offset, the reflected laser is closer to the tilt direction when the measured surface tilts clockwise around the y-axis. The reflected laser deviates from the objective lens more easily, resulting in a smaller maximum acceptable tilt angle. Under this tilt direction, the maximum acceptable tilt angles can reach 12.3°.

Figure 11 shows the measurement error when the measured surface tilts both clockwise and anti-clockwise around the x-axis for the same offset distances. Furthermore, the laser beam offset remains at 1 mm, 2 mm, 3 mm and 4 mm.

According to Formula (7), the theoretical maximum acceptable tilt angles A_3 with the four offset distances are 14.6°, 13.9°, 13.0° and 11.4°. Using Formula (9), we find the measured maximum acceptable tilt angles A_{m3} to be 14.5°, 13.8°, 13.1° and 11.7° for these distances. There is no offset in the projection of the incident laser and reflected laser on the yoz plane, so the maximum acceptable tilt angles are the same when the measured surface tilts clockwise and anti-clockwise around the x-axis. In this tilt direction, the maximum acceptable tilt angles can reach 14.5°.

The experimental findings reveal that the maximum acceptable tilt angle A_{m1} increases as the offset distance increases. Conversely, A_{m2} and A_{m3} decrease with a growing offset distance. Interestingly, there is minimal difference between the theoretical and measured values of the maximum acceptable tilt angles. In practical applications of point autofocus microscopy, the maximum acceptable tilt angles should be the lesser of the theoretical and measured values. Moreover, the surface being measured is recommended to move in the direction of the laser beam offset when it has a large inclination.

5. Conclusions

This article constructs a point autofocus microscope, for which a ray-tracing model is established. The functional relationship between the laser beam offset and the maximum acceptable tilt angles with different inclined directions of the measured surface is provided theoretically. To verify the accuracy of theoretical analysis, a novel experimental scheme was proposed. This scheme utilises the geometric features of a precision reference sphere to simulate the deflection angle and displacement of the measured surface. By maintaining the measurement accuracy of point autofocus microscopy within a maximum error of 1 μm, we were able to find the measurement values of the maximum acceptable tilt angles in different directions. The difference between the experiment results and the theoretical results is less than 0.5°. Therefore, the functional relationship formula proposed in this paper can effectively describe the relationship between the maximum acceptable tilt angles and the beam offset accepted when using point autofocus microscopy. This research demonstrates that the laser beam offset affects the maximum acceptable tilt angles differently in various directions for point autofocus microscopy. The microscope's maximum acceptable tilt angle reaches a peak of 20.3° at a laser beam offset of 4mm, when the tilt direction of the surface measured aligns with the direction of beam offset. Conversely, the microscope's maximum acceptable tilt angle reaches a maximum value of 12.3° or 14.5° at a laser beam offset 1mm when the tilt direction of the surface measured is opposite or perpendicular to the direction of beam offset. The device demonstrates its strongest ability to measure an inclined surface when measuring in the offset direction of the laser beam.

Author Contributions: Conceptualization, H.S.; methodology, H.S.; software, Q.L.; validation, H.S. and Z.S.; formal analysis, H.S.; investigation, Q.L.; resources, H.S.; data curation, Q.L.; writing—original draft preparation, Q.L.; writing—review and editing, H.S.; visualization, Q.L.; supervision, H.S.; project administration, Z.S.; funding acquisition, H.S. All authors have read and agreed to the published version of the manuscript.

Funding: The work was supported by the National Natural Science Foundation of China (52105043), R&D Program of Beijing Municipal Education Commission (KM202210005034), State Key Laboratory of Precision Measuring Technology and Instruments (Tianjin University) (PILAB2105) and Young Elite Scientists Sponsorship Program by CAST (2021QNRC001).

Institutional Review Board Statement: Not applicable.

Informed Consent Statement: Not applicable.

Data Availability Statement: The data presented in this study are available on request from the corresponding author.

Conflicts of Interest: The authors declare no conflict of interest.

References

1. Yu, Z.; Li, D.; Yang, J.; Zeng, Z.; Yang, X.; Li, J. Fabrication of micro punching mold for micro complex shape part by micro EDM. *Int. J. Adv. Manuf. Technol.* **2018**, *100*, 743–749. [CrossRef]
2. Wang, D.; Yan, N.; Liu, H.; Li, C.; Yin, S.; Zhang, X. Study on microscopic fringe structured light measurement method for tiny and complex structural components. *Appl. Phys. B* **2022**, *128*, 207. [CrossRef]
3. Geiger, M.; Kleiner, M.; Eckstein, R.; Tiesler, N.; Engel, U. Microforming. *CIRP Ann.* **2001**, *50*, 445–462. [CrossRef]
4. Zeng, N.; Wu, P.; Wang, Z.; Li, H.; Liu, W.; Liu, X. A small-sized object detection oriented multi-scale feature fusion approach with application to defect detection. *IEEE Trans. Instrum. Meas.* **2022**, *71*, 3507014. [CrossRef]

5. Luo, M.; Zhong, S. Non-Contact Measurement of Small-Module Gears Using Optical Coherence Tomography. *Appl. Sci.* **2018**, *8*, 2490. [CrossRef]
6. Minetola, P.; Galati, M.; Calignano, F.; Iuliano, L.; Rizza, G.; Fontana, L. Comparison of dimensional tolerance grades for metal AM processes. *Procedia CIRP* **2020**, *88*, 399–404. [CrossRef]
7. Kitsakis, K.; Kechagias, J.; Vaxevanidis, N.; Giagkopoulos, D. Tolerance Analysis of 3d-MJM parts according to IT grade. *IOP Conf. Ser. Mater. Sci. Eng.* **2016**, *161*, 012024. [CrossRef]
8. Miura, K.; Tsukamoto, T.; Nose, A.; Takeda, R.; Ueda, S. Accuracy verification of gear measurement with a point autofocus probe. In Proceedings of the Euspen's 18th International Conference and Exhibition, Venice, Italy, 4–8 June 2018.
9. Hadian, H.; Piano, S.; Xiaobing, F.; Leach, R. Gear measurements using optical point autofocus profiling. In Proceedings of the EUSPEN's 19th International Conference & Exhibition, Bilbao, Spain, 3–7 June 2019.
10. Miura, K.; Nose, A. Accuracy verification of a point autofocus instrument for freeform 3D measurement. In Proceedings of the 20th International Symposium on Advances in Abrasive Technology, Okinawa, Japan, 3–6 December 2017; pp. 1080–1084.
11. Thomas, M.; Su, R.; de Groot, P.; Leach, R. Optical topography measurement of steeply-sloped surfaces beyond the specular numerical aperture limit. In Proceedings of the Optics and Photonics for Advanced Dimensional Metrology, Online, 6–10 April 2020; pp. 29–36.
12. Chen, S.; Lu, W.; Guo, J.; Zhai, D.; Chen, W. Flexible and high-resolution surface metrology based on stitching interference microscopy. *Opt. Lasers Eng.* **2022**, *151*, 106915. [CrossRef]
13. Liu, M.Y.; Cheung, C.F.; Feng, X.; Wang, C.J.; Leach, R.K. A self-calibration rotational stitching method for precision measurement of revolving surfaces. *Precis. Eng.* **2018**, *54*, 60–69. [CrossRef]
14. Liu, M.Y.; Cheung, C.F.; Cheng, C.H.; Su, R.; Leach, R.K. A Gaussian process and image registration based stitching method for high dynamic range measurement of precision surfaces. *Precis. Eng.* **2017**, *50*, 99–106. [CrossRef]
15. Keller, F.; Stein, M. A reduced self-calibrating method rotary table error motions. *Meas. Sci. Technol.* **2023**, *34*, 065015. [CrossRef]
16. Nikolaev, N.; Petzing, J.; Coupland, J. Focus variation microscope: Linear theory and surface tilt sensitivity. *Appl. Opt.* **2016**, *55*, 3555–3565. [CrossRef] [PubMed]
17. Thomas, M.; Su, R.; De Groot, P.; Coupland, J.; Leach, R. Surface measuring coherence scanning interferometry beyond the specular reflection limit. *Opt. Express* **2021**, *29*, 36121–36131. [CrossRef] [PubMed]
18. Gao, S.; Felgner, A.; Hueser, D.; Wyss, S.; Brand, U. Characterization of the maximum measurable slope of optical topography measuring instruments. In Proceedings of the Optical Measurement Systems for Industrial Inspection XIII, Munich, Germany, 26–29 June 2023; pp. 576–582.
19. Leach, R. *Optical Measurement of Surface Topography*; Springer: Berlin/Heidelberg, Germany, 2011; Volume 8.
20. Maculotti, G.; Feng, X.; Su, R.; Galetto, M.; Leach, R. Residual flatness and scale calibration for a point autofocus surface topography measuring instrument. *Meas. Sci. Technol.* **2019**, *30*, 075005. [CrossRef]
21. Başkal, S.; Kim, Y. Lens optics and the continuity problems of the ABCD matrix. *J. Mod. Opt.* **2014**, *61*, 161–166. [CrossRef]
22. Martin-Sanchez, D.; Li, J.; Marques, D.M.; Zhang, E.Z.; Munro, P.R.T.; Beard, P.C.; Guggenheim, J.A. ABCD transfer matrix model of Gaussian beam propagation in Fabry-Perot etalons. *Opt. Express* **2022**, *30*, 46404–46417. [CrossRef] [PubMed]

MDPI AG
Grosspeteranlage 5
4052 Basel
Switzerland
Tel.: +41 61 683 77 34

Sensors Editorial Office
E-mail: sensors@mdpi.com
www.mdpi.com/journal/sensors

www.ingramcontent.com/pod-product-compliance
Lightning Source LLC
Chambersburg PA
CBHW041357210926
51850CB00013B/3103
* 9 7 8 3 7 2 5 8 4 4 9 5 1 *